ANIL V. VIRKAR
DEPARTMENT OF
MATERIALS SCIENCE & ENGINEERING
UNIVERSITY OF UTAH.

AF890694

QUANTITATIVE STEREOLOGY

ERVIN E. UNDERWOOD
Lockheed-Georgia Company
Marietta, Georgia

QUANTITATIVE STEREOLOGY

ADDISON-WESLEY PUBLISHING COMPANY
Reading, Massachusetts
Menlo Park, California • London • Don Mills, Ontario

This book is in
the Addison-Wesley series in
METALLURGY AND MATERIALS

Consulting Editor
MORRIS COHEN

Copyright © 1970 by Addison-Wesley Publishing Company, Inc. Philippines copyright 1970 by Addison-Wesley Publishing Company, Inc.

All rights reserved. No part of this publication may be reproduced, stored in a retrieval system, or transmitted, in any form or by any means, electronic, mechanical, photocopying, recording, or otherwise, without the prior written permission of the publisher. Printed in the United States of America. Published simultaneously in Canada. Library of Congress Catalog Card No. 70-83465.

To Marlit

PREFACE

"*Stereology*, sensu stricto, *deals with a body of methods for the exploration of three-dimensional space, when only two-dimensional sections through solid bodies or their projections on a surface are available. Thus, stereology could also be called extrapolation from two- to three-dimensional space.*"
Hans Elias, President, International Society for Stereology

As frequently happens in the scientific world, the time becomes ripe for the advance of a particular scientific subject. Then, and not until then, does activity become discernible, in that development of the subject proceeds rapidly and independently at different places. This is the state of affairs in the field of quantitative stereology. Metallurgists, petrographers, anatomists, biologists, ceramicists, mathematicians, and many others concerned with the geometrical aspects of structures have begun an accelerated and productive interplay.

At the same time, some aspects of the American educational system are undergoing significant revisions. In many engineering schools, the emphasis is on the science of materials as a whole, rather than on separate scientific disciplines devoted only to metals or ceramics or plastics or concretes. Thus it is fortunate to find the modern developments of stereology available to supplement the broad concepts of materials science.

Suitable textbooks for materials science students that deal with the quantitative aspects of microstructures are nonexistent at the present time. Saltykov's *Stereometric Metallography* (in Russian), Chayes' *Petrographic Modal Analysis* (an elementary statistical appraisal), DeHoff and Rhines' *Quantitative Microscopy*, and Weibel and Elias' *Quantitative Methods in Morphology* are the only known published books on the subject. For a complete coverage of the literature and an up-to-date treatment of modern quantitative stereology, one must pore through the literature of a dozen countries and through journals from many scientific fields.

The need is evident, therefore, for a unified and comprehensive treatment of the essential elements of stereological analysis. This book attempts to fill such a need. It is written for the student with no previous knowledge of the subject, as well as for the researcher who cannot afford to undertake an independent literature search. Because of the emphasis on principles and methods, the book can be used independently, or as part of a laboratory course in microstructures, or as the basis for a formal study of the theory of the subject. By adopting a generalized, geometrical approach, we appeal equally to those engaged in metallurgy, petrology, mineralogy, and research in polymers or ceramics; to those studying microscopic anatomy, histopathology, botany, or biology; and to those concerned with advancing the mathematics of statistics, geometrical probabilities, the theory of curves and surfaces, topology, and projective geometry. However, little is demanded of the reader in the way of mathematics, statistics, or geometrical probabilities. Some helpful information is given in the various chapters, but it is not expected to supersede standard textbooks or original articles.

In deriving equations, more emphasis is placed on the physical understanding of a relationship than on the degree of mathematical rigor achieved. A consistent scheme, based on simple geometrical models, has been adopted in order to establish a common ground for all readers. For those desiring a greater degree of mathematical sophistication, or alternative derivations, the references supplied should be adequate for most purposes.

The text also represents a brief, though complete, practical guide to the procedures used in estimating the three-dimensional properties of the microstructure. Simple measuring and counting operations, carried out under the microscope, are described in detail. Examples are given of how complex parameters are built up from the basic microstructural quantities in order to relate to some functional, physical, or mechanical property. A convenient device for the reader are the tables throughout the text that classify and summarize related data in readily usable form.

There are seven chapters in this book. After discussing nomenclature and basic measurements, we give the statistically exact expressions for specific volume, area, and length of line. Next we proceed to structural analyses for which more and more assumptions are required, and to subjects about which less and less is known, such as particle-size distributions, nonrandom structures, and projected images. Then approximation methods are explored. These can consist either of inexact representations of real structures, or of exact representations of unreal structures (i.e., idealized geometrical bodies whose characteristics may be known rigorously). In any event, most approximations have only a fleeting existence until a better approximation is made or until the rigorous solution is achieved. Chapter 7 treats the most difficult and indeterminate subject of all: specifying "shape." In that chapter, the rather qualitative state of the art is described in terms of the various approaches that have been used.

The text gives selected, representative problems. In addition, students can easily obtain classroom problems from photomicrographs which they take in the course of their work. Occasionally the book brings to the reader's attention an unsolved relationship or unsatisfactory state of development. The instructor can use these problem areas as tests of the student's ingenuity and ability to apply the basic principles outlined previously. Since these problems are far from being all solved, solutions to some of them may actually represent meaningful contributions to the literature of stereology.

References are quoted liberally and should be sufficient for full insight into the details of quantitative stereological methods. Although complete tabulation of the rapidly expanding literature in this field is impossible, an attempt has been made to include all significant work that has been done up to the end of June, 1969.

This book would not be complete without a final word of appreciation for Professor Morris Cohen of the Massachusetts Institute of Technology, who introduced me to the subject, illustrated its powers by practical example, and guided this effort to its conclusion.

Marietta, Georgia E.E.U.
August 1969

CONTENTS

1

BASIC CONCEPTS, SYMBOLS, AND MEASUREMENTS

"In many fields of scientific investigation the structure of cellular aggregates or random arrays of discrete particles imbedded in some matrix is observed on a two-dimensional section and inferences are drawn therefrom as to the real structure in three dimensions. The biologist's microtome slice, the petrologist's thin section, and the metallurgist's plane polished and etched sections are common examples, although the problem is a general one."
Cyril Stanley Smith and Lester Guttman

1.1 INTRODUCTION

All matter surrounding us can be described in terms of zero, one, two, and three dimensions. The geometrical relations that pertain to structures in minerals and metals are equally valid for human fat cells and air bubbles in concrete. Similarly, the problems that arise in the interpretation of three-dimensional structures from their two-dimensional sections are just as general.

When we talk in this book of structures, or microstructures, we are referring to the phases and constituents of a material that are visible on a microscopic, rather than on an atomic or macroscopic scale. Thus the magnifications that are usually required are those associated with light- and electron-microscopy. If a sectioning plane cuts a three-dimensional aggregate of space-filling polyhedrons, we observe a two-dimensional "structure" that consists of area-filling polygons. A two-dimensional structure is also obtained from the projected images of features in thin sections or foils. The problem is to relate our observations on the section or projection plane to the true three-dimensional microstructure.

Quantitative stereology attempts to characterize numerically the geometrical aspects of those features of the microstructure that interest us. Previous work in this area, labeled by such terms as morphometry [1], quantitative microscopy [2] or metallography [3], stereometric metallography [4], and micrometric [5] or modal analysis [6], has been applied in a relatively restricted sense. In its broadest context, stereology includes not only the quantitative study and characterization of any spatial structure, but also its qualitative interpretation, since exact representations are not yet possible in all cases.

Perhaps it would be useful to clarify what stereology is not [7]. Since stereology is based on uniaxial viewing, it is to be distinguished from stereoscopy, which is based on biaxial viewing. In a sense, stereology is the opposite of photogrammetry, since the latter utilizes three-dimensional images in order to construct flat maps.

Stereometry, or solid geometry, deals with the properties of solids of known shapes. Thus stereology is an extension of stereometry in that objects with more complex shapes are included in the former. Descriptive geometry, of which perspective is an integral part, treats the representation of projections from objects of known three-dimensional shapes.

There are various approaches to stereological problems. The statistico-geometrical approach depends on measuring and classifying a large number of two-dimensional figures or images. It is applicable when objects are randomly distributed in space. In such cases, a single section or projection, if extensive enough to contain a statistically significant number of features, may suffice to obtain valid results.

If objects of similar shape are arranged regularly in space, many section planes or visual angles may be necessary. Whether a section plane must be placed randomly throughout the material or just displaced systematically parallel to itself depends on the type of structural orientation involved.

In the case of objects of complicated shapes, reconstruction from serial sections may be the only desirable method to use. The same technique may also be required if only one or a few objects are available.

Finally, there are cases in which a small number of similar objects of complicated shape are available, but serial sectioning is impracticable. One depends, in these cases, on visual examination of many random sections in order to build up an actual or mental picture of the average three-dimensional shape. Random projections might also be used here and instead of sections, we would use projected images. The resulting average figure, reconstructed under the conditions just described, is called a "stereogram."

In this book we deal with the numerical or quantitative characterization of points, lines, surfaces, and volumes. Basic, exact expressions are known which relate measurements on two-dimensional sections to the three-dimensional structure. Regardless of whether the microstructural features occur randomly or in some preferential arrangement, the basic equations still apply. However, when a random, representative sampling of the microstructure is not attained on one sectioning plane, then multiple (and statistically random) sections must be employed.

Stated in other words, average values can be obtained from either random or oriented structures. In fact, in one general treatment of oriented structures [8], equations applicable to random structures are derived as special cases of the analysis. Additional information about oriented structures may be obtained by special methods. It may be that the extent of departure from the average value is the important consideration, or perhaps the directional nature of some measurement is desired.

1.2 BASIC SYMBOLS

Before we proceed to a discussion of the measurements that are most frequently employed in quantitative stereology, we shall present a list of the basic symbols used in this book. This list is based on the notation most frequently found in modern stereological literature and represents a systematic, self-consistent set of symbols for the basic geometrical elements [30]. For special requirements, we shall modify the basic symbols appropriately with subscripts or superscripts, parenthetical modifiers, etc. No attempt is made to give symbols for all quantities; as the need arises in the text, other terms will be introduced.

When one uses these symbols, one should take care to recognize clearly the possible different origins of points, lines, and areas. Points, for example, can exist as such in space, or can be formed by the intersection of a line with a line, plane, or curved surface. Figure 1.1 illustrates some relationships among surfaces, lines, and points in three dimensions and on a two-dimensional section.

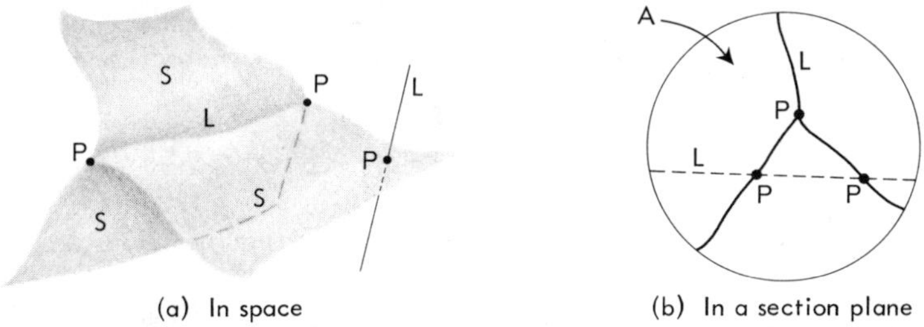

Fig. 1.1. Surfaces, lines, and points in three dimensions and in a two-dimensional section.

We should distinguish among the geometrical features that exist in space, appear on the section plane, or are generated by the test line or test plane. We must understand these distinctions clearly if we are to visualize correctly in three dimensions. In order to keep our system of notation as simple as possible, we shall make no distinction symbolically among the various origins of P, L, A, S, or V, unless the nature of the symbol is not clearly evident. The basic symbols, their definitions, and their interrelationships appear in Table 1.1.

Note the dual usage of symbols for both microstructural features and test quantities. For example, P stands for point elements in the microstructure, or for test points. The test quantities, or probes, could be points, lines, surfaces, or volumes.

The compound symbols represent a fraction in which the numerator refers to the microstructural quantity and the denominator pertains to the test or reference quantity. Consider S_V, which is equivalent to S/V. The symbol S refers to the area of surfaces in the microstructure contained within a test or reference volume V. P_P is the same as P/P, where the upper P refers to the grid points that fall in the microstructural areas of interest and the lower P represents the total number of grid points laid down (a test quantity).

TABLE 1.1

List of basic symbols and their definitions

Symbol	Dimensions*	Definition
P		Number of point elements, or test points
P_P		Point fraction. Number of points (in areal features) per test point
P_L	mm^{-1}	Number of point intersections per unit length of test line
P_A	mm^{-2}	Number of points per unit test area
P_V	mm^{-3}	Number of points per unit test volume
L	mm	Length of lineal elements, or test line length
L_L	mm/mm	Lineal fraction. Length of lineal intercepts per unit length of test line
L_A	mm/mm^2	Length of lineal elements per unit test area
L_V	mm/mm^3	Length of lineal elements per unit test volume
A	mm^2	Planar area of intercepted features, or test area
S	mm^2	Surface or interface area (not necessarily planar)
A_A	mm^2/mm^2	Area fraction. Area of intercepted features per unit test area
S_V	mm^2/mm^3	Surface area per unit test volume
V	mm^3	Volume of three-dimensional features, or test volume
V_V	mm^3/mm^3	Volume fraction. Volume of features per unit test volume
N		Number of features (as opposed to points)
N_L	mm^{-1}	Number of interceptions of features per unit length of test line
N_A	mm^{-2}	Number of interceptions of features per unit test area
N_V	mm^{-3}	Number of features per unit test volume
$\bar{L}$	mm	Average lineal intercept, L_L/N_L
$\bar{A}$	mm^2	Average areal intercept, A_A/N_A
$\bar{S}$	mm^2	Average surface area, S_V/N_V
$\bar{V}$	mm^3	Average volume, V_V/N_V

*Arbitrarily shown in millimeters.

Two symbols are used for surfaces: A and S. In deference to widespread usage, we have reserved A primarily for flat surfaces and S for curved surfaces. Also note that the number of points, P, is distinct from the number of features, N.

1.3 BASIC MEASUREMENTS

Certain measurements are required repeatedly in quantitative stereological work. These basic operations are all performed on two-dimensional sections or projections, and involve simple point, lineal, and areal measurements. Only the mechanics of the measurements are described in this chapter. Their applications to microstructures and equations will be elaborated on in later chapters.

1.3.1 P_P

"Point counting" (called the *Treffermethode* in German) is one of the simplest operations of quantitative stereology [3, 9, 10, 11]. The term refers to *test points* (for example, the intersections of a test grid or the end points of short test lines) that are counted with respect to some areal feature of the microstructure on the section plane. The method is especially effective for dealing with finely dispersed phases, such as small particles. One can use a grid inserted in the eyepiece of the microscope [4, 12, 13] or superimposed on a photomicrograph [3], or use the crosshairs in the microscope [14, 15]. Examples of typical point-counting grids designed for the microscope are shown in Fig. 1.2 and in the article by Sitte [27].

In these methods, a network of points is disposed uniformly over a sample area of the microstructure. The number of points that falls over the areas of interest (say a certain phase α) is counted for the first position of the grid. Then the grid is repositioned randomly, or translated in a particular direction, and the number of points falling over the α phase is counted again. This process is repeated until a sufficient number of measurements has been obtained.* The total number of points P_α falling over the phase α is then divided by the total number of grid points P_T, to give the ratio P_α/P_T, denoted simply by P_P (if additional identification is not required). Hilliard [12] recommends that those points that appear to fall on a boundary be counted as $\frac{1}{2}$. This gives the operator a guide to the magnitude of the experimental error.

If the grid is located in the microscope eyepiece, we work directly with the specimen; to obtain new grid positions we move the specimen on the microscope stage. An automatic click-stop mechanism can be employed to successively displace the specimen under the crosshairs of the microscope [14]. If photomicrographs are used instead, we may superimpose a grid marked on clear plastic sheets, or the grid can be printed photographically on each photomicrograph (in several random positions on duplicate pictures). In taking the photomicrographs, one must take care to select

*See Section 1.4 for a discussion of methods for calculating the number of measurements required for a certain accuracy.

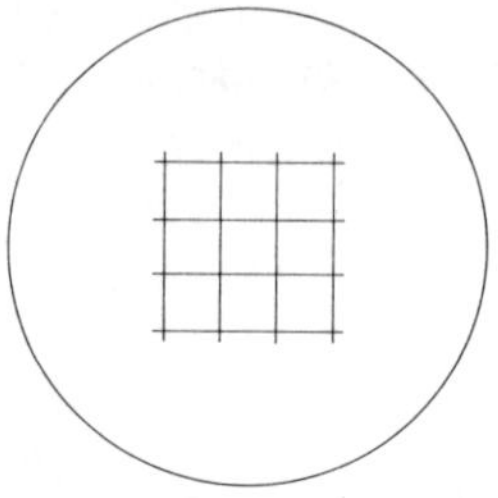

(a)

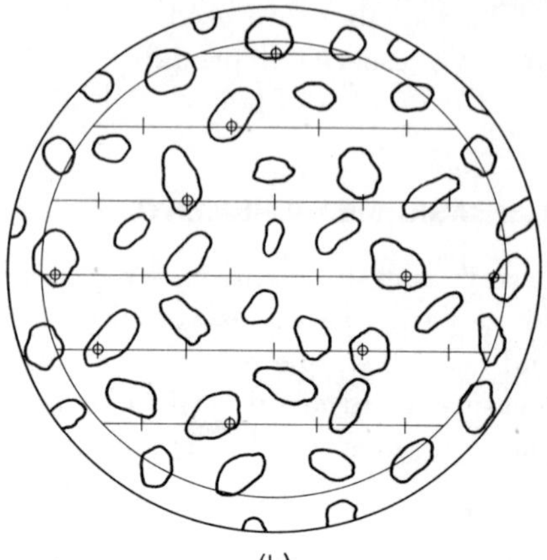

(b)

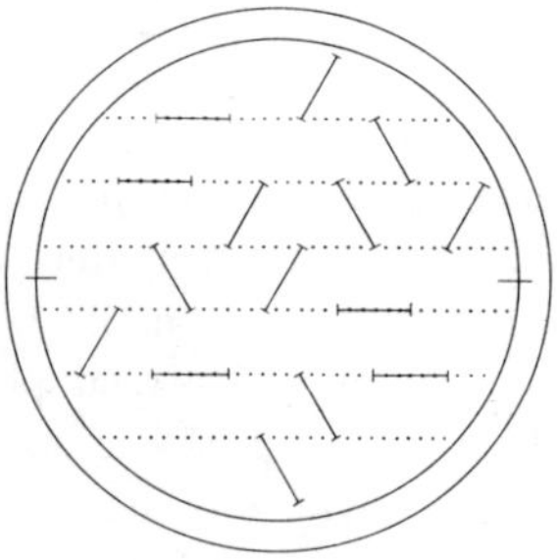

(c)

Fig. 1.2. Typical point-counting grids for the microscope eyepiece. (a) 16-point grid according to Hilliard [12]. 9- and 25-point grids are similar. (b) 25-point grid according to Hennig [13]. (c) Grid devised by Weibel [1] for point-counting and Chalkley-type measurements.

random positions, since there is always the tendency to take pictures where something of interest can be seen.

The operation of point counting is readily performed with automatic or mechanical equipment [14]. The Swift Automatic Point Counter [16] is one example of an automatic point-counting system. The specimen stage is moved under the crosshairs of the microscope by a ratchet mechanism with regularly spaced stops (the equivalent of the regularly spaced intersections of a grid). The total number of times that the crosshairs stop over the phase(s) of interest is counted and compared with the total number of stops. Numbers are tallied automatically. A practical disadvantage to this method is that only one test point is registered per position, whereas a grid gives many more test points per position.

There are substantial practical advantages to using the specimen directly. The labor of taking and preparing numerous photomicrographs is circumvented and a much greater area is available for measurements. On the other hand, if photographs are used, a permanent record is available.

1.3.2 P_L

Another important measurement frequently required in quantitative stereological work is the number of points (intersections) generated per unit length of test lines, P_L. A linear or circular test array is applied randomly to the microstructure in the section plane. The "points" usually consist of intersections made by test lines with traces of surfaces on the plane of polish, although systems of true lines are sometimes seen on projection planes. The test arrays may be applied randomly or displaced systematically over the entire microstructure until a sufficient number of intersections have been counted.

A P_L count can be performed in several ways. When the grid is inscribed on a glass disk inserted in the microscope eyepiece, measurements are made directly on the specimen as viewed under the microscope. The mechanics of performing a P_L count under these circumstances are illustrated in Fig. 1.2(b). The image of a linear grid of parallel lines appears over several particle sections. (For this operation, we ignore the tic marks on this particular grid, since they are used only for point-counting.) The number of intersections made by grid lines with the particle circumferences are counted systematically, giving a total of 27 intersections over a total grid length of L mm.* If the true grid length on the microstructure is 1.0 cm, then P_L would equal 2.7 mm^{-1}. The counting of actual objects in the microstructure, such as the particle traces shown in Fig. 1.2(b), will be described later.

Figure 1.3 reproduces linear and circular test arrays that are useful when one is working with photomicrographs or tracings. The total length of the test arrays may be fixed in advance to some convenient value in order to facilitate subsequent calculations. The linear grid in Fig. 1.3(a) has two major applications. It can be used when the microstructure is essentially random, in which case it is merely placed in various locations on the microstructure; P_L is determined for each location; then an average is calculated over all locations. On the other hand, the linear test array may be used for special measurements on an oriented structure. The test array is applied to one location, then rotated systematically at regular angular intervals with respect to some orientation axis. This procedure is repeated at other locations and the results are averaged, permitting one to assess the directional attributes of the total structure. This subject will be discussed in detail in Chapter 3, which deals with oriented structures.

If the structure is highly oriented in specific directions, a circular test array, such as that shown in Fig. 1.3(b), may be more suitable for obtaining average values than a linear test array. With the circular array, one is more certain of meeting the requirement of randomness of intersecting angle between test array and linear features on the microstructure.

Another scheme for determining P_L utilizes a single test line of fixed length, which is set down randomly or uniformly over the photomicrograph or microstructure. If the specimen is viewed directly under the microscope, a "line" can be generated by the relative motion of the crosshairs past the specimen image. As the

---

*The actual length of L depends on the magnification of the microstructure. The value of L can be determined at a standard magnification, for example, by comparison under the microscope with a ruled diffraction grating of known spacing. The true value of L at some other magnification would vary inversely with the relative change in magnification.

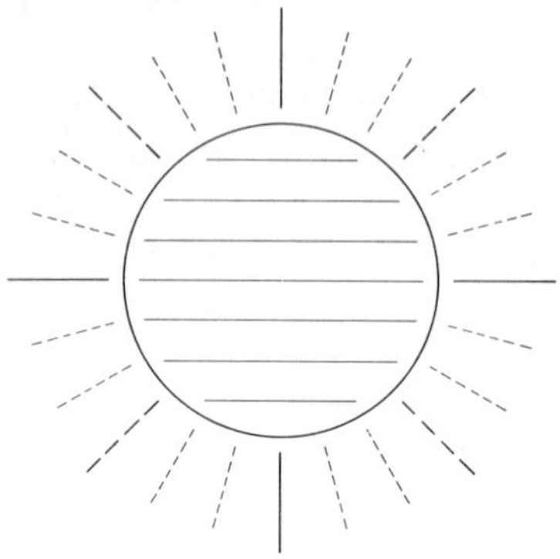

a) Parallel grid for random structures. Radii spaced at 15° angular intervals may be used for directional measurements on oriented structures.

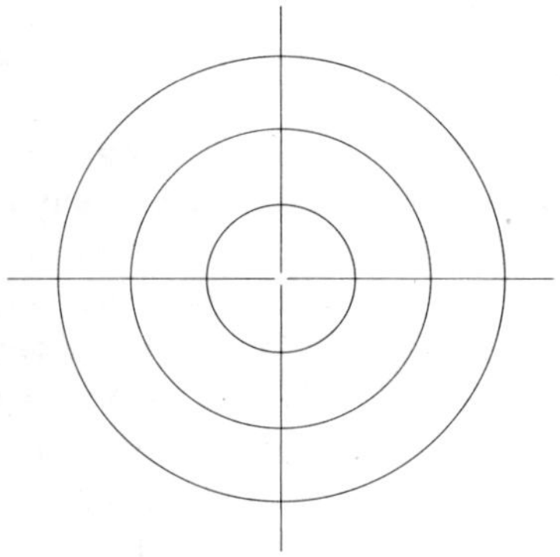

b) Circular grid for random intersections. May be used with random or oriented structures.

Fig. 1.3. Linear test arrays for measuring P_L, the number of intersections per unit length of test line.

specimen moves under the crosshairs, the number of intersections is recorded over the total traverse length. The latter quantity can then be obtained from the scale on the mechanical stage. Although distances obtained in this manner are not too accurate, they may be satisfactory for relatively coarse structures.

Usually linear traces in a microstructure are not mathematically thin, but possess a finite width. Thus there is a finite probability that tangent hits will be made, in addition to those situations in which a tangent hit *appears* to be made. In such cases of uncertainty, the P_L count can be recorded as $\frac{1}{2}$. Similarly, if a triple junction appears to be hit, it can be counted as equal to $1\frac{1}{2}$.

1.3.3 N_L

A measurement similar to P_L is N_L, defined as the number of interceptions of *objects* or *features* of the microstructure (rather than true points) per unit length of the test lines. Examples of N_L counts are the number of interceptions of second-phase areas, twin or martensite plates, or graphite nodules in cast iron per unit length of test line. A linear test array is laid down in a random manner, or better, it is displaced uniformly parallel to itself across the microstructure. Figure 1.2(b) shows parallel grid lines across a particulate structure. If the total grid length L equals 1.0 cm,

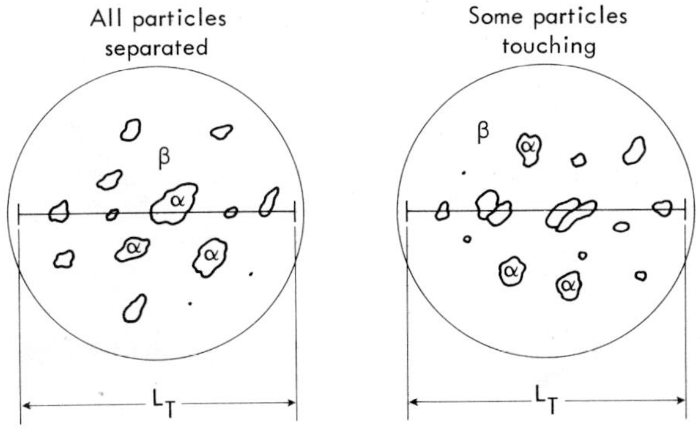

	All particles separated	Some particles touching
$(P_L)_{\alpha\alpha}$	0	2
$(P_L)_{\alpha\beta}$	10	8
$(N_L)_{\alpha}$	5	6

Fig. 1.4. Counting relationships between N_L and P_L for α-particles in a β-matrix (see Eq. 1.1).

N_L would equal 1.6 mm^{-1}, since 16 interceptions of particle sections are made. A test line of fixed length, applied randomly, could also be used here.

For separated convex figures (such as the nearly spherical graphite particles embedded in a cast iron), the N_L count is related simply to the P_L by a factor of two. In general, for every particle intercepted by the test line, we count *one* for the particle intercepted (N_L count), but *two* for the interceptions with the particle interfaces (P_L count); that is, $2N_L = P_L$. This simple relationship does not always hold, since, as we pointed out previously, the test line may appear to fall tangent to a particle trace, giving a P_L count of only *one* for that particle. If the particle configurations tend to promote tangent hits too frequently, separate P_L and N_L traverses may have to be run. When the test line in an N_L count appears to touch tangent to a boundary, it can be counted as $\frac{1}{2}$. The number of apparent tangent touches should help one to estimate the experimental error.

When the α-particles are not always separated, the N_L and P_L relationship must take into account the different types of interfaces that are intersected (see Fig. 1.4). That is, there is now the possibility of intersecting $\alpha\alpha$- as well as $\alpha\beta$-interfaces. Then, for the α-particles, we have

$$(N_L)_\alpha = \frac{2(P_L)_{\alpha\alpha} + (P_L)_{\alpha\beta}}{2}\ \text{mm}^{-1}. \tag{1.1}$$

When all particles are separated, $(P_L)_{\alpha\alpha}$ is zero and the relationship given previously is obtained. When the structure consists only of α-phase, $(P_L)_{\alpha\beta}$ is zero and $N_L = P_L$, as for space-filling grains or cells. This subject of contiguity is treated in greater detail in Chapter 4.

1.3.4 P_A

There is usually no great difficulty in a count of the number of points on a microstructure per unit area, P_A. Examples of true points on the section plane are "triple points" where the traces of three cell or grain boundaries meet at a point. Figure 1.5 represents a typical metallic structure [17] with grain-boundary traces intersecting at triple points. The dislocation etch pits commonly seen on metallic surfaces also represent true points, since they are formed at the intersections of dislocation lines with the surface. Although these etch pits have finite size, they frequently can be treated as true points, since their origin is known.

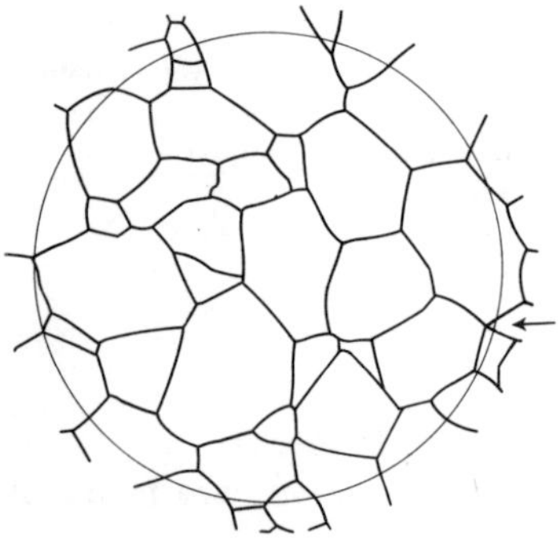

Fig. 1.5. Typical structure of grain boundary traces on the section plane [17]. Arrow indicates 4-rayed junction point.

When one is making the count, one may subdivide the microstructure into squares to help avoid counting the same point twice. Automatic counting or tallying devices are helpful here, especially if the numbers involved are very large. If desired, photomicrographs can be used directly, permitting each point to be marked off as it is counted. The total number of such points counted within the chosen area gives the ratio, P/A, or P_A, according to the convention adopted in this book.

1.3.5 N_A

The same techniques described above may be employed to count the number of *objects* in a certain area of the microstructure. These objects may be sections through precipitate particles, voids, or a second phase. Again the main problem is to avoid repetition in the count of any one object.

Two categories of N_A measurements are important. In one case, the features to be counted are space-filling (like cells or metal grains in a single-phase alloy) and $V_V = 1$. In the other case, the particles or other objects are not space-filling, and $V_V < 1$. For space-filling cells, the mean cell area is equal simply to $1/N_A$, while for a particulate structure, the mean area per particle is equal to A_A/N_A.

For the case in which $V_V = 1$, there are two straightforward methods for obtaining N_A. The first, due to Jeffries [18], is based on a count of those grains N_w lying wholly within a selected area A, and those grains N_i intercepted by the area perimeter. If enough perimeter grains are intercepted, on the average, they will be divided in half. Thus we obtain the total number of grains

$$N_T = N_w + \tfrac{1}{2}N_i, \tag{1.2}$$

with $N_A = N_T/A$. Referring to Fig. 1.5, we obtain $N_w = 20$ and $N_i = 21$. Thus

$N_T = 20 + 10.5 = 30.5$. The area A, in this case, is equal to 0.5 mm^2, and $N_A = 30.5/0.5 = 61$ mm^{-2}. The above procedure may also be applied to a particle count, i.e., when $V_V < 1$.

A second method for obtaining N_A when $V_V = 1$ is based on a count of grain corners (or "triple points") within a certain area. Smith [19] has pointed out how Euler's equation* relating the number of polygons, edges, and corners on a two-dimensional section may be applied to the usual equilibrium cell structure. Since, on a section through a cellular structure, the boundary traces generally meet three at a point, P, we find that

$$N_T = \tfrac{1}{2}P + 1, \qquad (1.3)$$

and again $N_A = N_T/A$. The occasional nonequilibrium 4-rayed junctions should be counted as two junction points. No special allowance need be made here for the border grains, since there is a correspondingly smaller number of junction points. Referring again to Fig. 1.5, we find that $P = 59$, and $N_T = 29.5 + 1 = 30.5$, which gives the same result we obtained when we used Jeffries' method. Note the 4-rayed junction point indicated by the arrow in Fig. 1.5.

The remaining two basic measurements to be discussed here are called *areal* and *lineal analysis*.

1.3.6 A_A

Usually areal analysis involves the determination of the relative area of a phase or constituent per unit area of the microstructure, and is denoted by A_A. Early workers [20, 21] frequently employed areal measurements, but such methods were superseded because methods that were faster and generally more accurate became available for estimating the same ratio. However, modern developments in electronic scanning instrumentation promise a return to area estimation methods [22].

Relative areas can be determined from tracings or photomicrographs by a planimeter, or by cutting out the phase of interest and weighing all the pieces, then comparing to the total weight. Areas can also be estimated by means of a square grid superimposed over the microstructure. The number of complete grid squares that fall over the phase of interest are added to the fractional areas of grid squares that are cut by the phase boundary, giving the total phase area. One can see that the determination of areas is more time-consuming than, say, point-counting or lineal analysis. Fortunately the latter measurements are equivalent to the areal method, so the slower measurement is not always necessary.

1.3.7 L_L

Lineal analysis [3, 23] is better employed with microstructural features that are aggregated into relatively large areas, rather than those which are finely dispersed. The term "lineal analysis" refers to the fact that intercepted lengths are measured along randomly applied straight lines across the microstructure. The individual lengths intercepted by particles or scond-phase regions are measured separately, added up,

*See Chapter 7, Section 7.5.5, for a discussion of Euler's equations.

and compared to the total length of traverse. The ratio gives L_L for the particular phase selected.

There are several instruments for measuring and adding up the individual intercepted lengths and the total length of the test line. Detailed descriptions of these instruments are presented by Fischmeister [14]. One of the better-known mechanical devices for lineal analysis is the Hurlbut Counter [24]. Let us describe its operation briefly. A constant-speed motor drives a shaft which moves the stage and specimen under the crosshairs of the microscope. Several stop-start buttons engage gears that connect the drive shaft with the stage. Veeder Root counters are actuated by each button and record the number of revolutions of the shaft that takes place while that button is depressed. There is also a total counter actuated by all buttons. A particular button is assigned to each phase of interest, and when the crosshairs traverse a particular phase, only the assigned button is depressed. Referring to Fig. 1.6, the button for the α-phase would be depressed at point 1 and released at point 2; then depressed at point 3 and released at point 4; etc. In between, of course, the button assigned to the matrix would be actuated. After one completes the traverses required to attain the desired accuracy, one reads the number of revolutions from the counter for the α-phase and for the total traverse length. Since the number of revolutions is proportional to line lengths, the simple numerical ratio of L_α to L_T gives the desired value of $(L_L)_\alpha$, or simply L_L, if no ambiguity is introduced thereby.

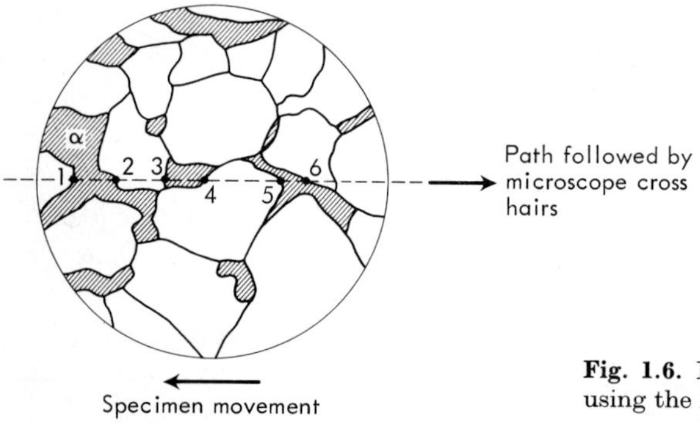

Fig. 1.6. Illustration of the measurement of L_L using the Hurlbut counter for lineal analysis.

Other modifications of the basic device employ, for example, ratios of the traversing times [25]. Constant speed of traverse is required in the latter case, and this may prove to be unduly restrictive. Another version utilizes a variable-speed motor, controlled with a foot pedal, in order to pass the crosshair through large empty areas of the microstructure quickly [26].

1.4 STATISTICAL ANALYSIS AND ESTIMATION OF ERRORS

An estimate of microstructural parameters depends on the reliability of the measurements. It is just as important to be able to estimate the errors in an analysis as it is to perform the analysis itself.

There are three general types of errors to guard against [28]. First are the errors arising from the experimental limitations. These include resolution of the microscope, etching effects, uncertainties in the length of the test figure, incorrect counting, and so on. In most cases, experimental errors will not be the limiting factor in the accuracy of the analysis.

The second major type of error is caused by improper sampling techniques [33]. During analysis, it is essential that all randomization processes be executed in a truly random fashion. The test area should be selected at random and the test figure applied at random, preferably without looking. If the observer must make a subjective decision, then some degree of bias will almost always be introduced. Bias is also introduced when the structural property being determined is not linearly related to the experimental measurement. For example, grain size numbers found by either the ASTM [17] or the intercept methods are representative of this type of bias. The magnitude of the bias is small, however, and can usually be neglected.

A third error results from nonrepresentative samples, whereby the areas selected for analysis are not exactly representative of the specimen as a whole. These sampling errors usually constitute the bulk of the total error; consequently, we shall treat them in some detail.

It is to be expected that the greater the number of observations involved in a measurement, the greater will be its reliability. In fact, it is a characteristic of sampling errors that the probability of exceeding any specified error can be made as small as desired if one makes a sufficient number of observations. Obviously, however, a compromise must be achieved between effort and accuracy, so that the investigator will neither make too few observations nor make more observations than necessary.

Sampling errors are important in another respect. Since there are alternative methods of measuring a particular property, the question arises as to the best method to use. The choice must then be made on the basis of the relative efficiencies; i.e., that method requiring the least effort for a given accuracy. We shall amplify this concept further below.

1.4.1 Statistical Relationships

Before we consider the individual analyses and their associated sampling problems, we shall collect, in this section, some basic statistical equations that we shall need.

The most common and important measure of the central value of a set of n measurements is the arithmetic mean, or average. The arithmetic mean of $x_1, x_2, \ldots, x_n$ observations is denoted by $\bar{x}$ and is given exactly by

$$\bar{x} = (1/n) \sum_{i=1}^{n} x_i \tag{1.4}$$

or approximately by

$$\bar{x} = \sum_{i=1}^{k} f_i x_i \Big/ \sum_{i=1}^{k} f_i, \tag{1.5}$$

where the f_i are the frequencies, or the number of observations having the same value, or the number of observations within a class interval.

A measure of the expected precision of a single measurement is afforded by the standard deviation $\sigma(x)$ of a population. In practice, we deal with the standard deviation $s(x)$ of the sample(s). It is usually more convenient to work with the variance, which is the square of the standard deviation.

The variance of a sample is given by

$$s^2(x) = \left(\frac{1}{n-1}\right) \sum_{i=1}^{n} (x_i - \bar{x})^2, \tag{1.6}$$

where n is the number of measurements of x. The terms $(x_i - \bar{x})$ are the deviations of each observation x_i from the arithmetic mean $\bar{x}$ of the n observations. A more convenient form for calculation of the variance is

$$s^2(x) = \left(\frac{1}{n-1}\right)\left[\left(\sum_{i=1}^{n} x_i^2\right) - n\bar{x}^2\right]. \tag{1.7}$$

Variances have a very useful additive property. Consider the parameter X which is a linear function of k independent random variables $x_1, x_2, x_3, \ldots, x_k$; that is,

$$X = \sum_{i=1}^{k} c_i x_i, \tag{1.8}$$

in which the c_i's are constants. The variance of X is related to the variances of the x_i's by

$$\sigma^2(X) = \sum_{i=1}^{k} c_i^2 \sigma^2(x_i). \tag{1.9}$$

From this result it is possible to obtain the standard deviation of a sample mean [28, 35]. That is, we can relate $\sigma^2(x)$, the variance in an individual observation, to $\sigma^2(\bar{x})$, the variance in the arithmetical mean of a sample. Consider the sum X of n individual observations x_1. If the sampling is random, the variables are independent of one another and their variances are equal. Thus Eq. (1.9) reduces to

$$\sigma^2(X) = n\sigma^2(x). \tag{1.10}$$

We next take a random sample of size n from the same population. Since $\bar{x} = X/n$, we have from Eq. (1.9)

$$\sigma(\bar{x}) = (1/n)\,\sigma(X), \tag{1.11}$$

which, in combination with Eq. (1.10), yields

$$\boxed{\sigma(\bar{x}) = \sigma(x)/\sqrt{n}\,.} \tag{1.12}$$

This result shows that the standard deviation of the mean varies inversely as the square root of the number of measurements entering into the mean [35].

For a nonlinear function,

$$\sigma^2(X) \approx \sum_i \left(\frac{\partial X}{\partial x_i}\right)^2_{x_i = \bar{x}_i} \sigma^2(x_i) \tag{1.13}$$

which is a satisfactory approximation provided that the condition $[\sigma(x_i)/x_i] \ll 1$ is met, and that the x's are independent [11, 28]. Usually we are more concerned about the relative value of the error than its absolute value. The common relative quantity is called the *coefficient of variation*, and is defined by

$$CV = \sigma(x)/\bar{x}. \tag{1.14}$$

The binomial frequency function has useful properties when one is dealing with the probability of occurrence of an event. If $\mathscr{P}$ is the probability that an event will occur in one trial, then the probability that it will occur exactly k times in N independent trials is given by the kth term of the binomial frequency function. In particular, the expected number of successes N_s in N trials is

$$\bar{N}_s = N\mathscr{P}, \tag{1.15}$$

and the variance in N_s is

$$\sigma^2(N_s) = N\mathscr{P}(1 - \mathscr{P}). \tag{1.16}$$

If $\mathscr{P}$ approaches zero as the product $N\mathscr{P}$ approaches a finite limit, then the binomial function tends to a Poisson function for which $\bar{N}_s$ in Eq. (1.15) is the same, but the variance becomes

$$\sigma^2(N_s) = N\mathscr{P} = \bar{N}_s. \tag{1.17}$$

These results will be used later.

1.4.2 Volume-Fraction Measurements

Of the three different methods for estimating the volume fraction—i.e., by areal, lineal, and point analyses—point-counting is preferred. It has been shown [11] that the systematic point count is markedly superior to the other methods in terms of efficiency, in that it requires the least amount of effort to obtain an estimate with a given sampling error. This is shown in Fig. 1.7 by Hilliard [12] in which he compares the values of $\sigma(V_V)/V_V$ for several different experimental procedures.

We now proceed to a brief consideration of the variances of the principal procedures used for estimating the volume fraction.

Areal analysis. It is assumed that the major contribution to the variance of an areal measurement arises in the selection of the subareas for analysis in the plane of polish. We further suppose that the subarea selected contains various sizes and numbers of α-features, such that there are n_1 features of area a_1, n_2 of area a_2, etc. Then the expected areal fraction A_A that would be obtained from the average of a very large

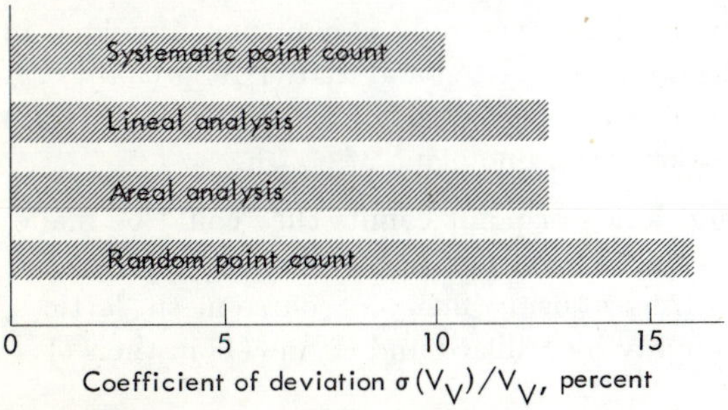

Fig. 1.7. Calculated coefficients of deviation for 100 observations on a structure composed of a random dispersion of equal-sized spherical particles [12].

number of measurements on equal-size α subareas selected at random from the specimen is

$$A_A = \sum_i \left(\frac{a_i}{A}\right)\bar{n}_i \tag{1.18}$$

where A is the area of the subarea. According to Eq. (1.17), we can write $\sigma^2(n_i) = \bar{n}_i$, since the expected number of randomly distributed α-features within a given area will follow a Poisson distribution. From Eqs. (1.9) and (1.18), it follows that

$$\sigma^2(A_A) = \sum_i \left(\frac{a_i}{A}\right)^2 \sigma^2(n_i) = \frac{1}{A^2}\sum_i a_i^2\bar{n}_i, \tag{1.19}$$

which can be rewritten (see Eq. 1.7 for the required functional relationship)

$$\sigma^2(A_A) = [N_a/A^2][\sigma^2(a) + \bar{a}^2], \tag{1.20}$$

where $N_a(= \sum_i \bar{n}_i)$ is the expected total number of α-features and $\bar{a}$ is their average area. Since we are usually more interested in the relative value of the error than in its absolute magnitude, we divide Eq. (1.20) through by $(A_A)^2$ $[= (N_a\bar{a}/A)^2]$, giving

$$\boxed{[\sigma(A_A)/A_A]^2 = (1/N_a)\{[\sigma(a)/\bar{a}]^2 + 1\}.} \tag{1.21}$$

The ratio $\sigma(a)/\bar{a}$ (see the coefficient of variation in Eq. 1.14) would equal zero if all α-features had the same area. In such a case, the right-hand side of Eq. (1.21) reduces to $1/N_a$. Normally, we will replace $\bar{a}$, $\sigma(a)$, and N_a by experimentally determined values from the sample(s).

Lineal analysis. If α-features are randomly dispersed on the plane of polish, then the intercepts l_i made on the random traverse line will also be randomly dispersed. The calculation of the variance in the lineal fraction L_L is then exactly analogous to that for the area analysis. From Eq. (1.21), we can write, for small $(V_V)_\alpha$,

$$\boxed{[\sigma(L_L)/L_L]^2 = (1/N_l)\{[\sigma(l)/\bar{l}\,]^2 + 1\},} \tag{1.22}$$

in which the average intercept length is $\bar{l}$ and the expected total number of intercepts is N_l. From Eq. (1.13) we obtain the more general expression

$$\boxed{[\sigma(L_L)/L_L]^2 = (1/N_l)(1 - L_L)^2\{[\sigma(l)/\bar{l}\,]_\alpha^2 + [\sigma(l)/\bar{l}\,]_\beta^2\},} \tag{1.23}$$

where β refers to all phases except the α-phase. The only assumptions involved in Eq. (1.23) are that the intercept lengths are uncorrelated and that $(1/N_l)[\sigma(l)/\bar{l}\,]^2 \ll 1$ for both phases. As long as the number of α-features greatly exceeds N_l, Eqs. (1.22) and (1.23) are valid for measurements taken from one plane of polish.

Systematic point count. There are several kinds of point counts that could be made, such as the systematic or random two-dimensional point counts, the systematic or random one-dimensional point counts, those using fine- or coarse-mesh lattices, etc. The possibilities are analyzed thoroughly by Hilliard and Cahn [11] in their classic paper. Here we shall outline the steps leading to the variance of a systematic two-dimensional point count using a coarse-mesh lattice.

The point array is defined by the intersections of a square grid, so that the total number of test points applied to the structure is some predetermined multiple of the points in the grid. The most critical item in determining the efficiency of the analysis is the grid spacing relative to the scale of the structure. For α-phase areas in an $\alpha + \beta$ phase mixture, with $(V_V)_\alpha < 0.5$, this spacing should be such that the average number of points falling in any α-phase area does not greatly exceed unity. If $(V_V)_\alpha > 0.5$, the same rule applies to the β-phase areas. With point densities exceeding this limit, there is a sharp increase in the sampling error (for a given number of points counted).

Provided that the α-features are randomly dispersed, the number $(p_i)_j$ of points in the jth subarea falling on features of area a_i will follow a binomial distribution. Therefore we obtain for the expected value

$$(p_i)_j = (n_i)_j a_i/d^2, \tag{1.24}$$

where d is the lattice spacing, and for the variance

$$\sigma^2(p_i)_j = (n_i)_j(a_i/d^2)[1 - (a_i/d^2)], \tag{1.25}$$

in which $(n_i)_j$ is the number of features of area a_i in the jth subarea. Averaging over all possible equal-size subareas, we obtain

$$\bar{p}_i = \bar{n}_i a_i/d^2$$

and

$$\sigma^2(p_i) = \bar{n}_i(a_i/d^2)[1 - (a_i/d^2)] + \sigma^2(n_i)[(a_i/d^2)]^2. \tag{1.26}$$

Because a random distribution of α-features was assumed, $\sigma^2(n_i) = \bar{n}_i$, and Eq. (1.26) becomes

$$\sigma^2(p_i) = \bar{n}_i a_i/d^2 = \bar{p}_i. \tag{1.27}$$

The above expressions relate only to α-features of one size. To obtain the expected total number of lattice points falling on the α-phase P_α and its variance $\sigma^2(P_\alpha)$, we must sum over all possible values of i. Thus

$$P_\alpha = \sum_i \bar{p}_i = \sum_i \bar{n}_i a_i/d^2 = N_a \cdot \bar{a}/d^2, \tag{1.28}$$

where N_a is the expected number of α-features of all sizes and $\bar{a}$ is their average area.

Summing the individual variances given by Eq. (1.27) results in

$$\sigma^2(P_\alpha) = \sum_i \sigma^2(p_i) = \sum_i \bar{p}_i = P_\alpha \tag{1.29}$$

or

$$\boxed{[\sigma(P_P)/P_P]^2 = 1/P_\alpha.} \tag{1.30}$$

Thus, for randomly dispersed features, the variance of the analysis is independent of their size distribution. This means that one can predict the variance of a systematic point count without making any additional measurements on the plane of polish.

Further analysis and comparison with the results for a random point count show that the variance of Eq. (1.30) is less than that for the random point count [11, 12].

1.4.3 Point-Count Procedure

To implement the preceding statistical analysis, we here outline a practical guide for conducting a systematic point count [12, 32, 33].

1) First: It is necessary to decide the maximum error that can be tolerated. In making this decision, the investigator should bear in mind that the effort required will vary inversely as the square of the error. Thus the difference between an accuracy of, say, 10% and 1% corresponds to an effort of 30 minutes and a week.
2) Second: One can determine an approximate value of the volume fraction by applying the grid a few times to the specimen to get P_P.
3) On the basis of this rough estimate, one can calculate a provisional value of the total number of lattice points P_T required to achieve the desired accuracy. For an accuracy with 95% confidence limits, DeHoff [34] gives

$$P_T = \left[\frac{200}{(\%\ \text{acc})}\,\frac{\sigma(P_P)}{P_P}\right]^2, \tag{1.31}$$

while Gladman and Woodhead [29] recommend

$$P_T = P_P(1 - P_P)/\sigma^2(P_P). \tag{1.32}$$

4) Next one selects a reticule of the type shown in Fig. 1.2(a) (with 9, 16, 25, . . . points), based on the shortest time required to count a given number of hits. Suppose that the grid giving the quickest count contains n points. Then, in order to lay down the required number of test points, one will have to examine P_T/n fields of view.
5) In addition to the speed of counting, there is also the requirement that only one point, on the average, falls on any one areal feature. Accordingly, the magnification is adjusted consistent with this goal.
6) The spacing between adjacent fields of view should be selected so that the entire specimen area is uniformly sampled.
7) One can record the number of hits per field of view automatically or manually, depending on personal preference and the magnitude of the undertaking. Points that fall (apparently) directly on a boundary should be counted as $\frac{1}{2}$. This procedure avoids introducing a subjective bias and indicates the magnitude of the experimental error.
8) If one requires an estimate of the standard deviation, one can keep a record of, say, ten subtotals, each having an equal number of applied test points.
9) About halfway through the analysis, the volume fraction should be recalculated in order to obtain a more accurate value of P_T.
10) When the measurements are completed, one can calculate the final value of P_P and obtain $\sigma^2(V_V)$ or $\sigma(V_V)$ from Eq. (1.32).

REFERENCES

1. E. R. Weibel, "Principles and Methods for the Morphometric Study of the Lung and Other Organs," *Lab. Investigation* **12,** 2 (1963) 131
2. *Quantitative Microscopy,* edited by R. T. DeHoff and F. N. Rhines. New York: McGraw-Hill (1968)
3. R. T. Howard and M. Cohen, "Quantitative Metallography by Point-Counting and Lineal Analysis," *Trans. AIME* **172** (1947) 413
4. S. A. Saltykov, *Stereometric Metallography,* second edition, *Metallurgizdat,* Moscow (1958)
5. F. Chayes, "The Relation Between Area and Volume in Micrometric Analysis," *Mineral. Mag.* (London) **30,** 221 (June 1953) 147
6. F. Chayes, *Petrographic Modal Analysis,* New York: John Wiley (1956)
7. H. Elias, "Stereology: Scope and Limits," *Stereologia* **1,** 1 (April 1962) 1
8. J. E. Hilliard, "Specification and Measurement of Microstructural Anisotropy," *Trans. AIME* **224** (1962) 1201
9. E. Thomson, "Quantitative Microscopic Analysis," *J. Geol.* **38,** 3 (April–May, 1930) 193–222
10. A. A. Glagolev, "Quantitative Analysis with the Microscope by the Point Method," *Eng. Min. J.* **135** (1934) 399
11. J. E. Hilliard and J. W. Cahn, "An Evaluation of Procedures in Quantitative Metallography for Volume-Fraction Analysis," *Trans. AIME* **221** (1961) 344
12. J. E. Hilliard, "Measurement of Volume in Volume," in Reference 2, page 45
13. A. Hennig, "A Critical Survey of Volume and Surface Measurement in Microscopy," *Zeiss Werkzeitschrift,* 30 (1958) 3
14. H. F. Fischmeister, "Scanning Methods in Quantitative Metallography," in Reference 2, page 336

    ———, "Automatic Measuring and Scanning Devices," *Stereology,* edited by H. Elias, *Proc. Second Int. Cong. for Stereology,* New York: Springer-Verlag (1967) 263

    ———, "Apparative Hilfsmittel der Stereologie," in *Quantitative Methods in Morphology,* edited by E. R. Weibel and H. Elias, New York: Springer-Verlag (1967) 221
15. F. Chayes, "A Simple Point Counter for Thin-Section Analysis," *Amer. Min.* **34** (1949) 1
16. James Swift and Sons, Ltd., London: "Swift Automatic Point Counter"
17. ASTM Standards, Part 31, May 1967, *Physical and Mechanical Testing of Metals* . . . " See article, "Estimating the Average Grain Size of Metals," Fig. 1, page 448, ASTM, Philadephia, Pa.
18. Z. Jeffries, *Chem. Met. Eng.* **18** (1918) 185
19. C. S. Smith, "Grain Shapes and Other Metallurgical Applications of Topology," *Metal Interfaces,* ASM, Cleveland, Ohio (1952) 65
20. A. Delesse, "Pour déterminer la composition des roches," *Annales des Mines* **13** (1848), fourth series, pages 379–388
21. A. K. Huntington and C. H. Desch, "The Planimetric Analysis of Alloys and the Structure of Phosphor-Copper," *Trans. Far. Soc.* **4** (1908) 51

22. A. Rose, L. Rademacher, and J. M. van Wyk, "Einfluss der Austenitisierungsbedingungen auf das Umwandlungsverhalten von Schnellarbeitsstählen," *Stahl u. Eisen* **79,** 18 (1959) 1243

23. A. Rosiwal, "Über Geometrische Gesteinsanalysen. Ein Einfacher Weg zur Ziffermässigen Feststellung des Quantitätsverhältnisses der Mineralbestandtheile Gemengter Gesteine," *Verhandl. der K.-K. geologischen Reichsanstalt* 5/6 (1898) 143

24. G. S. Hurlbut, "An Electric Counter for Thin Section Analysis," *Amer. J. Sci.* **237** (1939) 253

25. R. A. Huggins, "Quantitative Metallography by Lineal Analysis," *Met. Prog.* **69,** 6 (1956) 94

26. E. E. Underwood, unpublished research, Battelle Memorial Institute, Columbus, Ohio (1958)

27. H. Sitte, "Morphometrische Untersuchungen an Zellen," *Quantitative Methods in Morphology*, edited by E. R. Weibel and H. Elias, New York: Springer-Verlag (1967) 172

28. J. E. Hilliard, "Grain Size Estimation," G.E. Report 62-RL-3133M, December 1962, Schenectady, N. Y.

29. T. Gladman and J. H. Woodhead, "The Accuracy of Point Counting in Metallographic Investigations," *J. Iron and Steel Inst.* **194** (1960) 189

30. E. E. Underwood, "A Standardized System of Notation for Stereologists," *Stereologia* **3** (1964) 5

31. J. Silcox and M. J. Whelan, "Annealing of Dislocation Loops in Aluminum," in *Structure and Properties of Thin Films*, edited by C. A. Neugebauer, J. B. Newkirk, and D. A. Vermilyea, New York: John Wiley (1959) 164

32. R. T. DeHoff, "The Statistical Background to Quantitative Metallography," page 29 in Reference 2

33. J. E. Hilliard, "Applications of Quantitative Metallography in Recrystallization Studies," in *Recrystallization, Grain Growth and Textures*, Metals Park, Ohio: Amer. Soc. for Metals (1966) 267

34. R. T. DeHoff, "Quantitative Metallography," in *Techniques for the Direct Observation of Structure and Imperfections*, Vol. 2, Part 1, edited by R. F. Bunshah, New York: John Wiley (1968) 221

35. P. G. Hoel, *Introduction to Mathematical Statistics*, third edition, New York: John Wiley (1964) 138

PROBLEMS

1.1. Figure 1.8 gives a sequence of electron micrographs of dislocation line configurations. (a) Measure the average value of P_L for each photo. (b) Measure $P_L(\theta)$ as a function of direction (θ), using the vertical edge of the photos as reference direction. Plot the results on polar coordinates.

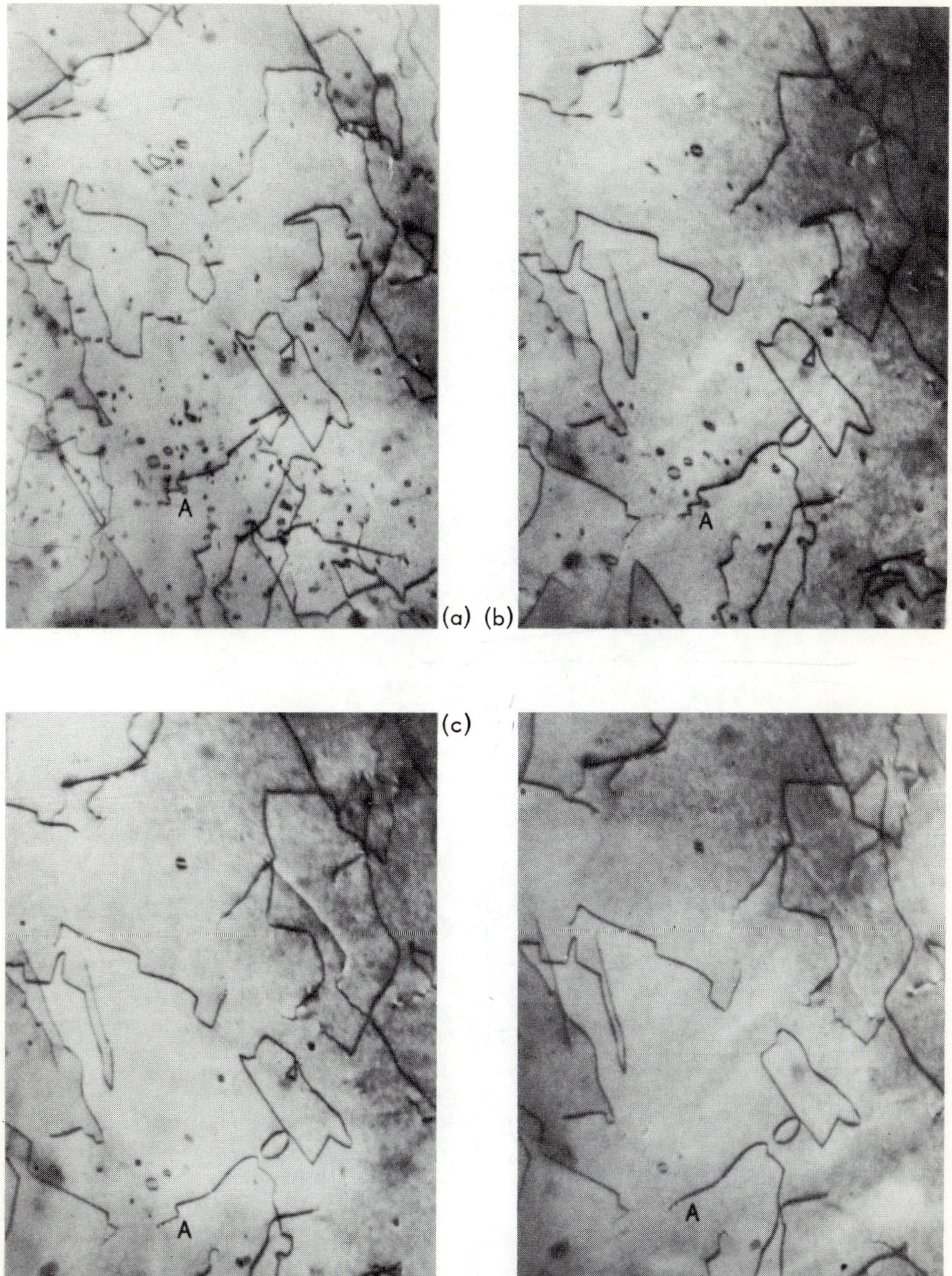

Fig. 1.8. Successive changes in the configuration of dislocation lines in a thin foil of Al [31].

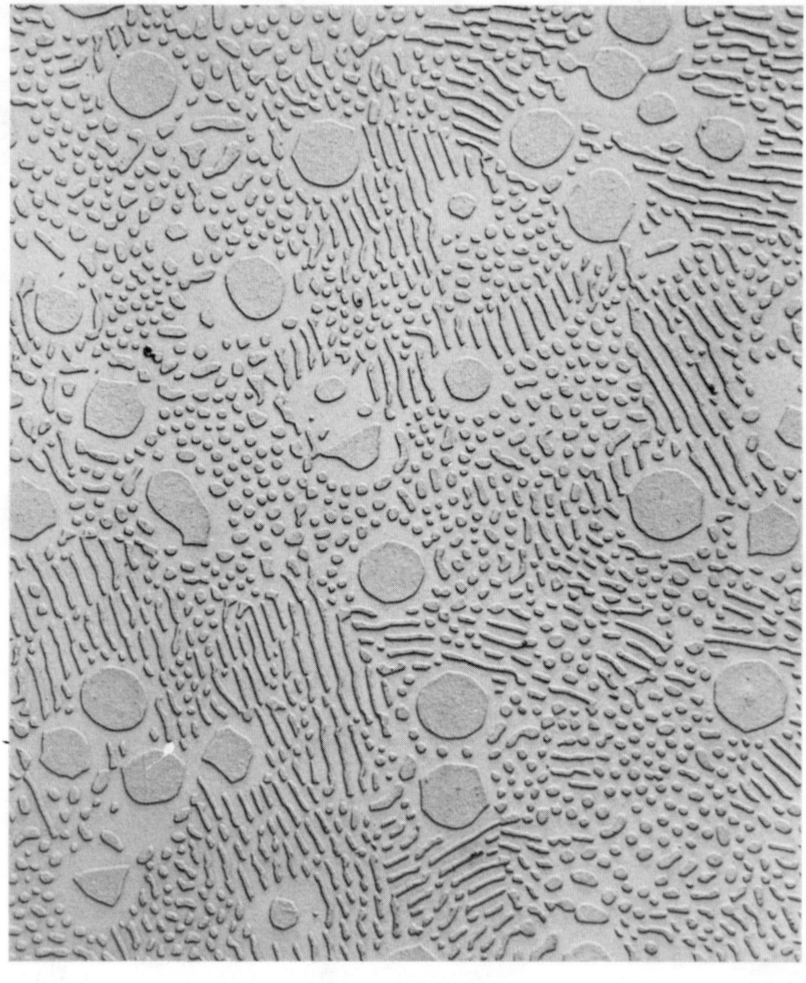

Fig. 1.9. Mg-Al eutectic alloy, unidirectionally solidified.

1.2. Measure P_L for the system of lines enclosed within the circle in Fig. 1.5. Using a single test line (length about equal to the circle radius), make (a) n systematic parallel translations of the test line from four directions, and (b) an equal number $4n$ of random applications of the test line. Which method gives the smaller standard deviation, $s(x)$? [*Note*: Use Eq. (1–6) or (1.7).]

1.3. Using the data obtained in Problem 1.2(a), calculate $s(\bar{x})$ for samples consisting of n, $2n$, $3n$, and $4n$ observations. How does the precision of the sample mean vary with sample size?

1.4. Figure 1.9 shows a eutectic alloy structure consisting of three types of features: large, nearly circular areas; small, nearly circular areas; and elongated areas. (a) Measure P_P, L_L, and A_A for the large circular-type areas. (b) How would you obtain the area fractions for each of these three types of features separately? Can measurements be made simultaneously or must they be made separately?

2

STATISTICALLY EXACT EXPRESSIONS FOR POINTS, LINES, SURFACES AND VOLUMES

". . . when you can measure what you are speaking about and express it in numbers, you know something about it; but when you cannot express it in numbers, your knowledge is of a meagre and unsatisfactory kind; it may be the beginning of knowledge, but you have scarcely in your thoughts advanced to the state of science, whatever the matter may be."

Lord Kelvin

2.1 INTRODUCTION

Exact treatment of all aspects of quantitative microstructural analysis is not yet possible. Among those subjects still requiring further development are size distributions of irregularly shaped particles or grains, projected images, and the specification of shape. These topics will be treated separately in Chapters 5, 6, and 7, respectively.

In this chapter, we deal with statistically exact expressions that relate points, lines, surfaces, and volumes in space to measurements made on two-dimensional sections through the space. Achieving the mathematically expected values does not depend, *per se*, on the type of microstructure involved, e.g., random, partially oriented, or completely oriented. However, if the structures do not fulfill the requirements of randomness, then randomly placed and/or randomly oriented section planes must be employed.*

The basic equations relate quantities measured on the two-dimensional section plane to the more inaccessible quantities located in three-dimensional space. The required level of accuracy in the expected values is attained by proper sampling and measuring techniques. We have considered these factors in Section 1.4. Here we

*By a "random" section plane (or test line), we mean that hypothetical plane (or line) having properties which equal the average or expected value of all possible sections (or lines) through the microstructure.

assume that a sufficient number of specimens are available, and that both the sampling from specimens and the number of measurements on the samples are statistically significant.

From a practical standpoint, only one section plane may be sufficient for many common microstructures, provided that there is approximate randomness of the features and that enough features appear on the plane. However, a knowledge of the history of the material, as well as the appearance of the microstructure, will help to assess the degree of statistical randomness actually attained.

2.2 MEASURED AND CALCULATED QUANTITIES

It may be helpful to compare the relationship of measured quantities in the two-dimensional section to calculated quantities in three-dimensional space. The triangular array of symbols in Table 2.1 is arranged systematically to emphasize this interrelationship of measured to calculated quantities. Although identified as measured quantities, A_A and L_A are usually calculated from other measurements. As we shall see, A_A is normally obtained from measurements of P_P or L_L, and L_A is calculated from measurements of P_L. The arrows indicate related quantities for which equations are developed in this chapter.

TABLE 2.1

Relationship of measured (○) *to calculated* (□) *quantities*

Microstructural feature	Dimensions of symbols (arbitrarily expressed in terms of millimeters)			
	mm^0	mm^{-1}	mm^{-2}	mm^{-3}
Points	P_P	P_L →	P_A →	P_V
Lines	L_L	L_A	L_V ↗	—
Surfaces	A_A	S_V ↗	—	—
Volumes	V_V	—	—	—

All symbols are interrelated by equations except between the first column and those on the right. By means of a rather simple, yet powerful, choice of particle or cell diameter, this gap can be bridged. This "diameter" is the mean intercept length; we shall discuss it later, in Section 4.2.

The equations relating the quantities in Table 2.1 may be expressed as follows:

$$V_V = A_A = L_L = P_P \qquad mm^0 \qquad (2.1)$$

$$S_V = (4/\pi)L_A = 2P_L \qquad mm^{-1} \qquad (2.2)$$

$$L_V = 2P_A \qquad mm^{-2} \qquad (2.3)$$

$$P_V = \tfrac{1}{2}L_V S_V = 2P_A P_L \qquad mm^{-3} \qquad (2.4)$$

These are the basic equations for determining the magnitude of lineal, areal, and volume elements from measurements made on a test section (the plane of polish).

It is important to note that all these equations are exact, in the sense that no simplifying assumptions are required as to size, shape, spacing, etc., of features in the microstructure. There is a requirement, however, that the measurements be made randomly or with statistical uniformity and that the test sections be representative of the entire sample. Then the calculated quantity can be obtained as precisely as desired, depending on the number of measurements taken.

Dimensions are indicated arbitrarily in millimeters in order to emphasize the dimensionality of the symbols used in the equations. By consistent adherence to these dimensions, we can ensure proper dimensions in the equations themselves.

2.3 VOLUME FRACTION V_V

The volume fraction is one of the most important quantities required in quantitative stereological analysis. Random two-dimensional section planes may be utilized in different ways [24] to obtain V_V. One method involves measurements of areas; another, the measurement of lineal intercepts; and a third, the counting of points. These three main methods are called *areal analysis*, *lineal analysis*, and *point counting*, respectively.

We wish to determine the volume fraction, V_V, from any of the other ratios A_A, L_L, or P_P. An example of the equivalence of the latter three ratios for a simple case [8] is given in Fig. 2.1. The areal ratio A_A of the randomly arranged squares to the total area is seen to be 20%. Also the lineal ratio L_L and the point ratio P_P approach 20%.

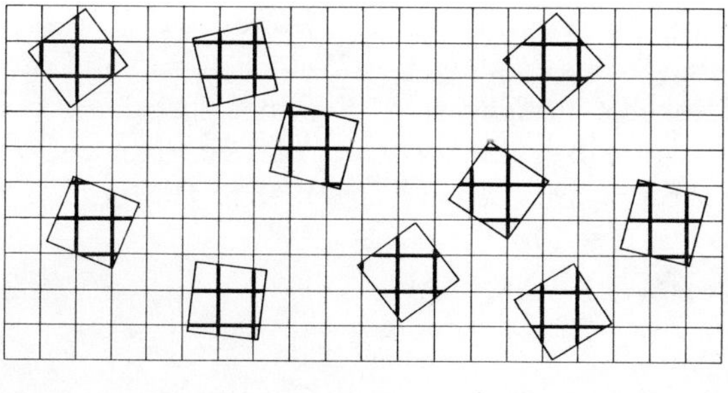

Fig. 2.1. Equivalence of areal, lineal, and point ratios (after Rosiwal [8]).

2.3.1 Areal Fraction A_A

The French geologist, Delesse (1), was the first to indicate the equivalence of V_V and A_A. To quote Delesse:

> *Let us assume that the volume occupied by the rock is related to a system of coordinates, and let p be the area occupied by one of the constituent minerals in a section formed by a plane parallel to xy. To obtain the exact volume occupied by this mineral in the rock, it would be necessary to know the successive values of p for a series of infinitely close planes parallel to xy. The integral $\int p\, dz$ would then give the volume sought.*

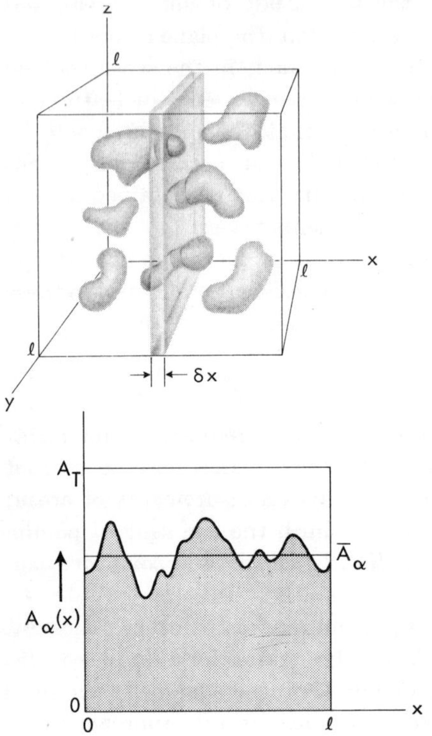

a) Cube containing particles of α-phase cut by a thin slice.

b) Variation of area of α-phase in thin slice, $A_\alpha(x)$, as a function of slice position x.

Fig. 2.2. Model for deriving the relationship $V_V = A_A$ (after Weibel [4]).

Delesse indicated the principle involved; others have given us detailed derivations [2, 3, 4, 5]. Here we borrow from the derivations employed by Weibel [4] and McLean [25] because of their directness and ease of visualization. Figure 2.2(a) depicts a test cube of volume $V_T = l^3$ which contains irregularly shaped particles (e.g., of α-phase), and a thin slice of thickness δx and area $A_T = l^2$ parallel to the zy-plane. The volume of α-phase in the thin slice is given by

$$\delta V_\alpha = l^2 \delta x (V_V)_\alpha \text{ mm}^3 \tag{2.5}$$

and, for δx sufficiently small, by

$$\delta V_\alpha = A_\alpha(x) \delta x \text{ mm}^3 \tag{2.6}$$

where $A_\alpha(x)$ is the area of α-phase on the slice surface as a function of slice position x. $A_\alpha(x)$ varies with x, as indicated in Fig. 2.2(b), and the average value between 0 and l is

$$\bar{A}_\alpha = (1/l) \int_0^l A_\alpha(x)\, dx \text{ mm}^2. \tag{2.7}$$

In the limit, the total volume of α-phase in the cube is given by

$$V_\alpha = \int_0^l dV_\alpha = \int_0^l A_\alpha(x)\, dx, \tag{2.8}$$

which becomes

$$V_\alpha = l\bar{A}_\alpha \tag{2.9}$$

when one introduces Eq. (2.7). Dividing both sides by V_T, we obtain

$$V_\alpha / V_T = \bar{A}_\alpha / A_T \tag{2.10}$$

or in general,

$$\boxed{V_V = \bar{A}_A = A_A} \tag{2.11}$$

where, as a rule, we shall omit bars in order to simplify the notation. Equation (2.11) states that the average areal fraction A_A, determined on sections through a volume, represents an estimate of the volume fraction V_V of the phase under investigation.

Although only one coordinate was used to specify the location of the section plane, the final form of Eq. (2.11) is independent of any such requirement. Some authors postulate the need for randomly oriented section planes [5, 6], while others claim that the section plane need merely be displaced parallel to itself [2], i.e., that the section plane need not be randomized with respect to angle. It appears that either procedure may be chosen for random structures. However, in oriented structures, parallel test sections may introduce special effects [7], depending on the particular structure under consideration. Indeed, random traverse intervals have been recommended for some oriented structures [3] to avoid special problems that could arise with equispaced parallel traverses.

For most purposes, systematic parallel sectioning (or traverses) will usually be more precise than "random" choices. In other words, for the same number of sections, systematic sectioning assures a more even distribution of information throughout the sample space. Saltykov [5] expresses the requirement in much the same way, calling for randomness, but with "statistical uniformity."

2.3.2 Lineal Fraction L_L

Delesse [1] also mentioned the possibility of determining volume fractions from lineal ratios, but felt that the latter would be less accurate than areal ratios in nonhomogeneous rocks. For this reason, perhaps, half a century elapsed before the subject was given further serious scrutiny. In 1898, the German geologist Rosiwal [8] published his classic paper on the equivalence of lineal and volume fractions. We quote only the qualitative idea expressed by Rosiwal:

> *[if] . . . a rock "fiber," an extremely thin prism, is introduced . . . the measured longitudinal lengths of the individual components show the same relative amounts as in the composition of the rock in question . . . . This "material measuring line" . . . shows immediately, by the ratio of the cumulative lengths across a selected mineral to the total line length over the cross section, that the numerical quantitative ratio of the third dimension is directly reduced to the first dimension.*

The equivalence of lineal fraction and volume fraction has been developed mathematically in several ways [2, 4, 5]. The derivation of this relationship follows the same

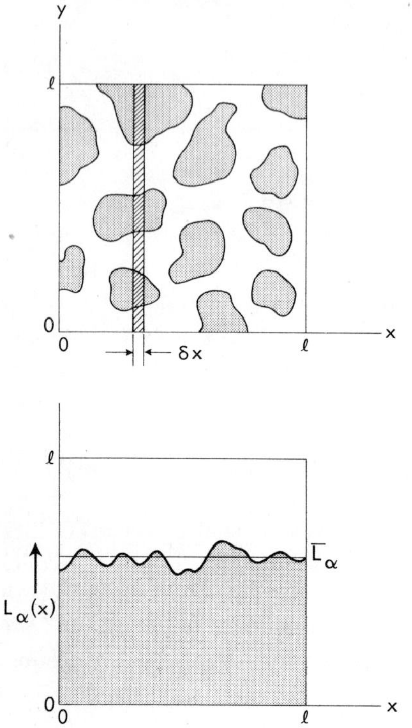

a) Section plane through particles of α-phase cut by thin strip.

b) Variation of length of α-phase in thin strip, $L_\alpha(x)$, as a function of strip position x.

Fig. 2.3. Model for deriving the relationship $V_V = L_L$ (after Weibel [4]).

scheme as we adopted in the derivation of Eq. (2-11). Instead of a cube, consider a section plane through an aggregate, such as was shown in Fig. 2.2(a). Figure 2.3(a) represents a square test section of area $A_T = l^2$ parallel to the xy-plane which contains irregularly shaped areas (e.g., of α-phase), and a thin strip of width δx across the section parallel to the y-axis. The area of α-phase in the thin strip is given by

$$\delta A_\alpha = l\, \delta x\, (A_A)_\alpha \text{ mm}^2 \tag{2.12}$$

and, for δx sufficiently small, by

$$\delta A_\alpha = L_\alpha(x)\, \delta x \text{ mm}^2, \tag{2.13}$$

where $L_\alpha(x)$ is the length of α-phase intercepted by the strip as a function of strip position x. $L_\alpha(x)$ may vary with x, as indicated in Fig. 2.3(b), and the average value between 0 and l is

$$\bar{L}_\alpha = (1/l) \int_0^l L_\alpha(x)\, dx \text{ mm}. \tag{2.14}$$

In the limit, the total area of α-phase in the test section is given by

$$A_\alpha = \int_0^l dA_\alpha = \int_0^l L_\alpha(x)\, dx, \tag{2.15}$$

which becomes

$$A_\alpha = l\bar{L}_\alpha \tag{2.16}$$

when we introduce Eq. (2.14). Dividing both sides by A_T, we obtain

$$A_\alpha/A_T = \bar{L}_\alpha/l \tag{2.17}$$

or, in general,

$$\boxed{A_A = \bar{L}_L = L_L.} \tag{2.18}$$

Thus we see that the fractional length of lines through a phase in an aggregate structure gives an estimate of the areal fraction of the same phase. The question of bias and consistency in a line-area analysis has been treated in detail by Chayes [3]. Fortunately, the line-area bias disappears if the total length of each individual traverse is held constant.

The equivalence of length and volume ratios

$$\boxed{V_V = L_L} \tag{2.19}$$

follows from Eqs. (2.11) and (2.18), or can readily be demonstrated by placing a test line randomly within the cube [2]. The proof is similar to those given above and will be omitted here.

2.3.3 Point Fraction P_P

The previous sections treated the equalities of V_V, A_A, and L_L. We now wish to derive the equations relating these three quantities to the ratio P_P, which is the ratio of the number of points that fall in the phase of interest to the total number of test points.

The earliest applications of point ratios as estimators of areal ratios appear to have been made by Thomson [9] in 1930 and Glagolev [10] in 1933. They used one- and two-dimensional point grids in order to determine the fractional number of points falling within the boundaries of a constituent in the section plane. We shall next demonstrate the equivalence of the point fraction to V_V, A_A, and L_L.

If a test point is placed at random within a test volume V_T containing a constituent α of volume V_α, the probability of the point falling within the constituent is equal to the ratio of the two volumes, or

$$\mathscr{P} = V_\alpha/V_T. \tag{2.20}$$

For many randomly placed test points, P_T, the expected value $E(P_\alpha)$ of the number of points lying within the constituent is given by

$$E(P_\alpha) = P_T\mathscr{P} = P_T(V_\alpha/V_T). \tag{2.21}$$

Therefore

$$V_\alpha/V_T = E(P_\alpha)/P_T \tag{2.22}$$

or, in the present notation,

$$\boxed{V_V = P_P.} \tag{2.23}$$

If a test point is placed randomly in a planar region of area A_T, the probability of hitting an area A_α within that region is defined as

$$\mathscr{P} = A_\alpha / A_T. \tag{2.24}$$

For many randomly applied test points, we have the situation depicted in Fig. 2.4. Following the same procedure indicated above, we obtain

$$\boxed{A_A = P_P.} \tag{2.25}$$

For a random dispersion of points along a line, the result is

$$\boxed{L_L = P_P.} \tag{2.26}$$

Equations (2.23), (2.25), and (2.26) are also valid for nonrandom arrays of test points, such as the intersections of a square grid, or the equal spacings set off by an automatic point-counting stage. It is necessary, however, that the grid array be placed randomly on the section plane.

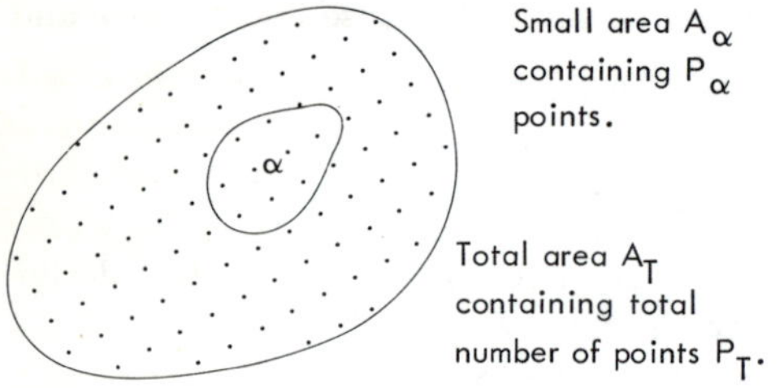

Fig. 2.4. Model for deriving the relationship $A_A = P_P$ (after Chayes [3]).

When the points are distributed in a regular fashion (e.g., at the intersections of a square lattice), the analysis is called a systematic point count. Hilliard and Cahn [11] have examined various point-counting procedures and concluded that a systematic point count (i.e., measurement of P_P) using a two-dimensional point grid should be the most efficient method of volume fraction analysis. However, to ensure optimum efficiency, the number of occupied points should about equal the number of structural features lying under the grid [11]. For further discussions, the reader may consult other papers [20, 21].

2.4 SURFACE-TO-VOLUME RATIO S_V

We turn now to methods for obtaining the surface-to-volume ratio of a system of surfaces in a volume. The surfaces may be true two-dimensional features unrelated to any other feature of the microstructure, or they may be the boundaries between space-filling cells or the interfaces between particles and matrix. In addition to these internal surfaces, we also consider external surfaces, such as the jagged irregular surface of an oxidized wire, a fracture surface, or the surface of an isolated particle. Methods that apply to external surfaces are not necessarily applicable to internal surfaces, and will be described separately.

First let us consider Eqs. (2.2), which relate the internal surface area in a volume to measurements on a test section. Next we shall introduce the equations of Tomkeieff [14, 17] and Chalkley [18, 23] for external as well as internal surfaces. Reference to the sketches in Table 2.2 will help to clarify the following discussion.

2.4.1 Two-Dimensional Internal Surfaces

The basic equation for obtaining the area of surfaces in a volume is

$$S_V = 2P_L \text{ mm}^2/\text{mm}^3, \tag{2.2}$$

which was derived by Saltykov [5] in 1945, and later by Smith and Guttman [6] in 1953. The importance of this equation can be appreciated from the fact that it has been rederived at least eight times [4, 5, 6, 12, 13, 17, 22, 26] since 1945.

Equation (2.2) applies to systems of surfaces with any configuration. It is as valid for systems of interconnected surfaces as for systems of discontinuous, separated, or bounded surfaces. In order to evaluate the extent of boundary areas in space, we need only perform a simple intersection (P_L) count with boundary traces on a section plane. If complete randomness between intersecting test lines and surfaces is not achieved in one test plane, then several test planes may be required.

In the derivation of Eq. (2.2) presented below, we attempt to eliminate the unsatisfactory aspects of previous derivations, while retaining the desirable features of simplicity, directness, and ease of visualization. Weibel's [4] treatment is especially good and we borrow heavily from his approach. We employ summations instead of integrals whenever we can obtain additional clarity thereby.

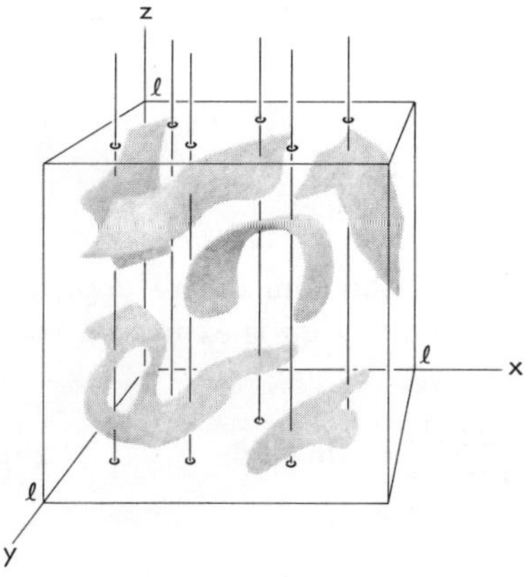

Fig. 2.5. Model for deriving the relationship $S_V = 2P_L$. This cube contains random surfaces cut by random vertical test lines.

Let us consider a test cube of edge lengths l and volume $V_T = l^3$ enclosing a system of surfaces oriented randomly throughout the cube. The surfaces may be planar or curved, continuous or interrupted, isolated or connected, as represented in Fig. 2.5. A set of N vertical test lines of total length $L_T = Nl$ is passed randomly through the cube, cutting horizontal planes through the cube with density N/l^2.

The total surface area S is divided into n elementary units of surface area δS so that $S = n\,\delta S$. The normals to each elementary area form angles θ_i to the vertical test lines, and the areas of the projections of the elementary areas on a horizontal plane equal $\cos \theta_i \, \delta S$. Thus the fraction of test lines intersecting the elementary areas is $\delta S \cos \theta_i / l^2$.

We set up an expression for the density of intersections of the test lines with the elementary areas, then add the individual contributions to obtain the total number of intersections with the total surface area. If P_i is the number of intersections associated with each elementary area, then the expected value of the total number of intersections with the entire surface is

$$E(P) = \sum^n P_i = N \sum^n \delta S \cos \theta_i / l^2 = \frac{N \, \delta S}{l^2} \sum^n \cos \theta_i. \tag{2.27}$$

Since the elementary units of surface are oriented randomly, every value of θ_i has equal likelihood and $(1/n) \sum^n \cos \theta_i$ equals the average value, $\overline{\cos \theta}$. Making appropriate substitutions in Eq. (2.27) gives us

$$E(P) = \left(\frac{N}{l^2}\right) (\delta S n) \, \overline{\cos \theta} = \left(\frac{L_T}{l^3}\right) S \, \overline{\cos \theta}, \tag{2.28}$$

the expected value of the total number of intersections of the test lines with the system of surfaces.

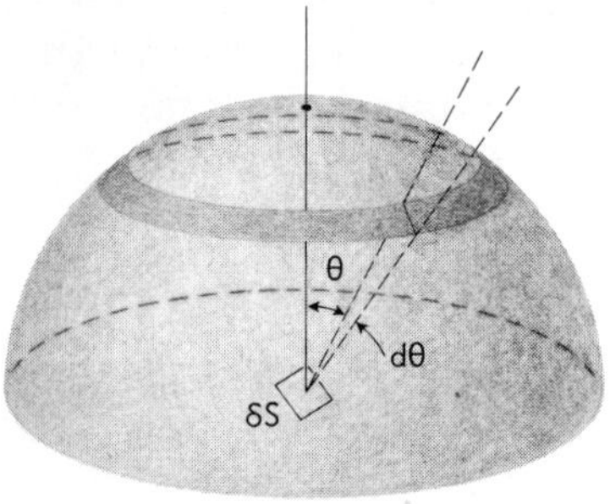

Fig. 2.6. Geometry involved in the determination of the probability that random normals lie between θ and $\theta + d\theta$.

The evaluation of $\overline{\cos \theta}$ can be visualized by means of the hemisphere depicted in Fig. 2.6. The probability that normals lie between θ and $\theta + d\theta$ is expressed [16] by

$$\mathscr{P}(\theta) \, d\theta = \frac{\text{area of zone}}{\text{area of hemisphere}} = \frac{2\pi r^2 \sin \theta \, d\theta}{2\pi r^2} = \sin \theta \, d\theta \tag{2.29}$$

and the average value of $\cos \theta$ is

$$\overline{\cos \theta} = \int_0^{\pi/2} \mathscr{P}(\theta) \cos \theta \, d\theta = \int_0^{\pi/2} \sin \theta \cos \theta \, d\theta = \tfrac{1}{2}. \tag{2.30}$$

Rearrangement of Eq. (2.28) gives

$$S/V_T = 2E(P)/L_T,$$

which yields the required relationship, in our notation,

$$\boxed{S_V = 2P_L} \ \text{mm}^2/\text{mm}^3. \tag{2.31}$$

The pertinent measurements and equations for determining the surface-to-volume ratios in different types of systems are summarized in Table 2.2.

2.4.2 Internal Surfaces of Finite Thickness

Instead of true two-dimensional surfaces separating space-filling cells or grains, there could be a thin boundary layer separating the cells. The surface-to-volume ratio in such a case can be treated in two ways: i.e., exactly, in the sense that a thin layer introduces double the interfacial area; or approximately, through the assumption that the thin layer is essentially two-dimensional.

If we make the latter assumption, only one intersection is counted each time the boundary layer is pierced by the test line. In such a case, $P_L = N_L$ and we must use the equations in part (a) of Table 2.2. This procedure is equivalent to measuring the area of an imaginary central plane through the boundary layer. If we treat the boundary as having thickness, then $P_L = 2N_L$ and we use equations similar to those given in part (b).

2.4.3 Interfaces of Particles Dispersed in a Matrix

The treatment accorded dispersed particles in a matrix, presented in part (b) of Table 2.2, is no different in principle from that given in part (a). The particle interfaces are usually taken to be simple two-dimensional surfaces, and one intersection is counted each time the particle boundary is penetrated by the test line. Alternatively, instead of counting intersections of interfaces, we can count the number of interceptions of particles per unit length of test line, N_L. If convex particles only are involved, this result can be doubled to get P_L.

When concave particles occur, the test line will occasionally penetrate a particle more than once, yielding more than one intercept length. In such cases, we merely count each discrete intercepted segment individually. This procedure maintains the validity of the relationship, $P_L = 2N_L$. Thus our definition of N_L is expressed as the number of *interceptions* of particles per unit length of a straight test line, rather than the number of *particles* intercepted per unit test line. Note that this definition is equally valid for convex particles as well.

In terms of N_L as defined above, Eq. (2.31) becomes

$$\boxed{(S_V)_\alpha = 4N_L} \quad \text{mm}^2/\text{mm}^3, \tag{2.32}$$

where $(S_V)_\alpha$ refers to the total particle interface area per unit volume of the total system (α-particles plus matrix).

When one is making measurements on real systems of particles in a matrix, using a test line of finite width, there is always a finite probability of getting tangent hits on particle perimeters. This is especially true in materials that are improperly sectioned or prepared. Over-etching the test section of particulate alloys is frequently responsible for enhanced tangent hits. This possibility obviously becomes less important with sharply defined interfaces on the test section.

TABLE 2.2 *Three-dimensional cells and particles*

(a) Space-filling contiguous cells of one phase ($V_V = 1$)	(b) Dispersed particles of α-phase $[(V_V)_\alpha < 1]$ in a matrix phase	(c) Isolated single particle (volume $= V$) not associated with a matrix
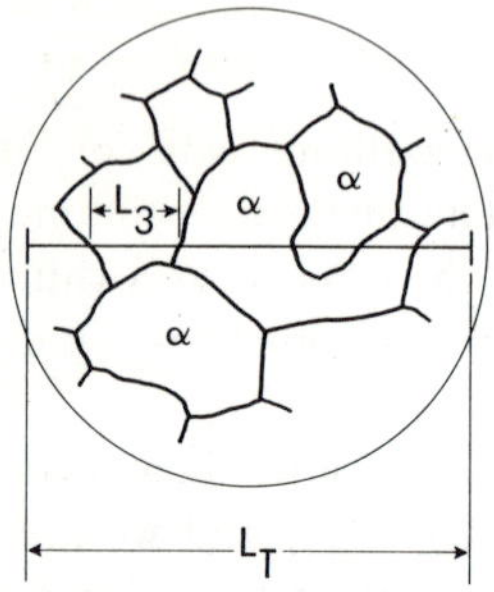	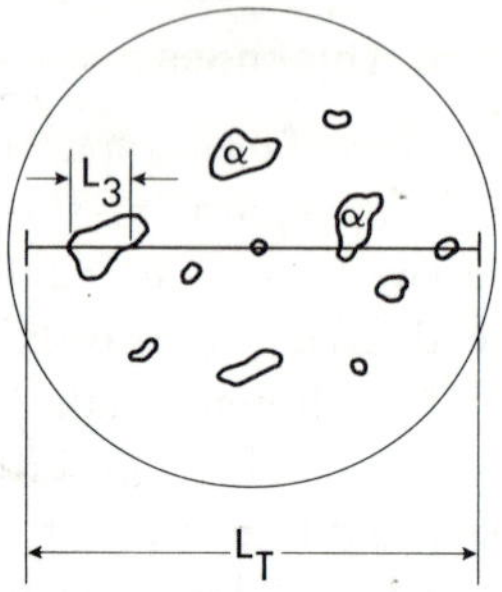	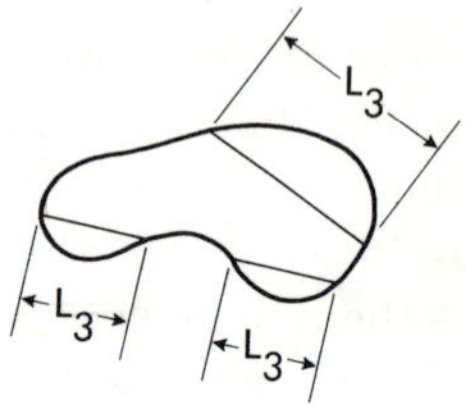
Two-dimensional internal surfaces* with sharing between contiguous cells. $N_L = \frac{1}{2} + 4 + \frac{1}{2} = 5$ $P_L = 5$ $P_L = N_L = 1/\bar{L}_3$	Two-dimensional interfaces† not shared with other α-particles. $N_L = 4$ $P_L = 8$ $P_L = 2N_L = 2(V_V)_\alpha/\bar{L}_3$	Two-dimensional external particle surface S.§ $N_L = 1, 2, \ldots$ $P_L = 2, 4, \ldots$ $P_L = 2N_L$
Saltykov [5,27] $S_V = 2P_L = 2N_L$	$(S_V)_\alpha = 2P_L = 4N_L$	$\dfrac{S}{V} = \dfrac{2P_L}{P_P}$
Tomkeieff [14] $\bar{L}_3 = 2\dfrac{\bar{V}}{\bar{S}} = \dfrac{2}{S_V}$	$\bar{L}_3 = \dfrac{4\bar{V}_\alpha}{\bar{S}_\alpha} = \dfrac{4(V_V)_\alpha}{(S_V)_\alpha}$	$\bar{L}_3 = \dfrac{4V}{S}$
Chalkley [18] $\bar{L}_3 = lp/2P$	$\bar{L}_3 = \dfrac{lp}{P} = \dfrac{4\bar{V}_\alpha}{\bar{S}_\alpha}$	$\bar{L}_3 = \dfrac{lp}{P} = \dfrac{4V}{S}$
*Internal surface area referred to test volume (S_V) or to mean cell volume ($\bar{S}/\bar{V}$).	†Interface area of α-particles referred to test volume $(S_V)_\alpha$ or to mean particle volume ($\bar{S}_\alpha/\bar{V}_\alpha$).	§External surface area referred to particle volume (S/V).

Explanation of symbols in Table 2.2

S_V = Total internal surface area per unit test volume
P_L = Number of intersections with surfaces per unit length of test line
N_L = Number of interceptions of cells or particles per unit length of test line
$\bar{L}_3$ = Mean intercept length of three-dimensional bodies $= L_L/N_L = A_A/N_L = V_V/N_L$
$\bar{V}_\alpha$ = Mean volume of α-particles $= (V_V)_\alpha/N_V$
$\bar{S}_\alpha$ = Mean surface area of α-particles $= (S_V)_\alpha/N_V$
V = Volume of individual particle
S = External surface area of individual particle
l = Constant length of short test lines, thrown randomly on microstructure
p = Number of (end) points of l-lines that fall in phase of interest
P = Number of intersections with surfaces made by l-lines
= Length of test line

2.4.4 Tomkeieff's Equation for Three-Dimensional Objects

In 1945, Tomkeieff proposed a relationship between the surface-to-volume ratio of an isolated, convex body and the mean intercept length through the body. It was later claimed, however, that the requirement of convexity is unduly restrictive [6].

Tomkeieff merely stated the final relationship without proof [14], so it will be derived here. Moreover, we shall generalize the derivation so as to include both concave and convex particles, and systems of such particles.

The mean intercept length is defined for a single particle as

$$\bar{L}_3 = \frac{1}{N}\sum^{N} (L_3)_i \text{ mm}, \tag{2.33}$$

where the subscript 3 refers to the dimensionality of the particle, and the $(L_3)_i$ are the intercepted lengths from N random penetrations of the particle by a straight test line. Note that this definition allows for the possibility of a particle's being intercepted more than once by a single penetration of the test line. Thus the value of $\bar{L}_3$ is applicable to concave bodies as well as to convex ones.

Referring the terms in the right-hand side of Eq. (2.33) to a test line of length L_T, we write

$$\bar{L}_3 = L_L/N_L \text{ mm}, \tag{2.34}$$

which applies equally to one particle intercepted many times, or to a system of many particles intercepted by a random test line. Substituting Eqs. (2.19) and (2.32), we obtain for many α-particles

$$\boxed{\bar{L}_3 = 4(V_V)_\alpha/(S_V)_\alpha} \text{ mm}. \tag{2.35}$$

It is seen that $\bar{L}_3$ has a unique value, regardless of convexity or concavity, since both $(S_V)_\alpha$ and $(V_V)_\alpha$ are fixed quantities for any given system of particles, independent of particle shape.

For one particle, we can write

$$\boxed{\bar{L}_3 = 4\frac{V}{S}} \text{ mm}, \tag{2.36}$$

which is the equation given by Tomkeieff. As an example of the application of Eq. (2.36), consider a sphere of diameter D for which $V/S = D/6$. If we know the diameter, we can calculate $\bar{L}_3$ from $\bar{L}_3 = 2D/3$ (see Table 4.1).

The average volume and surface area per particle may be required instead of the total quantities. Dividing both sides of Eq. (2.35) by N_V, the number of α-particles per unit volume, we obtain

$$\bar{L}_3 = 4\frac{(V_V)_\alpha/N_V}{(S_V)_\alpha/N_V} \tag{2.37}$$

or

$$\boxed{\frac{\bar{S}_\alpha}{\bar{V}_\alpha} = \frac{4}{\bar{L}_3}} \text{ mm}^2/\text{mm}^3, \tag{2.38}$$

where $\bar{S}_\alpha$ and $\bar{V}_\alpha$ refer, respectively, to the average surface area and the average volume per particle. $\bar{L}_3$ is averaged over all particles and is an experimentally obtainable quantity.

Note that we have shifted the base of reference from a test volume V_T in the total system [$(S_V)_\alpha$ in Eq. 2.32] to the average volume $\bar{V}_\alpha$ of the particles [$\bar{S}_\alpha/\bar{V}_\alpha$ in Eq. 2.38]. This distinction becomes important when we are dealing with separate particles or cells.

2.4.5 Chalkley's Equation for Three-Dimensional Objects

The method devised by Chalkley, *et al.* [18] for obtaining the average surface-to-volume ratio of a system of particles involves both a point count and an intersection count on random sections through the particles. Since the derivation of this equation appears in a rather inaccessible publication [23], we derive Chalkley's equation below.

Starting with Eq. (2.34), we have

$$\bar{L}_3 = L_L/N_L = P_P/\tfrac{1}{2}P_L \text{ mm}, \tag{2.39}$$

where P_P is the fractional number of points that hit the particle sections and P_L is the number of intersections with traces of the particle surfaces per unit length of test lines.

Equation (2.39) is further modified to

$$\bar{L}_3 = 2\left(\frac{p/P_T}{P/L_T}\right) \text{ mm}, \tag{2.40}$$

where p is the number of points that fall in the particulate phase and P is the number of intersections of particle interfaces by the test lines. P_T and L_T are the total number of test points and total length of test line, respectively.

The total test line length, L_T, is divided into m short segments of equal length l, and since each segment has two end points, there are $P_T = 2m$ end points available for point counting. The short segments are dropped randomly on the particle sections, intersecting the traces of the particle surfaces P times out of the total length of ml. The lower-case p represents the number of end points that hit within the particle sections.

Substituting these quantities into Eq. (2.40) gives us

$$\bar{L}_3 = 2\left(\frac{p/2m}{P/ml}\right) = pl/P \text{ mm}. \tag{2.41}$$

Simplifying, and introducing Eq. (2.38) for $\bar{L}_3$, we obtain Chalkley's equation for a system of particles, say, of α-phase

$$\boxed{\bar{S}_\alpha/\bar{V}_\alpha = 4P/pl} \text{ mm}^2/\text{mm}^3, \tag{2.42}$$

where P is the number of intersections of the line segments with the traces of α-particle surfaces, p is the number of (end) points that hit within the particle sections, and l is the length of the randomly thrown test line segments. Note that the mean

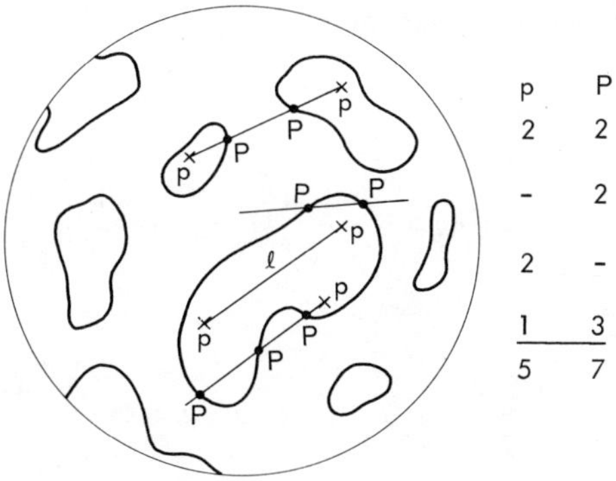

Fig. 2.7. Counting possibilities according to the Chalkley [18] method.

particle surface area is referred to the mean particle volume. Figure 2.7 illustrates various counting possibilities when test segments of length l are thrown randomly on a test plane containing particle sections. The grids in Figs. 1.2(c) and 2.8 are also adapted for simultaneous measurements of intersection and point counts.

The surface-to-volume ratio for a single particle is obtained by substituting Eq. (2.36) into Eq. (2.41). These relationships are tabulated in Table 2.2, and others may be formulated by combining appropriate terms in the vertical columns. Chalkley's equation for space-filling cells reduces to $\bar{L}_3 = 1/P_L$.

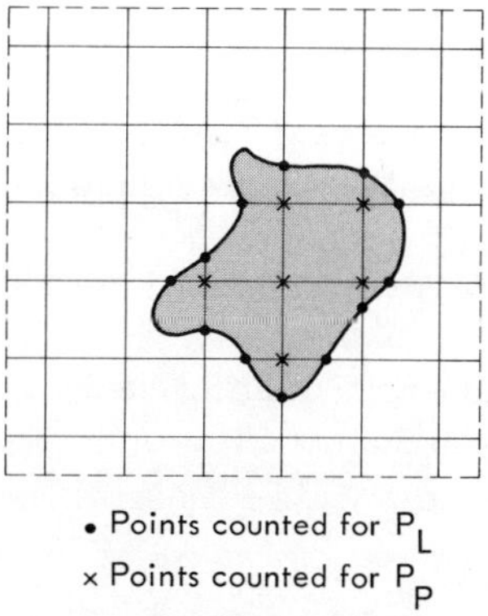

Fig. 2.8. Grid for simultaneous measurement of P_L and P_P [27].

Recently Saltykov [27] described a procedure for obtaining the surface-to-volume ratio of separate particles. The method is essentially the same as Chalkley's, in that both a point count and an intersection count are conducted simultaneously. The particles are embedded randomly in a resin matrix, then sectioned. A test grid, similar to that shown in Fig. 2.8, is applied randomly to the plane of polish. The values of P_P and P_L are determined by using the grid intersections for the P_P count and the grid lines for the P_L count. Note that this grid is constructed so that each grid intersection point occupies a central position in an area equal to the small squares outlined by the grid lines (see Problem 2.1).

If the sum of the surface areas and volumes of the particles are $\sum S_i$ and $\sum V_i$, respectively, we can write

$$\frac{\sum S_i}{\sum V_i} = \frac{\sum S_i/V_T}{\sum V_i/V_T} = \frac{S_V}{V_V} = \frac{2P_L}{P_P} \text{ mm}^{-1} \tag{2.43}$$

for a fixed number of loose particles embedded in an arbitrary material of total volume V_T.

For a single particle of surface S and volume V, we have

$$\boxed{\frac{S}{V} = \frac{2P_L}{P_P}} \text{ mm}^2/\text{mm}^3. \tag{2.44}$$

By combining the two measurements P_L and P_P, we avoid the problem that arises when one introduces an arbitrary test volume. It is readily apparent that with a fixed number of separate, isolated particles, there is a fixed value for their volume $\sum V_i$. However, if these particles are embedded in an arbitrary matrix, the ratio $\sum V_i/V_T$ can assume any magnitude, depending on our choice of V_T. For separate, isolated particles, consequently, we refer the surface area of the particles to the particle volume itself, that is, $\sum S_i/\sum V_i$, in order to obtain a fixed reference volume. For a system of α-particles uniformly dispersed in a matrix, on the other hand, we know (or assume) that the ratio $(S_V)_\alpha$ will remain constant no matter how great the test volume(s).

Note that Table 2.2 pertains especially to the two-dimensional surfaces of three-dimensional objects, such as cells, grains, and particles. The main difference among the equations in Table 2.2 is the factor of two, which arises because of sharing or nonsharing of surfaces between cells. Tomkeieff [14] and Chalkley, *et al.* [18] also give equivalent expressions for two-dimensional figures; we shall discuss these in the next section.

2.5 LENGTH OF LINE PER UNIT AREA L_A

Let us next consider the important equation that relates the length of lineal elements in a plane to their intersection with a test line,

$$L_A = \left(\frac{\pi}{2}\right) P_L \text{ mm/mm}^2, \tag{2.2b}$$

which has been listed previously as one of the basic equations. The quantity L_A is a basic microstructural parameter that serves usefully either as is or when manipulated into other forms. In corrosion studies, for example, one would want to know the length of grain-boundary traces exposed to the corrodent. The value of L_A might be required for the total or average length of crack traces in fracture studies, although a more significant parameter than L_A would be S_V, since cracks are essentially two-dimensional in nature.

2.5.1 Systems of Lines in a Plane

The derivation of Eq. (2.2b) follows much the same course as the derivation of Eq. (2.31). Again we utilize the scheme advanced by Weibel [4], although other derivations have been presented [5, 6, 19, 22, 24].

Given a randomly oriented system of lines in a plane, let us consider a square test area A_T of edge length l. A set of N vertical test lines of total length $L_T = Nl$ is passed randomly through the test area, cutting horizontal lines through the square with density N/l. The system of lines is divided into n straight elementary segments of length δL and the total line length in the system is $L = n\,\delta L$. The elementary segments form angles θ_i to the vertical test lines, and the lengths of the projections of the elementary segments on a horizontal line are $\delta L \sin\theta_i$. Thus the fraction of test lines intersecting the elementary segments is $\delta L \sin\theta_i/l$.

If P_i is the number of intersections of each elementary segment by the test lines, then the expected value of the total number of intersections with the entire system of lines is

$$E(P) = \sum^{n} P_i = N \sum^{n} \delta L \sin\theta_i/l = \frac{N\,\delta L}{l} \sum^{n} \sin\theta_i. \tag{2.45}$$

Since the elementary segments are oriented randomly, every value of the angles θ_i has equal likelihood of existence and $(1/n)\sum^n \sin\theta_i$ equals the average value, $\overline{\sin\theta}$. Making appropriate substitutions, we obtain

$$E(P) = \frac{N\,\delta L\, n}{l}\,\overline{\sin\theta} = \frac{L_T}{l^2}\, L\,\overline{\sin\theta}. \tag{2.46}$$

The probability that elementary segments are oriented between θ and $\theta + d\theta$ is equal to the fraction of the perimeter of a circle that is occupied by this orientation range. From the symmetry involved, only one quadrant of the circle need be considered, giving for the probability

$$\mathscr{P}(\theta)\,d\theta = \frac{r\,d\theta}{\pi r/2} = \frac{2}{\pi}\,d\theta.$$

Thus the average value of $\sin\theta$ is

$$\overline{\sin\theta} = \int_0^{\pi/2} \mathscr{P}(\theta)\sin\theta\,d\theta = \frac{2}{\pi}\int_0^{\pi/2} \sin\theta\,d\theta = \frac{2}{\pi}. \tag{2.47}$$

Rearrangement of Eq. (2.46) gives

$$\frac{L}{A_T} = \left(\frac{\pi}{2}\right)\frac{E(P)}{L_T}\ \text{mm/mm}^2 \tag{2.48}$$

or, in our notation,

$$\boxed{L_A = \left(\frac{\pi}{2}\right) P_L}\quad \text{mm/mm}^2. \tag{2.49}$$

The result is not changed if curved test lines are used instead of straight test lines, as long as randomness of intersecting angles is attained. Hence test lines may have any shape or curvature and the value of P_L is independent of the shape of the test line. In practice, in order to obtain all possible angles of intersection, one sometimes finds it convenient to use concentric circles or a spiral for the test line. When this is done, the test lines change direction and angle with respect to the system of lines in the plane, even when the lines are not random but have a preferred orientation.

Fig. 2.9. Section of a human oviduct with superimposed test grid [21].

An unusually complex system of lines in a plane is shown in Fig. 2.9, which represents a normal section through the tubular, cylindrical surfaces of a human oviduct. Hennig [21] has superimposed a parallel array of test lines on this structure and indicated the number of intersections. The length of the test lines depends on the magnification, of course. It should be noted that the test area is defined by the size of the test grid; however, a more meaningful test (or reference) area is that enclosed by the perimeter of the oviduct. In this connection, the comments made in the paragraph following Eq. (2.44) apply to this two-dimensional case as well.

The pertinent measurements and equations for determining the length-to-area ratio in different systems of lines in a plane are summarized in Table 2.3. In addition to the various forms of Eq. (2.49), we include the equations of Tomkeieff and of Chalkley; we shall discuss these next.

2.5.2 Tomkeieff's Equation for Two-Dimensional Figures

An equation for the ratio of perimeter length to area of isolated two-dimensional convex figures was given without derivation by Tomkeieff [14]. He expressed this ratio in terms of the mean intercept length, defined here as $\bar{L}_2 = (L_L)_\alpha / N_L$, where the subscript 2 refers to two-dimensional figures. In order to include concave figures in this procedure, we adopt the more general definition of N_L given previously in Sections 2.4.3 and 2.4.4. The results may also be extended to encompass systems of discrete figures as well as area-filling contiguous figures in a plane.

The Tomkeieff expression in the two-dimensional case [14] is derived much the same way as Eq. (2.36) for three-dimensional particles. For a system of separated closed loops, the mean intercept length is

$$\bar{L}_2 = \frac{(L_L)_\alpha}{N_L} = \frac{2(A_A)_\alpha}{P_L} \text{ mm}, \tag{2.50}$$

TABLE 2.3 *Two-dimensional planar figures and linear networks*

(a) Area-filling contiguous figures of one phase ($A_A = 1$)	(b) Separated figures of α-phase $[(A_A)_\alpha < 1]$ in a matrix phase	(c) Isolated single figure (area $= A$) not associated with a matrix
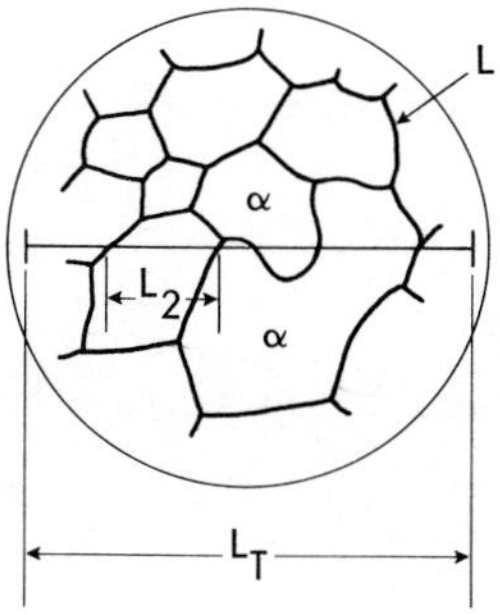	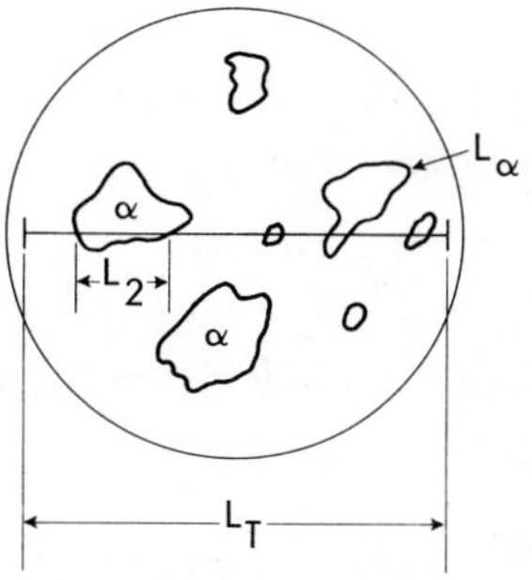	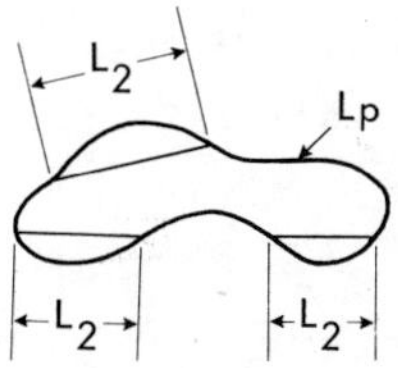
One-dimensional internal perimeters* with sharing between contiguous figures	One-dimensional perimeters† not shared with other α-regions	One-dimensional external perimeter§ length, L_p
$N_L = \frac{1}{2} + 4 + \frac{1}{2} = 5$ $P_L = 5$ $P_L = N_L = 1/\bar{L}_2$	$N_L = 4$ $P_L = 8$ $P_L = 2N_L = 2(A_A)_\alpha/\bar{L}_2$	$N_L = 1, 2, \ldots$ $P_L = 2, 4, \ldots$ $P_L = 2N_L$
Saltykov [5,27] $L_A = \frac{\pi}{2} P_L = \frac{\pi}{2} N_L$	$(L_A)_\alpha = \frac{\pi}{2} P_L = \pi N_L$	$\frac{L_p}{A} = \frac{\pi}{2} \frac{P_L}{P_P}$
Tomkeieff [14] $\bar{L}_2 = \pi \frac{\bar{A}}{\bar{L}_p} = \frac{\pi}{2L_A}$	$\bar{L}_2 = \frac{\pi \bar{A}_\alpha}{(\bar{L}_p)_\alpha} = \frac{\pi (A_A)_\alpha}{(L_A)_\alpha}$	$\bar{L}_2 = \frac{\pi A}{L_p}$
Chalkley [18] $\bar{L}_2 = lp/2P$	$\bar{L}_2 = \frac{lp}{P} = \frac{\pi \bar{A}_\alpha}{(\bar{L}_p)_\alpha}$	$\bar{L}_2 = \frac{lp}{P} = \frac{\pi A}{L_p}$
*Internal perimeter length referred to test area (L_A) or to mean figure area ($\bar{L}_p/\bar{A}$).	†Unshared perimeter length of α-regions referred to test area $(L_A)_\alpha$ or to mean figure area $(\bar{L}_p)_\alpha/\bar{A}_\alpha$).	§External perimeter length referred to figure area (L_p/A).

Explanation of symbols in Table 2.3

L_A = Total internal line length per unit test area
P_L = Number of intersections with perimeters per unit length of test line
N_L = Number of interceptions of planar figures per unit length of test line
$\bar{L}_2$ = Mean intercept length of planar figures $= L_L/N_L = A_A/N_L$
$\bar{A}_\alpha$ = Mean area of planar figures of α-phase $= (A_A)_\alpha/N_A$
$(\bar{L}_p)_\alpha$ = Mean perimeter length of planar figures of α-phase $= (L_A)_\alpha/N_A$
A = Area of single figure
L_p = External perimeter length of single figure
l = Constant length of short test lines, thrown randomly on microstructure
p = Number of (end) points of l-lines that fall on areas of interest
P = Number of intersections with perimeters made by l-lines
L_T = Length of test line

since $N_L = \frac{1}{2} P_L$. The mean perimeter length of α-phase areas is

$$(\bar{L}_p)_\alpha = \frac{(L_A)_\alpha}{N_A} = \frac{\pi P_L}{2N_A} \text{ mm}, \tag{2.51}$$

where $(L_A)_\alpha$ is given by Eq. (2.49) and N_A is the number of planar figures per unit test area. Moreover, the mean area of the figures is

$$\bar{A}_\alpha = \frac{(A_A)_\alpha}{N_A} \text{ mm}^2. \tag{2.52}$$

Substituting for P_L and $(A_A)_\alpha$ in Eq. (2.50) gives, for a system of isolated two-dimensional figures in a plane,

$$\boxed{\bar{L}_2 = \frac{\pi \bar{A}_\alpha}{(\bar{L}_p)_\alpha}} \text{ mm}, \tag{2.53}$$

where $\bar{A}_\alpha$ is the mean area of the planar figures and $(\bar{L}_p)_\alpha$ is their mean perimeter length.

For a single planar figure, the mean intercept length is

$$\boxed{\bar{L}_2 = \frac{\pi A}{L_p}} \text{ mm}, \tag{2.54}$$

where A is the area and L_p is the perimeter length. This is Tomkeieff's equation for the two-dimensional figure, analogous to Eq. (2.36) for the three-dimensional object.

For an area-filling system of contiguous planar figures, $N_L = P_L = 1/\bar{L}_2$ (see Table 2.3a). We start with

$$\bar{L}_p = 2L_A/N_A \text{ mm},$$

where L_A is doubled because each side is shared by two figures. By substituting $N_A = 1/\bar{A}$ and $L_A = (\pi/2)P_L$ and rearranging, we obtain

$$\bar{L}_p = 2\left(\frac{\pi}{2}\right)P_L\bar{A} = \pi N_L \bar{A} = \pi \bar{A}/\bar{L}_2 \text{ mm}. \tag{2.55}$$

In terms of the perimeter length to area ratio,

$$\boxed{\frac{\bar{L}_p}{\bar{A}} = \frac{\pi}{\bar{L}_2}} \text{ mm/mm}^2, \tag{2.56}$$

for the case of contiguous planar figures. The above results are summarized in Table 2.3.

2.5.3 Chalkley's Equation for Two-Dimensional Figures

Chalkley's [18] method applied to planar figures is similar to that described in Section 2.4.5. The equation for $\bar{L}_2$ is derived in the same manner as for $\bar{L}_3$ in Eq. (2.41). The ratio of perimeter length to area is

$$\boxed{\frac{L_p}{A} = \frac{\pi P}{lp}} \text{ mm/mm}^2 \tag{2.57}$$

for a single, isolated particle trace or two-dimensional closed loop in a plane. Appropriately modified equations for a system of separated figures are given in Table 2.3(b). Note also that the Chalkley equation for the case of area-filling figures reduces simply to $\bar{L}_2 = 1/P_L$. The derivation of Eq. (2.57) is given as a problem at the end of this chapter.

The two-dimensional analog to Saltykov's Eq. (2.44) is given in Table 2.3(c) for an isolated single figure (or closed loop in a plane). The ratio L_p/A is desired for a figure of perimeter length L_p and area A. A grid similar to that shown in Fig. 2.8 is used to obtain simultaneous measurements of P_L and P_P. We can then write

$$\frac{L_p}{A} = \frac{L_p/A_T}{A/A_T} = \frac{L_A}{A_A} = \frac{\pi P_L}{2P_P} \text{ mm/mm}^2, \tag{2.58}$$

where A_T is the total area occupied by the test grid.

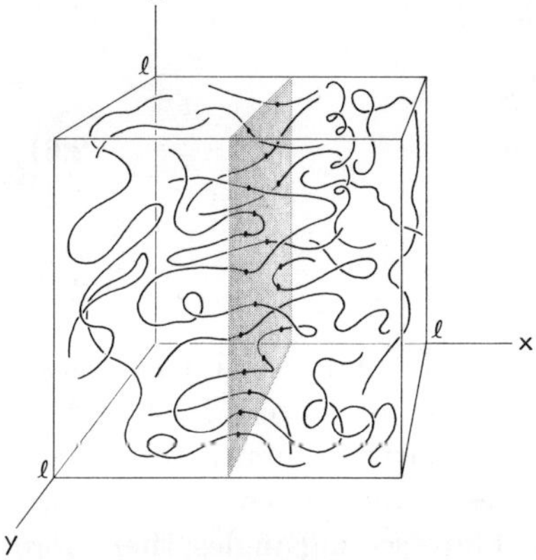

Fig. 2.10. Model for deriving the relationship [5] $L_V = 2P_A$. This cube contains random lines cut by a vertical test plane.

2.6 LENGTH OF LINE PER UNIT VOLUME L_V

The preceding equations have dealt with the length of lines in a plane. We now investigate the equation

$$L_V = 2P_A \text{ mm/mm}^3, \tag{2.3}$$

which concerns the total length of random lines in space. Derivations of this equation have been published by Saltykov [5] and Smith and Guttman [6], among others. The system of lines may consist of straight, curved, continuous or broken lines, oriented and located randomly throughout a certain volume. We intersect the lines with many test planes and the intersections of the lines with the planes appear as points on the plane, as shown in Fig. 2.10.

Consider a test cube of edge lengths l and volume $V_T = l^3$ enclosing a representative portion of the system of lines. A set of N parallel test planes of total area $A_T = Nl^2$ is passed vertically through the cube (say, parallel to the yz-plane), cutting a direction normal to the test planes (the x-direction) with a density of N/l.

We divide the total line length L into n elementary segments of length δL, such that $L = n\,\delta L$. Each elementary segment forms an angle θ_i with the normal to the test planes, and the lengths of the projections of the elementary segments along the x-direction are equal to $\cos\theta_i\,\delta L$. Thus the fraction of test planes intersecting the elementary segments is $\delta L \cos\theta_i/l$.

Given that P_i is the number of intersections of each elementary segment by the test planes, then the expected value of the total number of intersections with the entire system of lines is given by

$$E(P) = \sum^n P_i = N \sum^n \cos\theta_i\,\delta L/l = \frac{N\,\delta L}{l}\sum^n \cos\theta_i. \tag{2.59}$$

Since the elementary segments are oriented randomly, every value of θ_i has equal likelihood and $(1/n)\sum^n \cos\theta_i$ equals the average value, $\overline{\cos\theta}$. The expected value of the total number of intersections of the elementary segments with the set of test planes is then

$$E(P) = \frac{N\,\delta Ln}{l}\,\overline{\cos\theta} = \left(\frac{A_T}{l^3}\right) L\,\overline{\cos\theta}. \tag{2.60}$$

As shown previously, $\overline{\cos\theta}$ equals $\frac{1}{2}$; therefore we obtain

$$L/V_T = 2E(P)/A_T \text{ mm/mm}^3. \tag{2.61}$$

The required relationship, in our notation, is

$$\boxed{L_V = 2P_A} \text{ mm/mm}^3, \tag{2.62}$$

which states that the total length of random lines per unit volume is equal to twice the number of points of intersection with a test plane per unit area. It should be emphasized that Eq. (2.62) is applicable to any type of linear system in space, provided that all angles of intersection between lines and test planes are equally possible. If one section plane does not provide full randomness of intersecting angles, then more than one section plane (at different angles) will be required.

As a consequence of the latter requirement, when only a single section plane is used, the above equation may pertain to just the randomly oriented part of the system of lines. For nonrandom (oriented) systems of lines cut by just one section plane, different coefficients may apply. We shall take up these special matters in detail in Chapter 3.

2.7 NUMBER OF POINTS PER UNIT VOLUME P_V

Only under unusual circumstances would it be possible to observe true points in a volume. If the "points" became enlarged, perhaps as small precipitate particles, then they might be amenable to direct measurement. The intersections of dislocations in planar arrays are observed frequently in thin metallic foils by means of electron transmission. The intersections actually represent a set of points in a three-dimensional volume. If the grain edges of a polycrystalline aggregate were delineated, one could observe a three-dimensional network of lines (edges) terminating four at a grain corner (point) (see the corner points P in Fig. 1.1a). The latter experiment has been performed by Williams and Smith [15].

Points are also generated by lines intersecting surfaces. Hilliard [7] has developed an equation by which P_V for such points may be obtained. Two random arrays in space are considered: a lineal and an areal array. The intersections of the areal array by the lineal array produce the points per unit volume, P_V. The relationship is indicated in Table 2.1 and is given in Eqs. (2.4) as

$$P_V = \tfrac{1}{2} L_V S_V \text{ mm}^{-3} \tag{2.63}$$

or, by simple substitution, as

$$\boxed{P_V = 2P_A P_L} \text{ mm}^{-3}. \tag{2.64}$$

Both points and lines appear on the plane of polish. The points result from intersections of the lineal array with the section plane, and the linear traces arise from intersection of the areal array with the section plane. P_A is counted as described previously and P_L is obtained from intersections of a random test line. The values so obtained permit P_V to be calculated according to Eq. (2.64).

Equation (2.63) was developed from a rather complicated mathematical framework and will not be derived here. The interested reader is referred to Hilliard's original paper for details of the derivation.

If Eq. (2.63) is divided through by L_V, Eq. (2.31) ($P_L = \frac{1}{2}S_V$) results. Alternatively, division by S_V gives Eq. (2.62) ($P_A = \frac{1}{2}L_V$). Note that Hilliard's generalized formulation equates $\frac{1}{2}L_V$ to the point density (P_A) on either a curved or planar surface. This interesting result should prove useful when we are dealing with transparent structures (see Chapter 6).

REFERENCES

1. A. Delesse, "Pour déterminer la composition des roches," *Ann. des Mines* **13** (1848), fourth series, pages 379–388
2. J. E. Hilliard, "Volume Fraction Analysis by Quantitative Metallography," Rep. No. 61-RL-2652 M, General Electric Res. Lab., March 1961
3. F. Chayes, *Petrographic Modal Analysis*, New York: John Wiley, (1956) 5, 14, 29
4. E. R. Weibel, *Morphometry of the Human Lung*, Berlin: Springer-Verlag (1963) 12, 16, 31
5. S. A. Saltykov, *Stereometric Metallography*, second edition, *Metallurgizdat*, Moscow (1958)
6. C. S. Smith and L. Guttman, "Measurement of Internal Boundaries in Three-Dimensional Structures by Random Sectioning," *Trans. AIME* **197** (1953) 81
7. J. E. Hilliard, "Specification and Measurement of Microstructural Anisotropy," *Trans. AIME* **224** (1962) 1201
8. A. Rosiwal, "Über geometrische Gesteinsanalysen, usw.," *Verhandl. der K.-K. geologischen Reichanstalt* 5/6 (1898) 143
9. E. Thomson, "Quantitative Microscopic Analysis," *J. Geol.* **38,** 3 (April–May 1930) 193
10. A. A. Glagolev, "On the Geometrical Methods of Quantitative Mineralogic Analysis of Rocks," *Trans. Inst. Econ. Min., Moscow* **59** (1933)

11. J. E. Hilliard and J. W. Cahn, "An Evaluation of Procedures in Quantitative Metallography for Volume-Fraction Analysis," *Trans. AIME* **221** (1961) 344
12. R. J. Duffin, R. A. Meussner, and F. N. Rhines, "Statistics of Particle Measurement and of Particle Growth," Carnegie Inst. Tech., Tech. Rep. No. 32 (Dept. of Math.) C.I.T.-AF8A-TR32, April 1, 1953
13. E. Horikawa, "On a New Method of Representation of the Mixture of Several Austenite Grain Sizes," *Tetsu to Hagane* **40,** 10 (1954) 991
14. S. I. Tomkeieff, "Linear Intercepts, Areas and Volumes," *Nature* **155** (Jan. 6, 1945) 24; correction, page 107
15. W. M. Williams and C. S. Smith, "A Study of Grain Shape in an Aluminum Alloy and Other Applications of Stereomicroradiography," *Trans. AIME* **194** (1952) 755
16. G. E. Pellissier, M. F. Hawkes, W. A. Johnson, and R. F. Mehl, "The Interlamellar Spacing of Pearlite," *Trans. ASM* **30** (1942) 1049
17. H. Campbell and S. I. Tomkeieff, "Calculation of the Internal Surface of a Lung," *Nature* **170** (July 19, 1952) 117
18. H. W. Chalkley, J. Cornfield, and H. Park, "A Method for Estimating Volume–Surface Ratios," *Science* **110** (Sept. 23, 1949) 295
19. G. Schoeck, "Correlation Between Dislocation Length and Density," *J. Appl. Phys.* **33** (1962) 1745
20. A. Hennig, "Das Problem der Kernmessung," *Mikroskopie* **12** (1957) 174
21. A. Hennig, "Kritische Betrachtungen zur Volumen- und Oberflächenmessung in der Mikroskopie," *Zeiss Werkzeitschrift* **6** (1958) 78 (also available in English)
22. S. Corrsin, "A Measure of the Area of a Homogeneous Random Surface in Space," *Quart. Appl. Math.* **12** (1954) 404
23. J. Cornfield and H. W. Chalkley, "A Problem in Geometric Probability," *J. Wash. Acad. Sci.* **41,** 7 (15 July 1951) 226
24. M. G. Kendall and P. A. P. Moran, *Geometrical Probability,* London: C. Griffin & Co. Ltd. (1963) 90
25. McLean, D., *Grain Boundaries in Metals,* London: Oxford University Press (1957) 110
26. A. Hennig, "Bestimmung der Oberfläche beliebig geformter Körper, usw.," *Mikroskopie* **11** (1956) 1
27. S. A. Saltykov, "A Stereological Method for Measuring the Specific Surface Area of Metallic Powders," *Stereology*, edited by H. Elias, *Proc. Second Int. Cong. for Stereology,* New York: Springer-Verlag (1967) 63

PROBLEMS

2.1. a) Check the relationships $A_A = L_L = P_P$ for the randomly arranged squares in Fig. 2.1. What are your values for L_T (in length units) and P_T when all lines and points in the grid are counted?
b) Repeat the evaluation of $A_A = L_L = P_P$, but re-count the values of L_T and P_P by assigning only *half*-values to the lines and points along the borders of the grid. Which procedure comes closer to the expected value? Why?

2.2. Derive the value of $\overline{\cos\theta}$ using a method different from that in Eq. (2.30). [*Hint*: Use the properties of a rectangular distribution; e.g., see Burington and May, *Handbook of Probability and Statistics*, Sandusky, Ohio: Handbook Publishers, Inc. (1958), page 57.]

2.3 a) Derive Chalkley's equation for a system of closed loops in a plane, Eq. (2.57) $(A_A < 1)$.

b) Do the same for the case of $A_A = 1$.

2.4. If you were using the equations in Eqs. (2.1) (that is, $V_V = A_A$, $V_V = L_L$, $V_V = P_P$), which of the three would you think was most efficient in terms of smallest standard deviation and least time required? (See Reference 11.)

2.5. Given that $V_V = N_V \overline{V} = N_A \overline{A} = N_L \overline{L}$, what are the next terms involving points and numbers? Define the terms.

2.6. a) What is the value of $\bar{L}_3$ obtained from the measurements made in Problem 1.2?

b) What is the mean surface-to-volume ratio, $\overline{S}/\overline{V}$?

2.7. a) Construct a grid on clear plastic similar to the one shown in Fig. 2.8. Determine S/V for the irregular section in Fig. 2.8, using Saltykov's Eq. (2.44).

b) Assume that the irregular section in Fig. 2.8 is a two-dimensional figure. Calculate L_p/A by means of Eq. (2.58).

3

ORIENTED STRUCTURES

". . . we come to the conclusion that the magnitude of specific surface alone does not sufficiently characterize a system of boundary surfaces. Quantitative evaluation of the degree of orientation . . . is also needed."
Sarkis A. Saltykov

3.1 INTRODUCTION

Oriented structures occur frequently in nature. Meteorites, twinned minerals or metals, banded rocks, dislocation arrays, nonmetallic "stringers" in extruded rods, and the lamellar eutectic alloys are a few of the examples that come to mind. Methods have been given for describing banded rocks [1, 2, 3] and metallic structures [4–8].

In this chapter we make a distinction between completely oriented (idealized) structures and the partially oriented structures that are found in nature. Of the possible types of oriented structures, there are systems of lines in a plane or in space, and systems of surfaces in space. These types of oriented structures can be described according to the classification proposed by Saltykov [5]. He gives methods for characterizing the angular variation of lines or surfaces with respect to an orientation axis (axes) or orientation plane(s). A simple numerical index, called the *degree of orientation*, can also be used for describing oriented systems. In order to preserve its unity, we shall treat the general mathematical analysis for oriented systems developed by Hilliard [8] separately.

3.2 SYSTEMS OF LINES IN A PLANE

Let us recall that we can calculate the length per unit area of any system of lines in a plane according to Eq. (2.49). Our aim in this section is to extract additional information, other than the average value, from oriented systems of lines. Some ways of accomplishing this purpose are outlined below.

3.2.1 Completely Oriented Systems

A convenient classification scheme for completely oriented systems of lines in a plane is given by Saltykov [5]. Idealized linear systems with one, two, and three orientation axes are shown in Fig. 3.1. In part (a), the lines are equally or irregularly spaced,

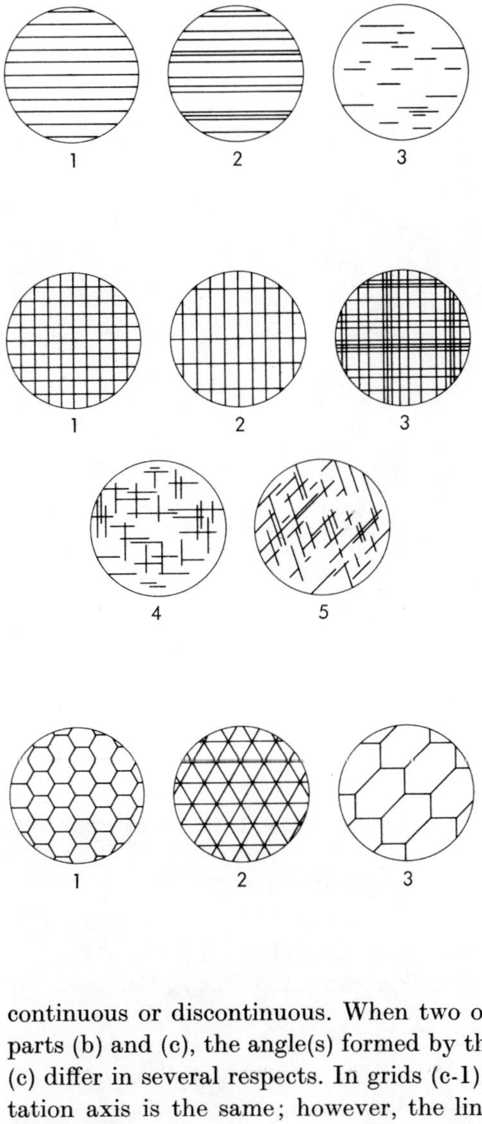

a) Systems of lines with one orientation axis.

b) Systems of lines with two orientation axes.

c) Systems of lines with three orientation axes.

Fig. 3.1. Examples of completely oriented systems of lines in a plane [5].

continuous or discontinuous. When two or more orientation axes are present, as in parts (b) and (c), the angle(s) formed by the axes must be specified. The grids in part (c) differ in several respects. In grids (c-1) and (c-2), the line length along any orientation axis is the same; however, the lines are continuously straight only in (c-2). The lengths of the lines in (c-3) are greater in one direction than another. A system of oriented lines in a plane is seen in Fig. 3.2 for a twinned single crystal of Mo-35 a/o Re. There are four main orientation axes in this case.

We saw in Chapter 2 that the length of linear elements in a plane is proportional to the number of intersections made with a test line. Using a test array of straight parallel lines, we expect the number of intersections with a randomly oriented (isometric) system of lines to be the same in any direction. For an oriented system of lines, however, the number of intersections will vary markedly with the direction of the test array.

Fig. 3.2. Twinned structure in a heavily etched Mo-35 a/o Re single crystal (128×).

The dependence of the number of intersections per unit length with the angle of the test array can be used to characterize the degree and types of orientation of a system of lines in a plane. Saltykov [5] proposes a polar plot of P_L with respect to the orientation axis (axes), and calls the resulting curve the *rose of the number of intersections*, or simply the *rose*.

We can readily obtain the rose for an oriented system of lines experimentally. We apply a test array to the system of lines at (equal) angular increments with respect to the orientation axis, and determine P_L separately at each angle θ. Then we draw radius vectors on polar graph paper, plotting P_L versus θ. We connect the ends of the radius vectors by a smooth curve, giving the rose.

We may also express the variation of P_L with the angle θ analytically for simple systems of lines. Let us consider a system of continuous lines with one orientation axis, as in Fig. 3.1, parts (a-1) and (a-2). A test array directed parallel to the orientation axis will not intersect any lines; thus the number of intersections per unit length in the parallel direction is zero. Conversely, a test array perpendicular to the orientation axis will intersect the greatest number of lines in the system per unit length of test array.

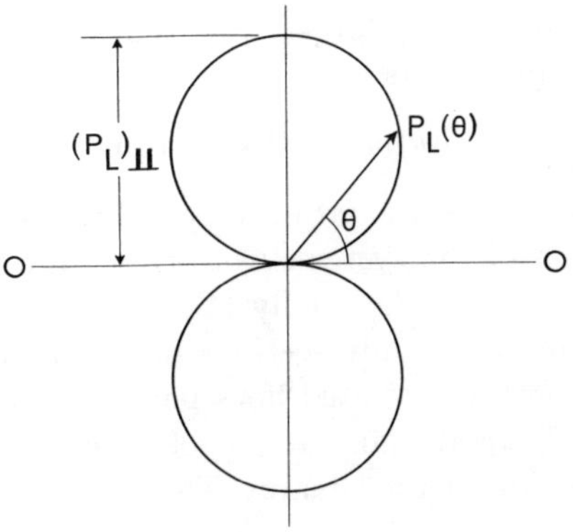

(a) From a grid of parallel lines

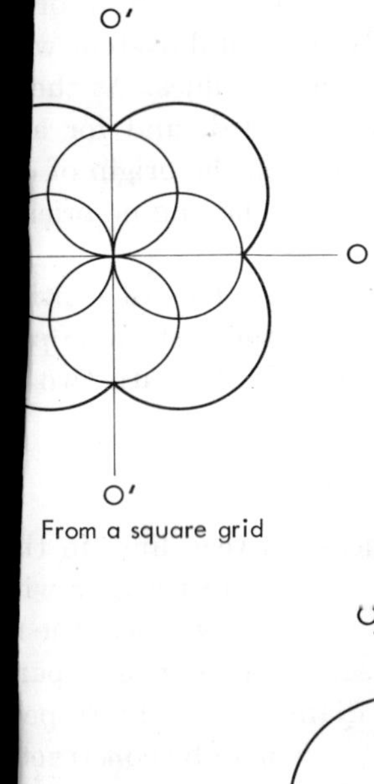

From a square grid

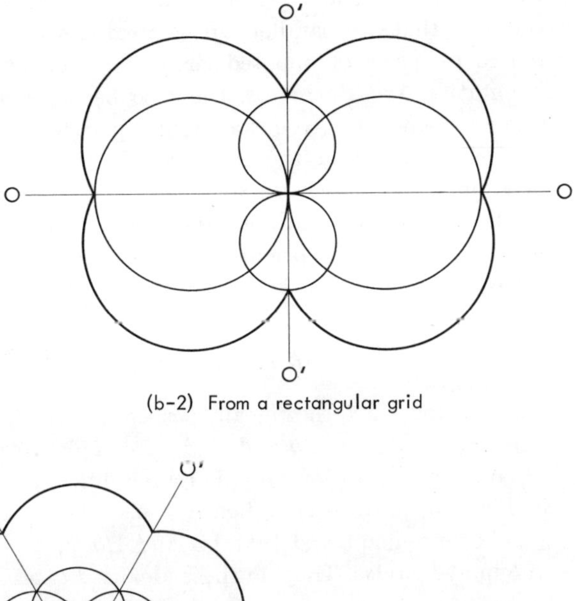

(b-2) From a rectangular grid

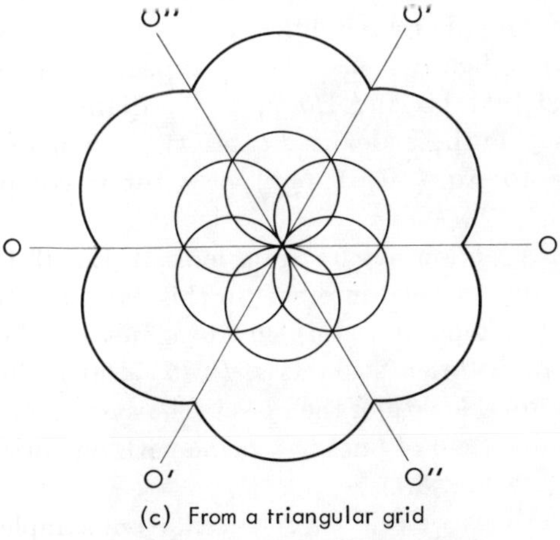

(c) From a triangular grid

Construction of the rose of the number of intersections for oriented systems of lines in]. (a) One orientation axis, (b) two orientation axes, (c) three orientation axes.

The angular dependence of P_L for an orie[n]
one orientation axis is expressed as

$$P_L(\theta) = (1/a) \sin \theta$$

where θ is the angle between the orientation axis
distance between the lines of the system. Note
the orientation axis, $1/a$ represents equally (1)
a unit length; (2) the number of intersections
test lines; and (3) the length of oriented lines p
consists of two circles of equal diameter oppos
orientation axis at the origin of coordinates. Th
value of $1/a$.

Figure 3.3 gives the polar plots and roses for so
In all cases there are opposed pairs of circles. Sin
the lines that are parallel can be treated separat
sented by pairs of opposed circles for each set o
orientation axes increases, the rose becomes mo
system of lines, the rose is a circular perimeter cen
An isometric system, therefore, may be regarded e
or an infinite number of axes.

The procedures outlined above can be extende
with more than one orientation axis, such as shov
3.1(b). The analytical expression for oriented syste
axes is

$$P_L(\theta, \omega) = (1/a) \sin \theta + (1/b) \text{s}$$

where a and b are the (mean) perpendicular distanc
of oriented lines, and θ and ω are the angles formed
responding orientation axes. For a rectilinear grid wit
in Fig. 3.3, part (b-1). When $a \neq b$, the rose appea
should be remembered that $1/a$ and $1/b$ equal the d
of opposed circles. Thus, in these idealized cases, the r
cally without recourse to Eq. (3.1) or (3.2), or witho
ments of P_L.

Before we leave this section, let us again look at
sented by $\theta = 90°$ assumes special interest. In this ca
ular to the lines of the system, and $(1/a) \sin \theta$ become
of lines in a plane with one orientation axis, P_L (θ =
but will depend on the total length of the lines in the sy
tionship for an oriented system of lines in a plane with
in the following way (see Fig. 3.4).

From the system of lines we select a representative
within a square test area $A_T = l^2$. One side lies parallel
we draw a network of equidistant, parallel test lines ov

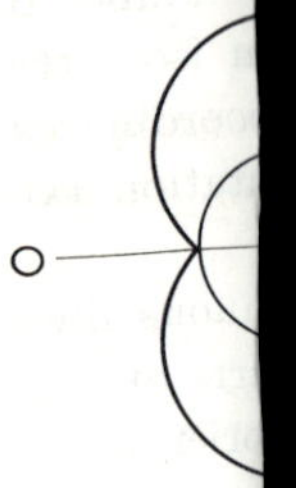

(b-1)

Fig. 3.3. (
a plane [5

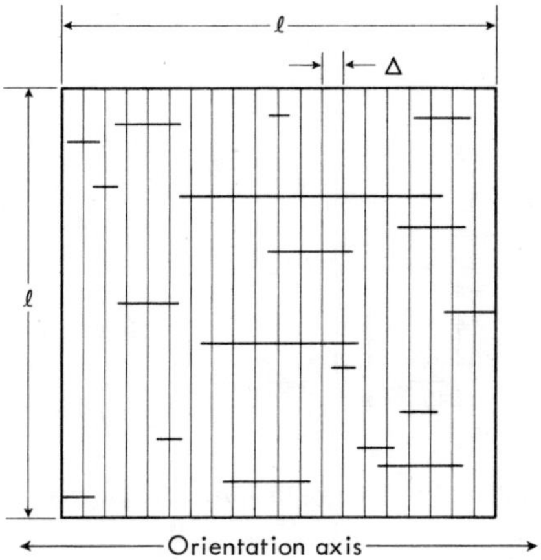

Fig. 3.4. Model for deriving the relationship $(L_A)_{or} = (P_L)_{\perp\!\perp}$ in an oriented system of lines with one orientation axis.

Δ apart, perpendicular to the orientation axis. The number of strips of width Δ is equal to l/Δ and the total length of strips is $(l/\Delta)l$.

We count the number of segments intercepted by each test strip of width Δ, ignoring those segments with lengths less than $\frac{1}{2}\Delta$. If the first strip has m_1 segments, the second strip m_2, the third strip m_3, etc., then the length of all intercepts in the first strip is $m_1\Delta$; in the second strip, $m_2\Delta$; etc. The total length of all segments in the test area is

$$L = \Delta(m_1 + m_2 + m_3 + \cdots) = \frac{m_1 + m_2 + m_3 + \cdots}{1/\Delta}. \tag{3.3}$$

Dividing both sides of Eq. (3.3) by l^2 gives us

$$\frac{L}{l^2} = \frac{(m_1 + m_2 + m_3 + \cdots)}{(1/\Delta)l^2}$$

where L/l^2 is equivalent to L_A.

For Δ sufficiently small, the strips approach test lines. Then $m_1 + m_2 + m_3 + \cdots$ represent the total number of intersections of the test lines, and l^2/Δ equals the total length of test lines. Thus the ratio of these two quantities equals the number of intersections per unit length of perpendicular test lines, P_L, and

$$\boxed{(L_A)_{or} = (P_L)_{\perp\!\perp}}\ \text{mm/mm}^2, \tag{3.4}$$

where $(L_A)_{or}$ represents the length per unit area of the parallel (oriented) lines of the system, and $(P_L)_{\perp\!\perp}$ is the number of intersections per unit length of the test array aligned in a direction perpendicular to the orientation axis.*

This result is equally valid for systems of equidistant, parallel straight lines and for arrays of parallel, linear segments with random lengths and locations. Equation

*The symbol "$\perp\!\perp$" applies only to measurements made on an oriented system of (parallel) lines, in a direction perpendicular to the orientation axis. The ordinary symbol "$\perp$" is used to identify measurements made in partially oriented systems of lines.

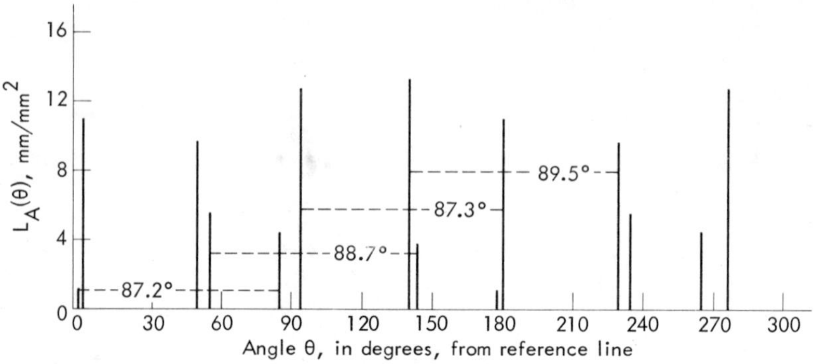

Fig. 3.5. Twin trace length per unit area versus grid angle θ for the Mo-35 a/o Re single crystal shown in Fig. 3.2.

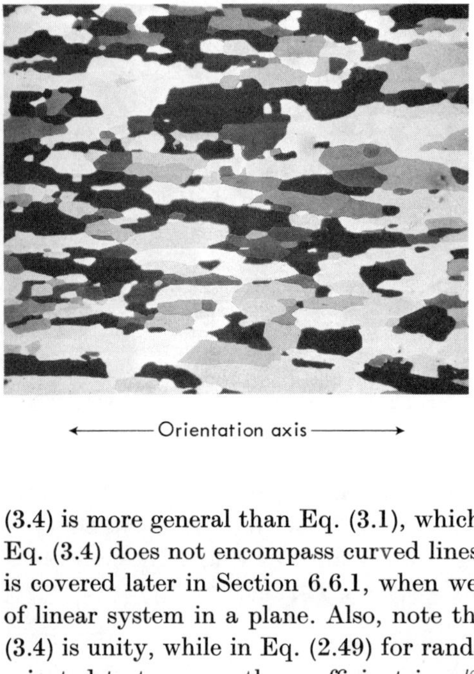

Fig. 3.6. Elongated grains in an extruded molybdenum alloy.

(3.4) is more general than Eq. (3.1), which applies only to continuous lines; however, Eq. (3.4) does not encompass curved lines or lines of all orientations. The latter case is covered later in Section 6.6.1, when we discuss the "total projection" of any type of linear system in a plane. Also, note that the coefficient relating the terms in Eq. (3.4) is unity, while in Eq. (2.49) for randomly oriented systems of lines or randomly oriented test arrays, the coefficient is $\pi/2$.

We see applications of Eq. (3.4) in Fig. 3.5, which is based on the twinned microstructure shown in Fig. 3.2, and in the relationships for partially oriented structures, developed in Section 3.2.2.

3.2.2 Partially Oriented Systems

In real structures the lines in a microsection are usually either isometric or partially oriented. Only in rare cases, such as the Widmanstätten patterns in meteorites, do we find a completely oriented system of lines. In a partially oriented system of

Fig. 3.7. Impurity segregation revealing dislocation lines in a silicon crystal [15].

lines in a plane, part of the total length of lines is oriented in a definite direction (or directions). The remaining segments may have essentially a random orientation.

Examples of partially oriented systems of lines in a plane are the traces of elongated grain boundaries of molybdenum (Fig. 3.6) and the linear patterns occurring in specially grown semiconductor crystals (Fig. 3.7) [15]. In the discussion that follows, the reader should keep in mind the characteristics of these photomicrographs. Let us now consider the derivation of the equation for partially oriented systems of lines in a plane, the graphical construction of the rose, and the calculation of the degree of orientation.

The lines of a partially oriented system of lines in a plane can be divided into elementary straight segments that are very small and of equal length. Some or all segments will lie parallel to one or more definite directions (the orientation axes). The remaining segments are assumed to be oriented randomly, i.e., isometrically. From this point of view, a partially oriented system of lines may be regarded as consisting of two superimposed systems of lines, i.e., an oriented portion and a random portion.

To determine the specific lengths of the isometric and oriented portions of a partially oriented system of lines in a plane with one orientation axis requires only two measurements. These are $(P_L)_{||}$ and $(P_L)_{\perp}$, where the subscripts indicate the directions of the test array with respect to the orientation axis. Since the test array laid down parallel to the orientation axis will not intersect oriented line segments, the value of $(P_L)_{||}$ will depend solely on the isometric line segments. Thus the length of the isometric segments of the system per unit area is

$$(L_A)_{\text{is}} = \frac{\pi}{2}(P_L)_{||} \text{ mm/mm}^2. \tag{3.5}$$

Next the test array is laid down perpendicular to the orientation axis, and the number of intersections counted regardless of the type of linear system. The value of $(P_L)_{\perp}$ thus derives from both the isometric and the oriented line segments. Consequently, the number of intersections per unit length due only to the oriented segments of the system is $(P_L)_{\perp} - (P_L)_{||}$, and the length of oriented lines per unit area is

$$(L_A)_{\text{or}} = (P_L)_{\perp} - (P_L)_{||} \text{ mm/mm}^2, \tag{3.6}$$

according to the relationship expressed by Eq. (3.4).

The total length of all lines per unit area in a partially oriented system of lines is the sum of the two linear systems, or

$$L_A = (L_A)_{\text{is}} + (L_A)_{\text{or}}. \tag{3.7}$$

Substituting from Eqs. (3.5) and (3.6) yields

$$\boxed{L_A = (P_L)_{\perp} + 0.571(P_L)_{||}} \text{ mm/mm}^2, \tag{3.8}$$

which indicates that only two positions of the test array, one parallel and one perpendicular to the orientation axis, are needed to determine the total line length per unit area in a partially oriented system of lines in a plane.

The data can be presented in alternate forms to emphasize the visual or numerical characterization of the linear anisotropy. A visual device is the rose plot of $P_L(\theta)$, while a numerical representation is afforded by the fraction *the degree of orientation.*

The rose for partially oriented systems of lines possesses characteristics that combine those of a completely isometric system of lines and those of a completely oriented system of lines. Thus the rose is built up from a pair (or pairs) of opposing circles plus a circular circumference centered on the origin. As indicated in Fig. 3.8, the rose can be constructed graphically from the values of $(P_L)_{||}$ and $(P_L)_{\perp}$. The rose can also be obtained directly by measuring $P_L(\theta)$ at suitable angular intervals.

Schmidt [9], in 1917, suggested a rose plot to portray the results of statistical analyses of two-dimensional fabric diagrams in petrology. Instead of using rectangular

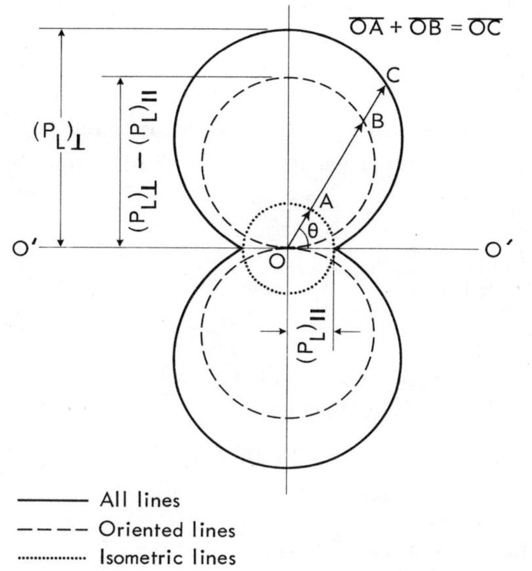

Fig. 3.8. Graphical construction of the rose for a system of partially oriented lines with one orientation axis O'–O'.

coordinates as in the conventional histogram, where, for example, the frequency of occurrence in certain directions is grouped in classes of 5°, he proposed a polar plot of the same data. Three examples from his original publication are shown in Fig. 3.9. The circle represents the limiting case of two-dimensional isotropy. Other applications of this idea may suggest themselves to the reader.

For practical applications, the complete rose may not be too convenient. A simpler method is available for characterizing the degree of orientation of a system of partially oriented lines in a plane. The length of oriented lines only per unit area is divided by the total length of all lines in the system per unit area. The resulting fraction gives an easily recognized, natural scale of anisotropy varying from 0 for no anisotropy to unity for a completely oriented system of lines. This ratio is called [5]

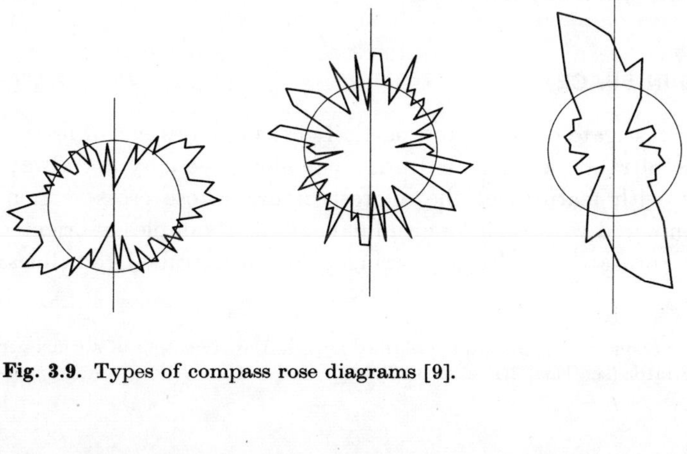

Fig. 3.9. Types of compass rose diagrams [9].

the *degree of orientation* Ω, and is defined as

$$\Omega_{12} = \frac{(L_A)_{\text{or}}}{(L_A)_{\text{is}} + (L_A)_{\text{or}}} \tag{3.9}$$

where the subscript 12 refers to lines (1) in a plane (2).* The values can vary from 0 to 1. Substituting experimentally measured quantities from Eqs. (3.5) and (3.6) into the above expression gives

$$\boxed{\Omega_{12} = \frac{(P_L)_\perp - (P_L)_{||}}{(P_L)_\perp + 0.571(P_L)_{||}}} \tag{3.10}$$

for a partially oriented system of lines in a plane.

An analysis of the partially oriented linear structure of a rolled metal sheet according to Eq. (3.10) is given by Saltykov [5]. Measured values are

$$(P_L)_\perp = 792 \text{ intersections/34 mm} = 23.3 \text{ mm}^{-1}$$

and

$$(P_L)_{||} = 378 \text{ intersections/50 mm} = 7.6 \text{ mm}^{-1}.$$

Substituting in Eq. (3.5) gives

$$(L_A)_{\text{is}} = 1.571(7.6) = 11.9 \text{ mm/mm}^2,$$

and in Eq. (3.6) gives

$$(L_A)_{\text{or}} = 23.3 - 7.6 = 15.7 \text{ mm/mm}^2.$$

The total length of grain boundary traces per unit area is

$$L_A = 11.9 + 15.7 = 27.6 \text{ mm/mm}^2,$$

and from Eq. (3.9), the degree of orientation is

$$\Omega_{12} = \frac{15.7}{27.6} = 0.57 \quad \text{or} \quad 57\%.$$

The equations discussed above, pertaining to systems of lines in a plane, are collected in Table 3.1 for comparison and ready reference.

3.3 SYSTEMS OF LINES IN SPACE

In this section we shall investigate oriented and partially oriented systems of lines in space. Just as in the case of randomly oriented lines in a plane, we may not always deal with true lines but with features of the microstructure whose cross sections are significantly small compared with their longitudinal extent. Examples of oriented linear features in a structure are parallel slag stringers in an extruded rod, linear

---

*The degree of orientation of a system of lines in a volume, Ω_{13}, and of a system of surfaces in a volume, Ω_{23}, are discussed later (see Eqs. 3.16 and 3.28).

TABLE 3.1

Equations for systems of lines in a plane

Type of system	Isometric lines $(L_A)_{is} =$	Oriented lines $(L_A)_{or} =$	Total specific line length $L_A =$	Degree of orientation, Ω_{12}
Isometric	$1.571P_L$	—	$1.571P_L$	$\Omega_{12} = 0$
Oriented*	—	$(P_L)_{\perp\!\!\perp}$	$(P_L)_{\perp\!\!\perp}$	$\Omega_{12} = 1.0$
Partially oriented†	$1.571(P_L)_{\parallel}$	$(P_L)_{\perp} - (P_L)_{\parallel}$	$(P_L)_{\perp} + 0.571(P_L)_{\parallel}$	$\Omega_{12} = \dfrac{(P_L)_{\perp} - (P_L)_{\parallel}}{(P_L)_{\perp} + 0.571(P_L)_{\parallel}}$

*The subscript $\perp\!\!\perp$ symbol refers to measurements perpendicular to the orientation axis of an oriented system of (parallel) lines.

†Subscript $\perp$ and $\parallel$ symbols refer to direction of test array with respect to the orientation axis in a partially oriented system of lines.

dislocation arrays in metal crystals, and linear elements in tree microstructures. The edges of elongated grains inside a rolled metal sheet can be considered as a partially oriented system of lines.

The detection of true lines or linear features in opaque materials offers more experimental difficulties than the linear traces dealt with in the preceding section. The probability of a plane intersecting a straight line in space along its length is vanishingly small. Consequently we observe the points of emergence of only those lines that intersect the section planes at an angle greater than zero. Several section planes may be required to ensure adequate randomness in the angles of intersection between lines and section planes.

3.3.1 Completely Oriented Systems

Completely oriented systems of lines in space with one orientation axis consist of straight lines that have equal or variable spacing, but are parallel to each other and to the orientation axis. The lines may be continuous or segmented. Section planes that are not parallel to the orientation axis will reveal points where the lines intersect the plane. When the section plane is perpendicular to the orientation axis, there is a simple relationship between the lengths of the oriented lines and the number of their intersections per unit area of the section plane. We shall derive this relationship next.

Let us consider an oriented system of lines in space, with one orientation axis, lying within a sample cube of edge lengths l. The orientation axis is parallel to an edge of the cube. We intersect the system of lines with many section planes, normal to the orientation axis, at a constant and very small distance Δ. The number of slices of width Δ is equal to l/Δ and the total area of the section planes is $(l/\Delta)l^2$.

We then count the number of segments intercepted by each slice of width Δ, ignoring those segments with lengths less than $\frac{1}{2}\Delta$. If the first slice has n_1 segments, the second slice n_2, the third slice n_3, etc., then the length of all intercepts in the first slice is $n_1\Delta$; in the second slice, $n_2\Delta$; etc. The total length of all segments in the test volume is

$$L = \Delta(n_1 + n_2 + n_3 + \cdots) = \frac{n_1 + n_2 + n_3 + \cdots}{1/\Delta}. \tag{3.11}$$

Dividing both sides of Eq. (3.11) by l^3 gives

$$\frac{L}{l^3} = \frac{n_1 + n_2 + n_3 + \cdots}{(1/\Delta)l^3},$$

where L/l^3 is equivalent to L_V.

For Δ sufficiently small, the slices approach test planes. Then $n_1 + n_2 + n_3 + \cdots$ represent the total number of intersections with test planes, and l^3/Δ equals the total area of the test planes. Thus the ratio of these two quantities equals the number of intersections per unit area of perpendicular test planes, P_A, and

$$\boxed{(L_V)_{\text{or}} = (P_A)_{\perp\!\perp}} \text{ mm/mm}^3, \tag{3.12}$$

where $(L_V)_{\text{or}}$ represents the length per unit volume of the parallel (oriented) system of lines, and $(P_A)_{\perp\!\perp}$ is the number of intersections per unit area of test planes erected perpendicular to the orientation axis. The similarity of Eq. (3.12) to Eq. (3.4) is noteworthy.

The coefficient in Eq. (3.12) is unity, whereas when the test planes are applied randomly to a system of lines in space (Eq. 2.62), the coefficient is two. Schoeck [10] investigated various planar arrays of dislocation lines in space and obtained coefficients with values between one and two, depending on the relative orientation of the section plane with respect to the dislocation arrays. Hence we must take care to specify the configuration of the linear system and the angular position of the sectioning plane(s), if they are other than random.

The rose for an oriented system of lines in space can be represented in two or three dimensions, depending on the particular requirements at hand and on the symmetry of the system. The path followed by the normal of the rotating test plane may be arbitrarily limited to a single great circle, or the normal may be permitted to rotate to any direction in space.

Consider a spherical coordinate system with an oriented system of lines lying parallel to the equatorial plane and to the orientation axis in this plane. If a test section lying originally in the equatorial plane is rotated so that its normal moves only in the great circle defined by the north pole and the orientation direction(s), we get two-dimensional figures completely analogous to the roses described previously in Fig. 3.3. The rose is still composed of pairs of opposed circles, but instead of P_L we plot P_A. The diameters of the rose circles are, of course, equal to $(P_A)_{\perp\!\perp}$, where the subscript $\perp\!\perp$ symbol refers to a section plane perpendicular to the orientation axis.

We obtain the three-dimensional rose for a system of lines in space with one orientation axis simply by allowing the test section to rotate over all directions of the reference sphere. Taking advantage of the symmetry involved, we see that a test plane normal to the orientation axis will have the same value of $(P_A)_{\perp\!\perp}$ in every position around the axis. Moreover, when this test plane is rotated into the direction parallel to the orientation axis, no intersections are possible and $(P_A)_{\parallel}$ is zero in every position around the axis. Thus we see that these conditions define a torus, whose equatorial plane lies normal to and centered about the orientation axis. We shall develop this subject further in subsequent sections.

3.3.2 Partially Oriented Systems

For the analysis of a system of lines in space with partial orientation, we apply a procedure similar to that described above for a partially oriented system of lines in a plane. For a system of lines in space with one orientation axis, two section planes are necessary: the first parallel and the second perpendicular to the orientation axis. The system of lines is assumed to be separable, as before, into isometric and oriented portions. We indicate briefly the derivation of the relationship between line length per unit volume and intersections per unit area.

A section plane parallel to the orientation axis will intersect isometric line segments but not oriented ones. Therefore we can write

$$(L_V)_{\text{is}} = 2(P_A)_{\parallel} \text{ mm/mm}^3, \tag{3.13}$$

where $(P_A)_{\parallel}$ is the number of points of intersection per unit area on parallel section planes. A perpendicular section plane will intersect both isometric and oriented line segments, and therefore

$$(L_V)_{\text{or}} = (P_A)_{\perp} - (P_A)_{\parallel} \text{ mm/mm}^3, \tag{3.14}$$

using the result expressed by Eq. (3.12). We can obtain the total line length of a partially oriented system of lines per unit volume by adding Eqs. (3.13) and (3.14), giving

$$\boxed{L_V = (P_A)_{\perp} + (P_A)_{\parallel}} \text{ mm/mm}^3, \tag{3.15}$$

where $(P_A)_{\perp}$ is measured on a test plane perpendicular to the orientation axis and $(P_A)_{\parallel}$ on a test plane parallel to it.

The space roses for partially oriented systems of lines in space are obtained by direct extensions of procedures outlined previously for partially oriented systems of lines in a surface. In the three-dimensional linear systems, however, we plot P_A instead of P_L. The three-dimensional space rose for a partially oriented system of lines in space would consist of a spherical surface centered on the origin of coordinates superimposed on a torus, as described in the preceding section [see also Fig. 3.12(b)]. The analogy to the two-dimensional case is evident.

The degree of orientation Ω_{13} of a partially oriented system of lines in space is defined by the general expression

$$\Omega_{13} = \frac{(L_V)_{\text{or}}}{(L_V)_{\text{is}} + (L_V)_{\text{or}}}. \tag{3.16}$$

By substituting from Eqs. (3.13) and (3.14), we see that the working formula is

$$\boxed{\Omega_{13} = \frac{(P_A)_{\perp} - (P_A)_{\parallel}}{(P_A)_{\perp} + (P_A)_{\parallel}}}. \tag{3.17}$$

The equations developed for systems of lines in space are summarized for convenience in Table 3.2.

TABLE 3.2

Equations for systems of lines in space

Type of system	Isometric lines $(L_V)_{is} =$	Oriented lines $(L_V)_{or} =$	Total specific line length $S_V =$	Degree of orientation, Ω_{13}
Isometric	$2P_A$	—	$2P_A$	$\Omega_{13} = 0$
Oriented*	—	$(P_A)_{\perp\!\perp}$	$(P_A)_{\perp\!\perp}$	$\Omega_{13} = 1.0$
Partially oriented†	$2(P_A)_{\|\|}$	$(P_A)_\perp - (P_A)_{\|\|}$	$(P_A)_\perp + (P_A)_{\|\|}$	$\Omega_{13} = \dfrac{(P_A)_\perp - (P_A)_{\|\|}}{(P_A)_\perp + (P_A)_{\|\|}}$

*The subscript $\perp\!\perp$ symbol refers to direction of section planes perpendicular to the orientation axis of an oriented system of (parallel) lines.

†Subscript $\perp$ and $||$ symbols refer to directions of section plane with respect to the orientation axis in a partially oriented system of lines.

3.4 SYSTEMS OF SURFACES IN SPACE

Even though sections through different planar structures appear to be identical, the spatial configurations may differ markedly. Surfaces play important roles in determining the behavior of materials [11]. The extent and special characteristics of surfaces, both internal and external, must frequently be known quantitatively. The extremely valuable and general relationship for determining the total surface area of any type of planar system has already been presented as Eq. (2.31). We also wish to investigate the additional relationships applicable to systems of oriented surfaces and their two-dimensional section planes.

If just the total surface area of an oriented system of surfaces were desired, we could use Eq. (2.31) provided, of course, that a sufficient number of section planes were used in order to satisfy the requirement of randomness. However, we are also interested in the *type* of orientation, and the *degree* of orientation Ω_{23} of partially oriented systems of surfaces in space. We shall treat these subjects next. We shall discuss completely oriented systems of surfaces first; then we shall classify and analyze partially oriented systems of surfaces in space.

3.4.1 Completely Oriented Systems

In completely oriented systems of surfaces with one orientation axis, two important types are possible. We say that these systems have *planar* orientation if the planes are perpendicular to the orientation axis, and that they have *linear* orientation if the planes are parallel to the orientation axis.

Imagine that the lines in Fig. 3.1 are traces of surfaces normal to the plane of the page. Then if the orientation axis is perpendicular to the traces in part (a), the traces represent systems of surfaces with planar orientation. If, on the other hand, the orientation axes are normal to the plane of the diagrams in parts (b) and (c), the traces

Fig. 3.10. System of lamellar surfaces with planar orientation in an austenitic Fe-30Ni-6Ti alloy. Thin-foil transmission electron micrograph, 31,000 × (courtesy of R. C. Glenn, United States Steel Corp.).

represent systems of planes with linear orientation. In both cases, however, the analysis plane (upon which we make our measurements) is parallel to the orientation axis.

An excellent example of a naturally occurring system of surfaces with planar orientation is shown in Fig. 3.10. The lamellar structure seen in this austenitic Fe-30Ni-6Ti alloy is actually a projection taken by thin foil transmission microscopy.

The surface area per unit volume of parallel plane surfaces bears a simple relationship to the number of intersections per unit length of a perpendicular test line or array. Consider the system of oriented plane surfaces within the cubical test volume of edge lengths l, shown in Fig. 3.11. The plane surfaces may have different shapes and sizes, and may be separated by fixed or variable distances. We select a position of the cube so that two sides are parallel to the system of surfaces. Next we stack identical vertical prisms with square bases equal to Δ^2 throughout the sample cube, perpendicular to the system of surfaces. The number of prisms of base area Δ^2 is equal to l^2/Δ^2, and their total length is $(l^2/\Delta^2)l$.

The number of surface elements carved out by each vertical prism of area Δ^2 is counted, except for those elements with area less than $\frac{1}{2}\Delta^2$. If the first prism has n_1 surface elements, the second prism, n_2, the third prism, n_3, etc., then the area of all surface elements in the first prism is $\Delta^2 n_1$; in the second prism, $\Delta^2 n_2$; etc. The total area of all surface elements in the test volume is

$$S = \Delta^2(n_1 + n_2 + n_3 + \cdots) = \frac{n_1 + n_2 + n_3 + \cdots}{1/\Delta^2}. \tag{3.18}$$

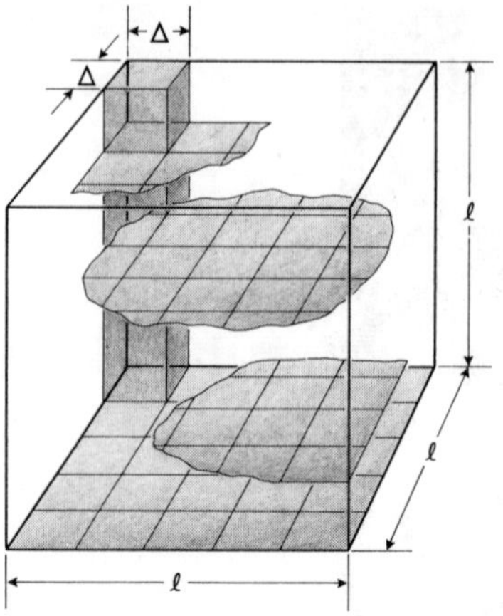

Fig. 3.11. Model for deriving the relationship $(S_V)_{\text{or}} = (P_L)_{\perp\!\perp}$ [5].

Dividing both sides of Eq. (3.18) by l^3 gives

$$\frac{S}{l^3} = \frac{n_1 + n_2 + n_3 + \cdots}{(1/\Delta^2)l^3},$$

where S/l^3 is equivalent to S_V.

In the limit, as the cross-sectional areas Δ^2 of the prisms approach zero, the prisms become test lines perpendicular to the system of surfaces. Then $n_1 + n_2 + n_3 + \cdots$ become the total number of intersections of the test lines with the surfaces, and l^3/Δ^2 equals the total length of the test lines. The ratio of these two quantities equals the number of intersections per unit length of perpendicular test lines, P_L, and

$$\boxed{(S_V)_{\text{or}} = (P_L)_{\perp\!\perp}} \quad \text{mm}^2/\text{mm}^3, \tag{3.19}$$

where $(S_V)_{\text{or}}$ represents the surface area per unit volume of the parallel (oriented) system of plane surfaces, and $(P_L)_{\perp\!\perp}$ is the number of intersections per unit length of a test array perpendicular to the surfaces.

It should be noted that Eq. (3.19) is valid for both planar- and linear-oriented systems of surfaces, provided that the $(P_L)_{\perp\!\perp}$ count is made perpendicular to the surfaces in question.

Comparison of Eqs. (3.4) and (3.19) yields the special result

$$\boxed{(L_A)_{\text{or}} = (S_V)_{\text{or}}} \quad \text{mm}^{-1}, \tag{3.20}$$

where $(L_A)_{\text{or}}$ refers to the parallel traces on section planes normal to the parallel surfaces. Another form of this equation will be used later in Section 3.4.2, when we deal with partially oriented systems of surfaces. Table 3.3A summarizes equations for systems of surfaces applicable to this point.

The space roses for completely oriented systems of surfaces with linear or planar orientation are constructed in a manner similar to that employed for lines in a plane.

TABLE 3.3A

Equations for systems of surfaces in space

Type of system	Isometric surfaces $(S_V)_{is} =$	Oriented surfaces $(S_V)_{or} =$	Total specific surface area $S_V =$	Degree of orientation, Ω_{23}
Isometric	$2P_L$	—	$2P_L$	$\Omega_{23} = 0$
Oriented*	—	$(P_L)_{\perp\perp}$†	$(P_L)_{\perp\perp}$	$\Omega_{23} = 1.0$

*Applies to either completely planar or completely linear systems of surfaces under appropriate conditions.

†The subscript $\perp\perp$ symbol refers to direction of section planes perpendicular to oriented system of (parallel) planes.

Test sections may be restricted to angular positions that correspond to a great circle through the orientation axis, giving a two-dimensional representation. Alternatively, test sections may be taken in every direction over the reference sphere.

The evolution of the space rose from that for an isometric system of surfaces to that for a system of surfaces with 100% planar orientation is portrayed in Fig. 3.12(a), due to Saltykov [5]. The orientation axis is vertical in each case and only axial sections of surfaces of revolution are shown. The space rose progresses from the isometric case (a spherical surface), through increasing percentages of planar orientation, to the completely oriented system of planar surfaces (two spherical surfaces tangent to each other). A similar sequence of space roses is given in Fig. 3.12(b) for an isometric system of surfaces evolving into a system of surfaces with linear orientation.

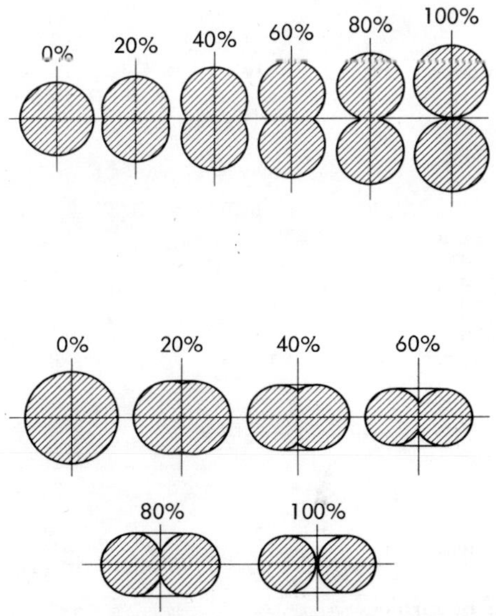

a) Planar systems of surfaces with various percentages of planar orientation.

b) Linear systems of surfaces with various percentages of linear orientation.

Fig. 3.12. Space roses for partially oriented systems of surfaces with planar and linear orientation [5]. (Orientation axes are vertical.)

It can be seen that a torus develops from the spherical surface as the degree of linear orientation increases to 100%. Because of the structural symmetry around the vertical orientation axis, all sections through the orientation axis will appear as tangent circles, which generate the torus surface in three dimensions. The space roses for the intermediate degrees of orientation shown in Fig. 3.12 apply to the partially oriented systems of surfaces discussed in the next section.

3.4.2 Partially Oriented Systems

Saltykov [5] proposes a convenient system of classification for partially oriented systems of surfaces. The three types considered, in addition to the isometric system of surfaces discussed previously, are the partially oriented systems of surfaces having linear, planar, and planar-linear orientations. Figure 3.13 illustrates this classification schematically and designates the various orientation planes, orientation axes, and "analysis planes." In the linear system, the surfaces tend to lie parallel to the orientation axis and the orientation plane; in the planar system the surfaces tend to align themselves perpendicular to the orientation axis and parallel to the orientation plane; and in the planar-linear system, the surfaces have one orientation plane and an orientation axis parallel to the orientation plane in the direction of the longest unit dimension.

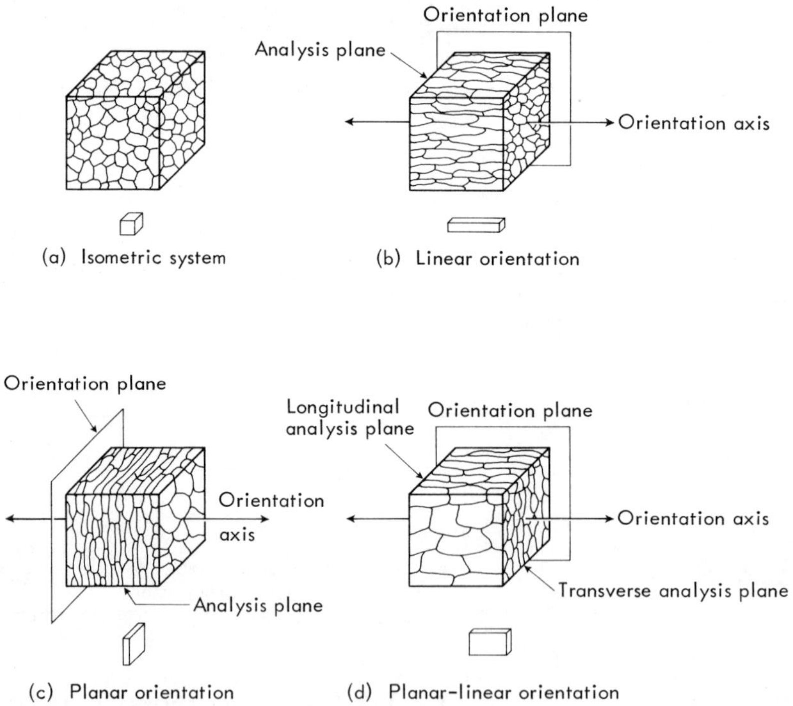

Fig. 3.13. Classification of partially oriented systems of surfaces [5].

The analysis planes are the test planes from which we obtain the experimental data: i.e., the number of intersections with the surface traces per unit length of a test array. After we have established the proper analysis planes, measurements are made parallel or perpendicular to the appropriate orientation axes or planes, similar to the procedure required for partially oriented systems of lines. Although this system of analysis, involving as it does only parallel and perpendicular measurements, is not as accurate as a complete angular analysis, the error is negligible (5).

If the surfaces of an isometric system are divided into small elementary areas of equal size, each of which are plane, then normals to the elementary areas will be oriented randomly in space. An isometric system of surfaces (see Fig. 3.13a) may be defined either as having no orientation axis or as having an infinite number of axes. A partially oriented linear system of surfaces (see Fig. 3.13b) has a major portion of its elementary areas disposed parallel to the orientation axis. The remaining elementary areas are assumed to be oriented isometrically. Although a partially oriented planar system of surfaces (see Fig. 3.13c) also has axial symmetry, the grains are flat instead of elongated. The elementary oriented areas lie parallel to the orientation plane and the remaining areas have isometric symmetry. In both the linear and planar oriented systems, cross sections perpendicular to the orientation axis reveal equiaxed grains.

Both planar and linear orientations, as well as a residue of isometric areas, are present in a partially oriented planar-linear system of surfaces (see Fig. 3.13d). Section planes cut in three mutually perpendicular directions all reveal different types of linear structures in a plane.

The equations for partially oriented surfaces are derived in much the same way as for partially oriented systems of lines in a plane. We shall not go through all these derivations here. In order to demonstrate the procedure, however, we shall derive the equations for partially oriented systems of surfaces with linear orientation.

As is seen in Fig. 3.13(b), only one analysis plane is required for linearly oriented surfaces, i.e., a longitudinal section. On this test section, a linear test array is applied in directions parallel and perpendicular to the orientation axis. Since the oriented boundary traces are parallel to the orientation axis, only the isometric portion of the boundary traces will be intersected by a parallel test array. We may then write

$$(S_V)_{\text{is}} = 2(P_L)_{\parallel} \text{ mm}^2/\text{mm}^3, \tag{3.21}$$

where $(S_V)_{\text{is}}$ is the surface area of the isometric portion of surfaces per unit volume, and $(P_L)_{\parallel}$ is the number of intersections per unit length of the test array in a direction parallel to the orientation axis.

The test array is next placed perpendicular to the orientation axis, so intersections measured in this direction, $(P_L)_{\perp}$, are composed of both isometric and linearly oriented components. The number of intersections per unit length attributable only to the linearly oriented surfaces is $(P_L)_{\perp} - (P_L)_{\parallel}$.

Our problem now is to relate this quantity to $(S_V)_{\text{lin}}$, the area per unit volume of the linearly oriented surfaces. To do this, we need to develop an intermediate expression first for just the linearly oriented portion of the surfaces. Consider a transverse section through a partially oriented linear system of surfaces. The traces of the surfaces appear as linear segments disposed and oriented in a random manner. Thus the length of the oriented traces per unit area is given by

$$(L_A)_{\text{lin}} = \frac{\pi}{2}[(P_L)_\perp - (P_L)_\parallel] \text{ mm/mm}^2, \tag{3.22}$$

where the subscript "lin" refers to the fact that only the traces from linearly oriented surfaces are considered.

Moreover, due to the fact that these traces originate from oriented surfaces that are all lying normal to the transverse section, the trace lengths per unit area are equal to the surface area per unit volume, or

$$(L_A)_{\text{lin}} = (S_V)_{\text{lin}} \text{ mm}^{-1}, \tag{3.23}$$

which is an alternative form of Eq. (3.20). Equating the last two expressions yields

$$(S_V)_{\text{lin}} = \frac{\pi}{2}[(P_L)_\perp - (P_L)_\parallel] \text{ mm}^2/\text{mm}^3, \tag{3.24}$$

which is the desired relationship for a linearly oriented system of surfaces or for the linearly oriented portion of a partially oriented system of surfaces.

The total surface area of a partially oriented system of surfaces with linear orientation per unit volume is

$$S_V = (S_V)_{\text{is}} + (S_V)_{\text{lin}}.$$

By substitution of Eqs. (3.21) and (3.24), we obtain the working equation

$$\boxed{S_V = 1.571(P_L)_\perp + 0.429(P_L)_\parallel} \text{ mm}^2/\text{mm}^3. \tag{3.25}$$

Note that $(P_L)_\perp$ and $(P_L)_\parallel$ are measurements made on a longitudinal section, perpendicular and parallel to the orientation axis, respectively. The derivations for the surface area per unit volume of partially oriented systems of surfaces with planar and planar-linear orientation follow much the same course as outlined above.

We have already shown, in Fig. 3.12, space roses for varying degrees of planar and linear orientation. Similar figures can be built up for specific cases by extension of methods described previously, so we shall not consider further details here. For additional discussion of space roses, see Saltykov [5].

The degree of linear orientation of the partially oriented system of surfaces is defined as

$$\Omega_{\text{lin}} = \frac{(S_V)_{\text{lin}}}{(S_V)_{\text{is}} + (S_V)_{\text{lin}}}.$$

By appropriate substitution, we obtain

$$\boxed{\Omega_{\text{lin}} = \frac{(P_L)_\perp - (P_L)_\parallel}{(P_L)_\perp + 0.273(P_L)_\parallel}} \tag{3.26}$$

in terms of the measurable quantities, $(P_L)_\perp$ and $(P_L)_\parallel$.

The equations for the degree of orientation of partially oriented systems of surfaces with planar or planar-linear orientation will not be derived here. However, the final equations for all three partially oriented systems of surfaces are summarized in Table 3.3B, which points out the similarities and differences. Two general relationships are implied in Table 3.3B. First, the total surface area per unit volume is

$$S_V = (S_V)_{\text{is}} + (S_V)_{\text{pl}} + (S_V)_{\text{lin}} \text{ mm}^2/\text{mm}^3, \tag{3.27}$$

where the subscripts "is", "pl", and "lin" refer respectively to isometric, planar, and linear components of the system of surfaces, if applicable. Secondly, the general equation for the degree of orientation of a system of partially oriented surfaces in space is

$$\Omega_{23} = \frac{(S_V)_{\text{or}}}{S_V}, \tag{3.28}$$

where the subscript "or" refers to the particular oriented surface(s) of interest, i.e., planar, linear, or planar-linear.

Let us consider an example of the application of the equations in Table 3.3B. We select six extruded beryllium specimens with grains in the planar-linear orientation. Measurements of $(P_L)_\perp$, $(P_L)_{||}$, and $(P_L)_|$ were made on transverse and longitudinal analysis planes, as indicated in Fig. 3.13(d). The extrusion direction is parallel to the orientation axis. The results of the basic measurements are as follows.

Specimen designation	Initial powder size, μ	$(P_L)_{\|\|}$, mm^{-1}	$(P_L)_{\|}$, mm^{-1}	$(P_L)_\perp$, mm^{-1}
QMV-1	250	56.8	57.5	69.0
-2	100	69.5	75.6	104.2
-3	44	115.0	127.0	164.8
PEC-1	300	82.0	94.5	124.7
-2	125	82.5	86.3	123.7
-3	74	89.2	121.5	152.1

Assuming planar-linear orientation, we obtain the following values for the QMV-1 sample.

$$(S_V)_{\text{is}} = 2(P_L)_{||} = 2 \times 56.8 = 113.6 \text{ mm}^{-1}$$

$$(S_V)_{\text{pl}} = (P_L)_\perp - (P_L)_| = 69.0 - 57.5 = 11.5 \text{ mm}^{-1}$$

$$(S_V)_{\text{lin}} = 1.571[(P_L)_| - (P_L)_{||}] = 1.571(57.5 - 56.8) = 1.1 \text{ mm}^{-1}$$

$$S_V = (S_V)_{\text{is}} + (S_V)_{\text{pl}} + (S_V)_{\text{lin}} = 113.6 + 11.5 + 1.1 = 126.2 \text{ mm}^{-1}$$

$$\Omega^*_{\text{pl}} = \frac{(S_V)_{\text{pl}}}{S_V} = \frac{11.5}{126.2} = 0.0911$$

$$\Omega^*_{\text{lin}} = \frac{(S_V)_{\text{lin}}}{S_V} = \frac{1.1}{126.2} = 0.0087$$

$$\Omega_{\text{pl-lin}} = \frac{(S_V)_{\text{pl}} + (S_V)_{\text{lin}}}{S_V} = \frac{11.5 + 1.1}{126.2} = 0.0998 = 9.98\% \text{ oriented}$$

TABLE 3.3B

Equations for partially oriented systems of surfaces in space

Type of orientation	Isometric surfaces $(S_V)_{is} =$	Planar surfaces $(S_V)_{pl} =$	Linear surfaces $(S_V)_{lin} =$	Total specific surface area $S_V =$	Degree of orientation, Ω_{23}
Planar†	$2(P_L)_{\parallel}$	$(P_L)_{\perp} - (P_L)_{\parallel}$	—	$(P_L)_{\perp} + (P_L)_{\parallel}$	$\Omega_{pl} = \frac{(P_L)_{\perp} - (P_L)_{\parallel}}{(P_L)_{\perp} + (P_L)_{\parallel}}$
Linear‡	$2(P_L)_{\parallel}$	—	$1.571[(P_L)_{\perp} - (P_L)_{\parallel}]$	$1.571(P_L)_{\perp} + 0.429(P_L)_{\parallel}$	$\Omega_{lin} = \frac{(P_L)_{\perp} - (P_L)_{\parallel}}{(P_L)_{\perp} + 0.273(P_L)_{\parallel}}$
Planar-linear§	$2(P_L)_{\parallel}$	$(P_L)_{\perp} - (P_L)_{\vert}$	$1.571[(P_L)_{\vert} - (P_L)_{\parallel}]$	$(P_L)_{\perp} + 0.429(P_L)_{\parallel} + 0.571(P_L)_{\vert}$	$\Omega^*_{pl} = \frac{(P_L)_{\perp} - (P_L)_{\vert}}{(P_L)_{\perp} + 0.429(P_L)_{\parallel} + 0.571(P_L)_{\vert}}$ ▲ $\Omega^*_{lin} = \frac{1.571[(P_L)_{\vert} - (P_L)_{\parallel}]}{(P_L)_{\perp} + 0.429(P_L)_{\parallel} + 0.571(P_L)_{\vert}}$ $\Omega_{pl\text{-}lin} = \frac{(P_L)_{\perp} - 1.571(P_L)_{\parallel} + 0.571(P_L)_{\vert}}{(P_L)_{\perp} + 0.429(P_L)_{\parallel} + 0.571(P_L)_{\vert}}$

All directions (perpendicular, $\perp$, and parallel, $\parallel$ and $\vert$) of test arrays are referred to orientation planes illustrated in Fig. 3.13.

†‡ Subscripts $\parallel$ and $\perp$ refer to direction of test array placed on the analysis plane.

§ Subscript $\parallel$ here refers to direction of test array placed on the longitudinal analysis plane.
Subscript $\vert$ here refers to direction of test array placed on the transverse analysis plane.
Subscript $\perp$ here refers to direction of test array placed on either the transverse or longitudinal analysis planes.

▲ The asterisks in Ω^*_{pl} and Ω^*_{lin} specify planar and linear orientations in the planar-linear system of surfaces.

Assuming planar orientation:

$$(S_V)_{\text{pl}} = (P_L)_\perp - (P_L)_{||} = 69.0 - 56.8 = 12.2\ \text{mm}^{-1},$$
$$S_V = (S_V)_{\text{is}} + (S_V)_{\text{pl}} = 113.6 + 12.2 = 125.8\ \text{mm}^{-1},$$
$$\Omega_{\text{pl}} = \frac{12.2}{125.8} = 0.097 = 9.7\%\ \text{oriented.}$$

It is apparent that $(P_L)_{||}$ and $(P_L)_{|}$ will have the same value in a system of surfaces with planar orientation. A comparison of the values in these two columns indicates that most samples belong in the planar-linear orientation.

3.5 HILLIARD'S ANALYSIS OF ORIENTED STRUCTURES

Recently, Hilliard [8, 22] has provided a new mathematical framework for the quantitative description of anisotropic microstructures. He has developed a self-consistent and generalized mathematical treatment for both specifying and determining anisotropy. His paper deals with the orientation distribution of lines in a plane, lines in space, and surfaces in space. Spektor [6] has also given an exact treatment for oriented systems of surfaces, but his method is restricted to symmetrical structures.

Hilliard's approach is a completely general one. Two systems of lines are superimposed on one another and their probabilities of intersection are considered. One system of lines is arbitrarily designated as the test array, and the general equation is set up relating the line length of the other system of lines to the number of intersections with the test array. Methods are then indicated for obtaining the specific line length as a function of direction. Similar procedures are followed for systems of surfaces. The advantages of his analysis are that the functions are easy to interpret and provide for direct correlation of structural anisotropy with other properties; disadvantages lie in the number of angular measurements required and in the mathematical complexity of the solutions. The details of the mathematical analysis are beyond the scope of this book, but we shall indicate Hilliard's procedure and give useful equations without derivation. For further details, the reader may consult the original paper.

3.5.1 Lines in a Plane

The orientation distribution of a system of lines in a plane is defined by a function $L_A(\theta)$. $L_A(\theta)\,d\theta$ is the total line length per unit area having tangents with angles in the range $\theta \pm (d\theta/2)$, where θ is measured from an arbitrary reference direction. The line length over any finite range of θ is given by the integral of $L_A(\theta)$ over that range. In particular, the total line length per unit area is

$$L_A = \int_0^\pi L_A(\theta)d\theta\ \text{mm/mm}^2, \tag{3.29}$$

and for a randomly oriented system of lines,

$$L_A = \pi L_A(\theta)\ \text{mm/mm}^2. \tag{3.30}$$

The equation relating $L_A(\theta)$ and $P_L(\omega)$, the expected number of intersections per unit length of the test array with the microstructure, is

$$P_L(\omega) = \int_{\omega-(\pi/2)}^{\omega+(\pi/2)} L_A(\theta) \cos(\theta - \omega)\, d\theta, \tag{3.31}$$

where the angles θ and ω refer to lines on the microstructure and the grid, respectively. The orientation of the test lines is defined by the angle ω that their perpendiculars make with respect to the orientation axis of the system of lines. Two methods for solving Eq. (3.31) for $L_A(\theta)$ are proposed: one is based on second differences and the other on a Fourier analysis.

Experimentally, one uses a linear grid similar to that shown in Fig. 1.3(a), and counts $P_L(\omega)$ at convenient angular increments. For a random array of dislocation lines, for example, $\Delta\omega$ of 15° is quite satisfactory. For a geometrically perfect set of twin traces [12], on the other hand, extremely close-spaced values of $\Delta\omega$ are required in order to obtain satisfactory resolution of the twin traces. Since, in this case, finite line lengths appear only at discrete orientations, the distribution function is the sum of a set of delta functions. In the limit, the function can be represented by a Fourier series with an infinite number of terms. For a finite number of terms, the discrete orientations along which the twin traces appear are represented by peaks of finite width. The widths of the peaks decrease with an increasing number of terms in the Fourier series. The number of terms that must be included depends on the resolution required, and this also determines the intervals $\Delta\omega$ at which $P_L(\omega)$ must be measured.

3.5.2 Lines in Space

Spherical coordinates are used to define the orientation of a three-dimensional system of lineal or areal features. The orientation distribution of a system of lines in space is specified by a function $L_{V}(\phi, \theta)$ defined in terms of a solid angle. Thus $L_{V}(\phi, \theta) \sin\phi \, d\phi \, d\theta$ is the line length per unit volume with orientations in the range $\phi \pm (d\phi/2)$ and $\theta \pm (d\theta/2)$. It follows that the total line length per unit volume is

$$L_V = \int_0^{2\pi} \int_0^{\pi/2} L_V(\phi, \theta) \sin\phi \, d\phi \, d\theta \tag{3.32}$$

where the angles ϕ and θ refer to the tangent to the line. The total length of a random system of lines in a unit volume is

$$L_V = 2\pi L_V(\phi, \theta). \tag{3.33}$$

To analyze the orientation distribution of the system of lines, one applies a test array of parallel planes randomly, but at a specified angular position. This gives

$$P_A(\psi, \omega) = \int_0^{2\pi} \int_0^{\pi/2} L_V(\phi, \theta) \,|\cos\gamma|\, \sin\phi \, d\phi \, d\theta, \tag{3.34}$$

where $P_A(\psi, \omega)$ is the expected number of intersections per unit area of the test planes, and γ is the angle between line segments and the normals to areal elements.

3.5.3 Surfaces in Space

Similarly, the orientation distribution of a system of surfaces in space is specified by $S_V(\psi, \omega)$, also with respect to a solid angle. The total area of the system of surfaces per unit volume is defined as

$$S_V = \int_0^{2\pi} \int_0^{\pi/2} S_V(\psi, \omega) \sin \psi \, d\psi \, d\omega, \tag{3.35}$$

where the angles ψ and ω refer to normals to the boundary elements. The total area of a random distribution of surfaces in a unit volume is

$$S_V = 2\pi S_V (\psi, \omega). \tag{3.36}$$

The analysis of the orientation distribution of a system of surfaces uses a test array of parallel straight lines whose orientation is referred to the same axes as for the system of surfaces. Thus we can write

$$P_L(\phi, \theta) = \int_0^{2\pi} \int_0^{\pi/2} S_V(\psi, \omega) \, |\cos \gamma| \sin \psi \, d\psi \, d\omega, \tag{3.37}$$

where $P_L(\phi, \theta)$ is the expected number of intersections per unit length of the test lines and γ is the same as in Eq. (3.34). Methods for the solution of Eqs. (3.34) and (3.37) are set forth in the original paper [8].

In closing this brief presentation of Hilliard's analysis of anisotropy, let us note that the important equations for S_V, L_A, and L_V in the case of random structures are also obtained as special cases. In addition to Eqs. (2.31), (2.49), and (2.62), Hilliard's analysis also yields the multiple relation

$$P_V = \tfrac{1}{2} S_V L_V \text{ mm}^{-3}, \tag{3.38}$$

where, for a random superimposition of lineal and areal systems, the number of points (intersections of surfaces and lines) per unit volume is expressed in terms of two, more accessible, quantities (see Section 2.7).

In further development of superimposed lineal and areal arrays, Philofsky and Hilliard [20] show how the quantity P_V can be used to determine the orientation distribution of either one of the arrays, provided the orientation of the other array is known.

3.6 LAMELLAR STRUCTURES

At this stage in our discussion, it may be appropriate to analyze in greater detail an oriented structure of considerable importance. These are the lamellar structures, found in many materials, in which alternating lamellae join to create a system of partially oriented surfaces. A well-known example of such a system of surfaces is that of pearlite in steels.

A measure of the fineness of a lamellar structure is the so-called *interlamellar* or *true spacing*, defined as the perpendicular distance across a single pair of contiguous lamellae. The specific interface area of a lamellar structure depends on the magnitude of the true spacing. In fact, both approximate and exact expressions for the interface area have been worked out in terms of various spacings, such as the true spacing, the apparent spacing, and the random spacing.

Because of the wide variation in "spacing" terminology in the literature, and because this terminology is defined inadequately, if at all, a more unified and logical system of nomenclature is suggested for lamellar structures. We define both *spacings* and *distances*, and list three types in each category, i.e., true, apparent, and random. The spacings between alternate lamellae are measured from center to center, while the distances are measured from edge to edge. Figure 3.14 defines the terms schematically, with corresponding symbols and descriptive titles.

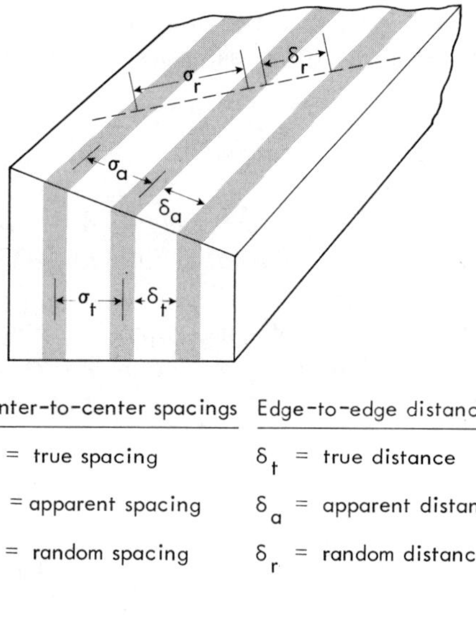

Fig. 3.14. Classification of spacings and distances in a geometrically ideal lamellar structure.

Center-to-center spacings	Edge-to-edge distances
σ_t = true spacing	δ_t = true distance
σ_a = apparent spacing	δ_a = apparent distance
σ_r = random spacing	δ_r = random distance

Both σ_t and σ_a are difficult to determine experimentally and require arduous measurements. A simpler and more readily obtained spacing is the mean random spacing, $\bar{\sigma}_r$, of Gensamer, *et al.* [13]. A test line is applied randomly to the microstructure, intersecting a great number of lamellae at all possible angles in the plane of polish. By dividing the total lengths of test lines by the total number of alternate lamellae intersected, we obtain the mean random spacing $\bar{\sigma}_r$.

The outstanding attribute of $\bar{\sigma}_r$ is that it is related simply to both the total interface area per unit volume, S_V, and the mean true spacing, $\bar{\sigma}_t$. If N_L alternate lamellae are intersected per unit length of test line, then the number of interfaces intersected per unit length is simply $P_L = 2N_L$ and $\bar{\sigma}_r = 1/N_L$. From Eq. (2.31) we obtain

$$S_V = 4/\bar{\sigma}_r \text{ mm}^2/\text{mm}^3. \tag{3.39}$$

This equation is generally valid, provided, of course, that adequate statistical coverage is obtained.

If a unit cube is cut out of an ideal lamellar structure such that two cube faces are parallel to the lamellae surfaces, the total number of pairs of lamellae is $1/\bar{\sigma}_t$ and the total interface area per unit volume is

$$(S_V)_{or} = 2/\bar{\sigma}_t \ \mathrm{mm^2/mm^3}, \tag{3.40}$$

where "or" specifies a completely oriented structure.

For this idealized structure, we see that $(S_V)_{or}$ can be obtained if $\bar{\sigma}_t$ is known, or vice versa. By equating S_V to $(S_V)_{or}$ in Eqs. (3.39) and (3.40) we obtain

$$\boxed{\bar{\sigma}_r = 2\bar{\sigma}_t} \ \mathrm{mm}. \tag{3.41}$$

This important equation is not generally true, but appears to represent real lamellar structures adequately. In fact, it has been found experimentally [13] in pearlitic steels that $\bar{\sigma}_r$ and $\bar{\sigma}_t$ are related by a constant with an average value between 1.9 and 2.0.

An analysis [14] of nonideal lamellar structures (i.e., of partially oriented surfaces with planar orientation) gives an additional term,

$$\boxed{\frac{2}{\bar{\sigma}_r} = \frac{1}{\bar{\sigma}_t} + (P_L)_{\parallel}} \ \mathrm{mm^{-1}}, \tag{3.42}$$

where $(P_L)_{\parallel}$ is the number of intersections per unit length of test array with interface traces of the lamellae parallel to the orientation plane. For a completely oriented lamellar system, $(P_L)_{\parallel}$ is zero, giving Eq. (3.41).

The interlamellar distances are related readily to the interlamellar spacings through the lineal intercept ratio, L_L. From Table 2.2, we see that

$$N_L - \frac{(V_V)_\alpha}{\bar{L}_3} = \frac{(L_L)_\alpha}{\bar{L}_3} \ \mathrm{mm^{-1}}. \tag{3.43}$$

Substituting $\bar{\sigma}_r = 1/N_L$ and $\bar{L}_3 = \bar{\delta}_r$, we obtain

$$\bar{\delta}_r = (L_L)_\alpha \, \bar{\sigma}_r, \tag{3.44}$$

where α represents the light lamellae in Fig. 3.14. The same quantity relates δ_a and σ_a, as well as δ_t and σ_t. It should be pointed out that the mean random distance $\bar{\delta}_r$ for lamellar structures has the same significance as λ, the mean free distance, for general two-phase structures. This important parameter will be discussed in detail in the next chapter.

REFERENCES

1. F. J. Turner and L. E. Weiss, *Structural Analysis of Metamorphic Tectonites*, New York: McGraw-Hill (1963)
2. H. W. Fairbairn, *Structural Petrology of Deformed Rocks*, Reading, Mass.: Addison-Wesley (1949) 304
3. F. Chayes, *Petrographic Modal Analysis*, New York: John Wiley (1956) 16
4. N. T. Belaiew, *J. Iron Steel Inst.* **105** (1922) 203; *Proc. Roy. Soc.* **108A** (1925) 295
5. S. A. Saltykov, *Stereometric Metallography*, second edition, Moscow: Metallurgizdat (1958)

6. A. G. Spektor, *Zavod. Lab.* **20** No. 5 (1954)
7. R. W. Kraft, F. D. Lemkey, and F. D. George, "Quantitative Metallographic Analysis of Linear Features in Anisotropic Structures. Substructure of Lamellar Eutectic Alloy," *Trans. AIME* **224** (1962) 1037
8. J. E. Hilliard, "Specification and Measurement of Microstructural Anisotropy," *Trans. AIME* **224** (1962) 1201
9. W. Schmidt, "Statistische Methoden beim Gefügestudium kristalliner Schiefer," *Sitz. Kaiserl. Akad. Wiss. Wien, Math.-nat. Kl.*, Abt. 1, **126** (1918) 515
10. G. Schoeck, "Correlation Between Dislocation Length and Density," *J. Appl. Phys.* **33** (1962) 1745
11. D. McLean, *Grain Boundaries in Metals*, London: Oxford University Press (1957) 81, 92, 110
12. E. E. Underwood and W. C. Coons, "The Role of Quantitative Stereology in Deformation Twinning," from *Deformation Twinning*, edited by R. E. Reed-Hill, J. P. Hirth, and H. C. Rogers. Published in AIME Metallurgical Society Conference Series, **25** New York: Gordon and Breach Sci. Publ. (1964) 405
13. M. Gensamer, E. B. Pearsall, W. S. Pellini, and J. R. Low, Jr., "The Tensile Properties of Pearlite, Bainite, and Spheroidite," *Trans. ASM* **30** (1942) 983
14. E. E. Underwood, "Surface Area and Length in Volume," in *Quantitative Microscopy*, edited by R. T. DeHoff and F. N. Rhines, New York: McGraw-Hill (1968) 77
15. S. O'Hara, "Dislocations in Webs of Germanium and Silicon," *J. Appl. Phys.* **35** (1964) 409
16. A. S. Yue, "Microstructure of Magnesium-Aluminum Eutectic," *Trans. AIME* **224** (1962) 1010
17. E. W. Sinnott, *The Problem of Organic Form*, New Haven: Yale University Press, (1963) 59
18. J. R. Low and A. M. Turkalo, "Slip Band Structure and Dislocation Multiplication in Silicon-Iron Crystals," *Acta Met.* **10** (1962) 215
19. J. E. Hilliard, personal communication
20. E. M. Philofsky and J. E. Hilliard, "On the Measurement of the Orientation Distribution of Lineal and Areal Arrays," *Quart. Appl. Math.* **27**, 1 (April 1969) 79
21. B. J. Bayles, J. A. Ford, and M. J. Salkind, "The Effect of Elevated-Temperature Exposure on the Microstructure and Tensile Strength of Al_3Ni Whisker-Reinforced Aluminum," *Trans. AIME* **239** (1967) 849
22. J. E. Hilliard, "Determination of Structural Anisotropy," *Stereology*, edited by H. Elias, *Proc. Second Int. Cong. for Stereology*, New York: Springer-Verlag (1967) 219

PROBLEMS

3.1. Construct the rose for the twinned structure in Fig. 3.2. Can the data in Fig. 3.5 be used?

3.2. Show qualitatively the relative appearance of the roses for the systems of lines No. 1 and No. 3 in Fig. 3.1(a).

3.3. a) What is the degree of orientation Ω_{12} for the system of grain boundary traces in Fig. 3.6? (b) Assuming a partially oriented system of surfaces with linear orientation, what is the value of Ω_{23}?

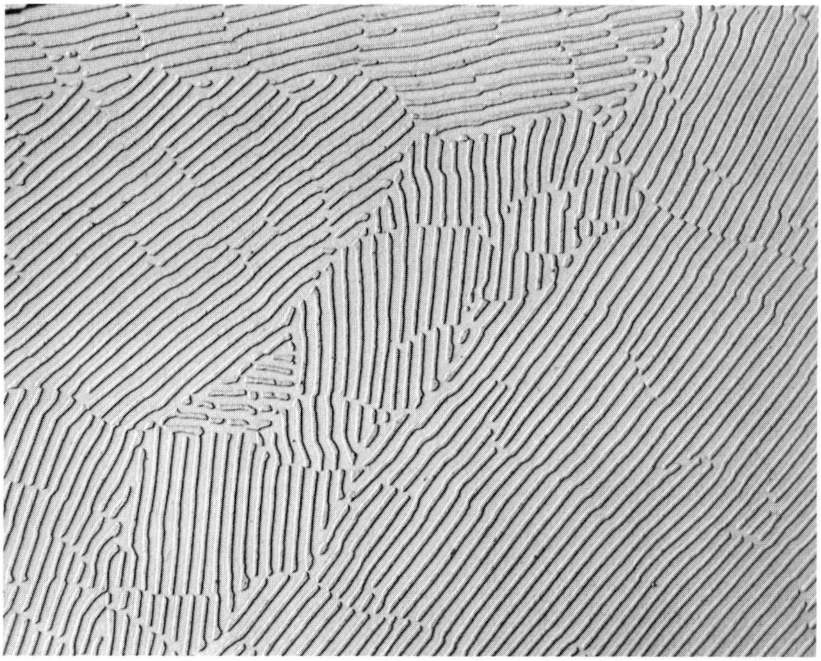

Fig. 3.15. Lamellar structure of a unidirectionally solidified Al-Cu eutectic alloy, 1200×.

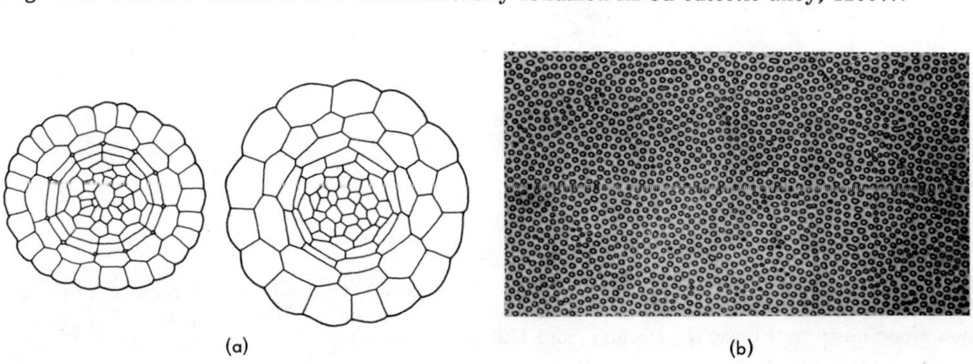

Fig. 3.16. Transverse sections through oriented structures. (a) Two roots. Left, *sporobolus*; right, *agrostis* [17]. (b) Rods formed in Mg-Al eutectic alloy [16].

3.4. Obtain $\bar{\sigma}_t$ for the microstructure in Fig. 3.15, using both Eq. (3.41) and (3.42). [*Note:* Assume that we have an adequate statistical sample.]

3.5. How would you characterize the degree of orientation of the structures shown in parts (a) and (b) of Fig. 3.16? [*Hint:* In (a), how about a radial grid and/or concentric test circles; a rose plot; a spiral grid; etc.? In (b), how about the number of rod sections aligned uniformly within a local area; the frequency of nearest neighbors—that is, 4, 5, 6, 7 . . .—for all rod sections taken one at a time in a chosen

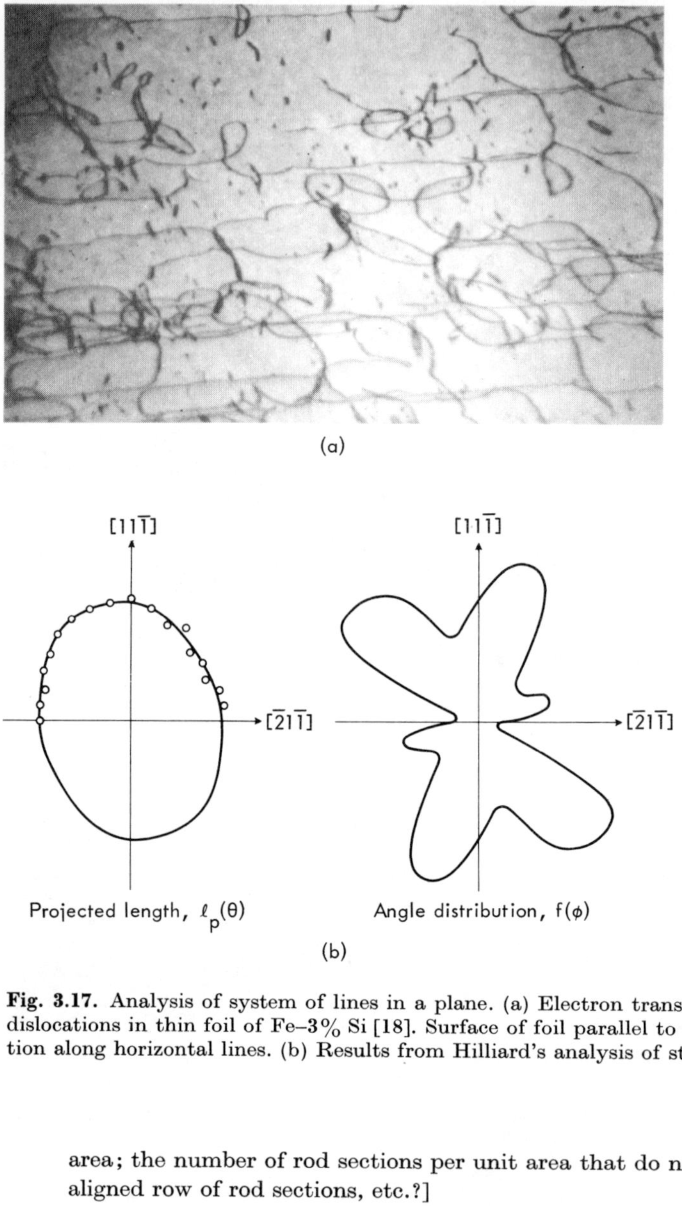

Fig. 3.17. Analysis of system of lines in a plane. (a) Electron transmission micrograph of screw dislocations in thin foil of Fe–3% Si [18]. Surface of foil parallel to (011) slip plane. ⟨111⟩ direction along horizontal lines. (b) Results from Hilliard's analysis of structure in part (a) [19].

area; the number of rod sections per unit area that do not lie along any uniformly aligned row of rod sections, etc.?]

3.6. Figure 3.17(a) is the projected image of an array of screw dislocations in a thin foil of Fe-3% Si single crystal. Assuming that this is a system of partially oriented lines in a plane, (a) measure the rose with respect to an orientation axis parallel to the (approximately) horizontal lines (the ⟨111⟩ direction). (b) Analyze the lines according to Hilliard's [8] method for lines in a plane. [*Note*: His projected length, $l_p(\theta)$, is equal to $P_L(\omega)$ in Eq. (3.31) when $\omega = \pi/2$.] Some of Hilliard's results for this system of lines are given in part (b) of Fig. 3.17.

3.7. In a composite of aligned continuous filaments, a transverse section has the appearance of Fig. 3.16(b). Given the following conditions, what is the relationship of the filament surface area per unit volume $(S_V)_{\text{lin}}$ to the number of filaments per unit area in a transverse plane $(N_A)_{\perp}$?

a) The filaments have circular cross sections of diameter d.

b) The filaments have convex cross sections of mean tangent diameter $\bar{d}$.

c) The filaments have unspecified cross-sectional shapes.

d) The test section normal lies at an angle θ to the orientation axis of filaments with circular cross sections.

[Compare the solution offered in Reference 21.]

4

PARTICLE AND GRAIN CHARACTERISTICS

"But although absolute dimension of particle size may not—because of the lack of uniqueness of that concept—be attainable, we may yet arrive at valid comparisons . . . by using the appropriate statistical technique, and so provide the type of information which is needed in particle technology."
Gustav Herdan

4.1 INTRODUCTION

There are many points of similarity in the geometrical properties of particles and grains. Both represent volume elements with bounded surfaces; the number per unit volume is calculated in much the same way; and similar methods are employed in obtaining mean diameters, size distributions and surface areas. The gross effects of grains or particles on the bulk properties are by no means the same; however, when we are considering their geometrical properties we can usefully discuss both particles and grains on a common basis.

We can approach an analysis of the geometrical properties of these volume elements in several ways. A grain or particle may be considered as a separate entity, and its average properties examined without regard to its surroundings. Alternatively, grains or particles and their surrounding matrix may be considered *in toto* and average properties obtained for the overall aggregate. Moreover, we can measure some properties precisely and without special assumptions. These properties include, for example, surface areas, volumes, and intercept lengths. Other properties cannot be obtained rigorously, especially when the shape of the particles must be specified or when there are many re-entrant angles. In such cases, it may be necessary to approximate the true shape by some geometrical figure and base the calculations on this approximation.

4.2 STATISTICALLY EXACT RELATIONSHIPS

First, let us see what particle and grain characteristics can be derived (without any restrictive assumptions). Foremost among the exact characteristics applicable to systems of particles in a matrix are the volume fraction, the surface area per unit volume, the mean free distance, and the mean random spacing. In addition, the mean volume, the mean surface area, and the mean intercept length have significance both for systems of particles in a matrix and for separate, isolated particles or grains. The appropriate application of the following relationships should be evident from the discussion.

4.2.1 Mean Intercept Length

The mean intercept length of space-filling cells or grains is defined by

$$\bar{L}_3 = \frac{1}{P_L} = \frac{1}{N_L} \text{ mm,} \tag{4.1}$$

where P_L and N_L are the number of intersections per unit length of test line with the particle outlines or with the particles, respectively. For isolated α-particles or cells in a matrix,

$$\bar{L}_3 = \frac{(V_V)_\alpha}{N_L} = \frac{(A_A)_\alpha}{N_L} = \frac{(L_L)_\alpha}{N_L} = \frac{(P_P)_\alpha}{N_L} \text{ mm.} \tag{4.2}$$

The mean intercept length of separated particles may be expressed [1] by

$$\bar{L}_3 = \frac{4\bar{V}}{\bar{S}} \text{ mm,} \tag{4.3}$$

while for space-filling cells and grains, we have

$$\bar{L}_3 = \frac{2\bar{V}}{\bar{S}} \text{ mm,} \tag{4.4}$$

where $\bar{V}$ and $\bar{S}$ refer to the mean volume and surface area, respectively, of space-filling grains or separate particles. These relationships were tabulated and discussed in other contexts in Chapter 2.

An important consequence of Eq. (4.1) is that the mean intercept length is defined for concave as well as convex particles. A simply measured, unique value of $\bar{L}_3$ is obtained for a single particle or aggregate of particles, and this value is directly related to the particle surface and volume. Other implications of this very general "size" parameter were touched upon in Chapter 2.

It might be well at this point to discuss the physical nature of the mean intercept length as opposed to other definitions of "diameter." If we imagine an irregularly shaped convex particle, through which lines are passed at all possible locations and angles, the intercept lengths will vary from zero to a maximum at a particular location and direction. The mean value of the intercept lengths will lie somewhere between these two extremes. In a sphere, for example, the mean intercept length is equal to two thirds of the diameter (see Eq. 4.3).

Other "diameters" are the caliper diameter, tangent diameter, equivalent sphere diameter, Feret's diameter, Martin's diameter, and the statistical diameters, to name a few [2, 3]. Some of these diameters depend on assumed spherical shapes; others on an intermediate linear dimension. Most are based on an arbitrary or maximum dimension.

When we consider the bulk properties of an aggregate containing grains, cells, or particles, it appears that some *average* distance across these randomly oriented volume elements is the important quantity, rather than a maximum dimension [42]. Such an average distance is $\bar{L}_3$, the *mean intercept length.*

The latter is also related directly to the specific surface area of a system of grains or particles. Since in a space-filling system of grains or cells $P_L = N_L$ and $\bar{L}_3 = 1/N_L$, we combine these identities with the general equation

$$S_V = 2P_L \text{ mm}^2/\text{mm}^3 \tag{4.5}$$

to obtain the significant relationship

$$\bar{L}_3 = 2/S_V \text{ mm}. \tag{4.6}$$

By this equation we emphasize the fundamental character of the mean intercept length. Since S_V is uniquely determined for any space-filling aggregate of cells, then $\bar{L}_3$ must also enjoy a unique status, regardless of grain convexity or concavity. It is as fundamental a concept as S_V, which has been reckoned one of the most general and important of the quantities for characterizing grains [4, 5, 6].

4.2.2 Mean Free Distance

A spatial parameter of great importance in particulate systems is the *mean free distance*, λ, between particles, say, of α-phase. As defined by Fullman [7],

$$\boxed{\lambda = \frac{1 - (V_V)_\alpha}{N_L}} \text{ mm}, \tag{4.7}$$

where $(V_V)_\alpha$ refers to the volume fraction of α-particles, or isolated cells, or second-phase regions in a two-phase aggregate, etc., and N_L is the number of interceptions per unit length of test lines with the particles, cells, etc. Equation (4.7) is valid regardless of size, shape, or distribution of particles or other separated volume elements.

The relation derives simply from the fact that the number of interceptions with α-particles is the same as the number of interceptions with the matrix. Thus the lineal fraction occupied by the matrix is λN_L, which is also equal to its volume fraction, $1 - (V_V)_\alpha$.

From the foregoing, it can be seen that the mean free distance, λ, is essentially a mean edge-to-edge distance. It represents the uninterrupted interparticle distance through the matrix averaged between all possible pairs of particles, as opposed, for example, to an interparticle distance between nearest neighbors only [8]. Moreover, contrary to previous statements in the literature, although λ is measured on random section planes, it gives the true three-dimensional distance between particles. A truly random distribution of particles has the same mean interparticle distance on a plane as in a volume, since this geometric property is the same in any direction [9].

The expression for mean free distance in Eq. (4.7) is quite simple, and because there are no special assumptions, possesses great generality. We can extend this general concept to redefine the particle mean intercept length in another way. Briefly, since N_L is the number of interceptions with particles per unit length of test line, the reciprocal, $1/N_L$, is the mean traverse length associated with each particle. This quantity, the *mean random spacing*, σ, represents the mean uninterrupted center-to-center length* between all possible pairs of particles. Therefore we can also define the particle mean intercept length as

$$\bar{L}_3 = \sigma - \lambda \text{ mm} \tag{4.8}$$

*Compare with $\bar{\sigma}_T$ for lamellar structures in Fig. 3.14.

without specification or dependence on particle shape. In essence, $\bar{L}_3$ is a diameter, because σ extends from center to center, and λ measures the edge-to-edge distance. Substitution for σ and λ in Eq. (4.8) gives Eq. (4.2) directly.

4.2.3 Particle Relationships

Several useful expressions for particulate systems can be obtained by combining Eqs. (4.5), (4.7), and (4.8) in various ways. Bearing in mind that $P_L = 2N_L$ in this case, we can write for a system of α-particles

$$\bar{L}_3 = 4\frac{(V_V)_\alpha}{(S_V)_\alpha}\ \text{mm}, \tag{4.9}$$

$$\lambda = 4\frac{1-(V_V)_\alpha}{(S_V)_\alpha}\ \text{mm}, \tag{4.10}$$

$$\boxed{\lambda = \bar{L}_3\frac{1-(V_V)_\alpha}{(V_V)_\alpha}}\ \text{mm}. \tag{4.11}$$

Equation (4.9) (for particles) and Eq. (4.6) (for space-filling cells) are analogous to Tomkeieff's expressions in Eqs. (4.3) and (4.4), respectively. For a constant volume fraction of particles, Eq. (4.11) predicts that the mean free distance varies directly as the particle mean intercept length. The constant of proportionality is related to the volume fraction of particles. Thus Eq. (4.11) has been used to verify the value of volume fraction in a two-phase alloy in which the particle size was varied by heat treatment [10]. The volume per cent determined by point counting (73.5 v/o) and from chemical analysis (71.4 v/o) corresponded well with the volume fraction (73.2 v/o) calculated from the slope of the curve of λ versus $\bar{L}_3$. It is of interest to note that the equivalent equation for spheres of constant diameter [7] includes a factor of two-thirds which converts the diameter to the mean intercept length.

As a final comment on Eq. (4.11), it should be said that, for constant volume fraction, either λ or L_3 could be used as a structural parameter. This point has not been fully appreciated in theories of particle hardening advanced in the past, probably because a shape-independent equation such as Eq. (4.11) was not generally known.

If desired, Eq. (2.49), which is exact,

$$L_A = \left(\frac{\pi}{2}\right)P_L\ \text{mm/mm}^2 \tag{4.12}$$

can be combined with Eqs. (4.7) and (4.8) to yield different combinations of variables. For example, in a particulate system of α-particles

$$\bar{L}_3 = \pi\frac{(V_V)_\alpha}{(L_A)_\alpha}\ \text{mm} \tag{4.13}$$

and

$$\lambda = \pi\frac{1-(V_V)_\alpha}{(L_A)_\alpha}\ \text{mm}, \tag{4.14}$$

where $(L_A)_\alpha$ is the perimeter length of the sectioned particles per unit test area. Substitution of $(A_A)_\alpha$ for $(V_V)_\alpha$ in Eq. (4.13) gives a variant of Tomkeieff's equation [1] (see Tables 2.2 and 2.3).

Inspection of Eqs. (4.9) and (4.10), and of Eqs. (4.13) and (4.14), reveals an interesting result: that is, S_V and V_V as well as L_A and V_V are related in a general way without special assumptions or dependence on shape. By means of the four equations just quoted, the gap in Table 2.1 can be bridged. Specifically, the terms on the right side of the table are linked to the column of terms on the left. The additional parameter required to effect this transition is either λ or $\bar{L}_3$, as defined above.

4.2.4 Nearest-Neighbor Distances

Other distance and spacing parameters are available [11, 12, 13, 14, 15] for expressing separation of particles. Depending on the desired application of the particular "spacing," the definitions are based on one, two, or more "nearest"-neighbor particles. The spacings are called planar or volumetric, regardless of whether the data are taken from sections or from projected images, or whether the particles are randomly distributed or not. Other spacings are calculated from square arrays in a plane, or cubical arrays in a volume.

Two precisely defined nearest-neighbor distances are used by Gurland [8, 11], based on the derivation given by Chandrasekhar [12]. These two distances are denoted by Δ_2 and Δ_3, and are defined as the average distance between nearest-neighbor pairs of point particles in a plane and in a volume, respectively. Strictly speaking, the derived expressions pertain to point particles only, but should apply to aggregates of small, randomly dispersed particles.

The derivations of Δ_2 and Δ_3 run rather similar courses, so we shall employ parallel columns to emphasize their similarities.

Nearest Neighbors in a Volume

We wish to find Δ_3, the mean distance between nearest neighbors in a random distribution of point particles in a volume. The probability that a point particle exists in the spherical shell between R and $R + dR$ is

$$\mathscr{P}(R)\,dR = \left[1 - \int_0^R \mathscr{P}(R)\,dR\right] 4\pi R^2 P_V\,dR, \tag{4.15}$$

where P_V is the mean number of point particles per unit volume and the function $\mathscr{P}(R)$ that satisfies Eq. (4.15) is found to be

$$\mathscr{P}(R) = 4\pi R^2 P_V e^{-(4/3)\pi R^3 P_V}. \tag{4.16}$$

By the definition of average value, we have

$$\Delta_3 = \int_0^\infty R\mathscr{P}(R)\,dR \tag{4.17}$$

Nearest Neighbors in a Plane

We can derive the mean distance of separation Δ_2 of nearest-neighbor pairs of point particles in a plane in a way that is similar to the way we derive Δ_3. The probability of finding a nearest neighbor within the circular band between r and $r + dr$ is

$$\mathscr{P}(r)\,dr = \left[1 - \int_0^r \mathscr{P}(r)\,dr\right] 2\pi r P_A\,dr, \tag{4.15a}$$

where P_A is the mean number of point particles per unit area of the test plane. The function $\mathscr{P}(r)$ that satisfies Eq. (4.15a) is found to be

$$\mathscr{P}(r) = 2\pi r P_A e^{-\pi r^2 P_A}. \tag{4.16a}$$

Equation (4.16a) is substituted into the expression that defines the average distance of separation of nearest-neighbor point pairs,

$$\Delta_2 = \int_0^\infty r\mathscr{P}(r)\,dr, \tag{4.17a}$$

which, when Eq. (4.16) is substituted therein, gives

$$\boxed{\Delta_3 = 0.554\, P_V^{-1/3}}. \tag{4.18}$$

Thus the quantity Δ_3 depends simply on the spatial point density, and appropriate substitutions for P_V (or N_V) can be made from Tables 4.1 or 4.2.

which gives

$$\boxed{\Delta_2 = 0.500\, P_A^{-1/2}}. \tag{4.18a}$$

This expression for Δ_2 can be evaluated by direct substitution of a measurable quantity, P_A (or N_A).

It is evident that the above expressions in Eqs. (4.18) and (4.18a) for point particles are not the same as for the mean free distance λ, in Eq. (4.7) for real particles. The latter represents the average uninterrupted surface-to-surface distance between all possible particle pairs, while Δ_3 and Δ_2 refer to the average distance between nearest-neighbor pairs of point particles.

We can demonstrate approximate relationships between the Δ-parameters and λ by assuming that the points represent a dilute dispersion of randomly distributed spheres of radius r. Using expressions for N_V and r from Table 4.1 for spheres and assuming $\lambda \approx 1/N_L$ for small V_V, we substitute in Eqs. (4.18) and (4.18a), which gives, respectively,

$$\Delta_3 \approx 0.554(\pi r^2 \lambda)^{1/3} \tag{4.19}$$

and

$$\Delta_2 \approx 0.500\left(\frac{\pi}{2} r\lambda\right)^{1/2}. \tag{4.20}$$

We can obtain other approximate relations for Δ_3 and Δ_2 by appropriate substitutions, depending on the actual particle shapes observed. The Δ-parameters should be used only for particulate systems in which either the particle radius and/or the volume fraction are small.

4.3 EXACT RELATIONSHIPS FOR PARTICLES OF REGULAR SHAPES

Frequently the properties of a particulate or cellular structure can be represented satisfactorily by calculations based on a regular polyhedron or volume of revolution. Many types of particles have been employed in these approximations. To name a few, spheres, cubes, ellipsoids, truncated octahedrons, and pentagonal dodecahedrons have been utilized for the above purpose.

It should be emphasized in this regard that by "approximation" we refer to the substitution of a regular solid of simple shape for the real particles of relatively more complicated shapes. Of course, the calculations based on these known bodies can be exact. Average values obtained from an idealized system have agreed quite well with data obtained from the actual aggregate of cells or particles, so the results have some applicability to real structures, within limits.

In this section we attempt to collect and present useful information about regular polyhedrons and volumes of revolution in a systematic fashion. One can readily appreciate why assumptions of regular shape are so attractive. If one can thereby avoid the labor of measuring section sizes and calculating particle sizes, much time and effort can be saved. Ideally, we would like to confine our measurements to simple counting measurements, such as P_P, P_L, N_L, or N_A counts.

Many scientists and mathematicians have contributed to the geometrical analysis of regular polyhedrons [16, 17]. Others have dealt with general relationships for convex particles and figures [1, 18, 19, 20] of unspecified shape that are basic to a geometrical analysis. Among those whose work assumes significance in the present context are Smith and Guttman [21], who present general relationships for two-dimensional figures and three-dimensional objects. Rutherford, Aborn, and Bain [22] made calculations on the truncated octahedron* that form the basis for the ASM Metals Handbook tabulation [23] of spatial and planar grain properties. Kaiser [24] tabulated properties of four basic space-filling regular polyhedrons with plane faces: i.e., the cube, hexagonal prism, rhombic dodecahedron, and the truncated octahedron. Hull and Houk [25] selected four regular polyhedrons for an analysis of intercept area distribution curves. Saltykov [6] reviews work in this field, and contributes some interesting analyses for spheres.

The sphere has been substituted for particles or cells more than any other volume of simple shape. If it can properly be assumed that spheres approximate the real structure satisfactorily, one can readily obtain spatial parameters from simple planar measurements alone. The problem is still simpler if a constant sphere diameter can be assumed.

It remained for Fullman [7], however, to show the power of the statistical-geometrical probability approach to the analysis of simple particles. Let us start this discussion with a review of his important contributions.

4.3.1 General Relationships for Convex Bodies

In two articles appearing in 1953, Fullman [7] develops a technique for determining certain properties of simple solids of revolution. The bodies selected include the sphere, disk, rod, and cylinder. Quantitative relationships are derived between various average quantities obtained from a section plane and the actual particle dimensions. By assuming that the particles in each particulate system have the same size and shape, and by using a combination of geometry and probability theory, one can solve the relationships with mathematical rigor.

In this section, we derive the basic equations applicable to any type of convex particle when intersected by a line or a plane. This may be done by means of the general principles of geometrical probability [19, 28], but we shall adopt a simplified, more intuitive approach here.

There are some obvious but rather basic truisms involved in these derivations that bear repeating. They are:

1) The mean projected area of one particle averaged over all orientations is the same as the mean of the projected areas of a system of identical, randomly oriented particles.

---

*Variously called the (regular) tetrakaidecahedron, cubic octahedron, or cuboctahedron, although among mathematicians, truncated octahedron appears to be preferred.

2) If a single test probe pierces the test volume many (n) times, or if many (n) identical test probes pierce the test volume once, the results are statistically the same (for random penetrations).
3) The fraction of parallel, uniformly distributed test lines (or planes) intersecting a convex particle in a volume equals the ratio of projected area (or height) of particle to projected area (or height) of the test volume.

According to these statements, the properties of systems of particles are closely related to those of a single, randomly oriented particle. The probabilities of intersection are expressed differently, however, depending on the nature of the test probe. In the ensuing discussion, we shall investigate the geometrical relationships of a general convex figure intersected by a test line or a test plane. Note that special attention is given to the question of dimensions and to the definitions of the probabilities (or fractions) involved.

Penetration by Test Lines

We start with a system of polydispersed convex particles, randomly positioned and oriented in space, from which is cut a sample cube of edge lengths l and volume $V_T = l^3$.

A parallel array of n_T test lines of total length $L_T = n_T l$ penetrates the cube uniformly in a direction parallel to one edge. A projection plane is erected perpendicular to the direction of the test lines.

Considering first the relationships applicable to one particle, we write

$$n_x = f_1 n_T, \tag{4.21}$$

which states that the number of test lines n_x intersecting the particle is equal to a fraction f_1 of the total number n_T of test lines. The fraction f_1 is an areal ratio defined by

$$f_1 = \frac{A'}{l^2}, \tag{4.22}$$

where the subscript 1 refers to the dimensionality of the test probes. As seen in Fig. 4.1(a), A' is the projected area of the particle and l^2 that of the test cube on the projection plane.

Substituting for f_1 in Eq. (4.21) and dividing through by L_T gives

$$\frac{n_x}{L_T} = \frac{n_T A'}{L_T l^2} = \frac{A'}{l^3}. \tag{4.23}$$

Penetration by Test Planes

A sample cube of edge lengths l and volume $V_T = l^3$ is cut from a polydispersed system of convex particles that are randomly positioned and oriented in space.

A parallel array of n_T test planes of total area $A_T = n_T l^2$ penetrates the cube uniformly, with the test planes parallel to one side of the cube. A projection plane is erected perpendicular to the test planes and parallel to one of the cube faces.

For one particle, we can write

$$n_x = f_2 n_T, \tag{4.21a}$$

which states that the number of test planes n_x intersecting the particle is equal to a fraction f_2 of the total number n_T of test planes. The fraction f_2 is a lineal ratio defined by

$$f_2 = \frac{H'}{l}, \tag{4.22a}$$

where the subscript 2 refers to the dimensionality of the test probes. As depicted in Fig. 4.1(b), H' is the projected height of the particle and l that of the test cube on the projection plane.

Substituting for f_2 in Eq. (4.21a) and dividing through by A_T gives

$$\frac{n_x}{A_T} = \frac{n_T H'}{A_T l} = \frac{H'}{l^3}. \tag{4.23a}$$

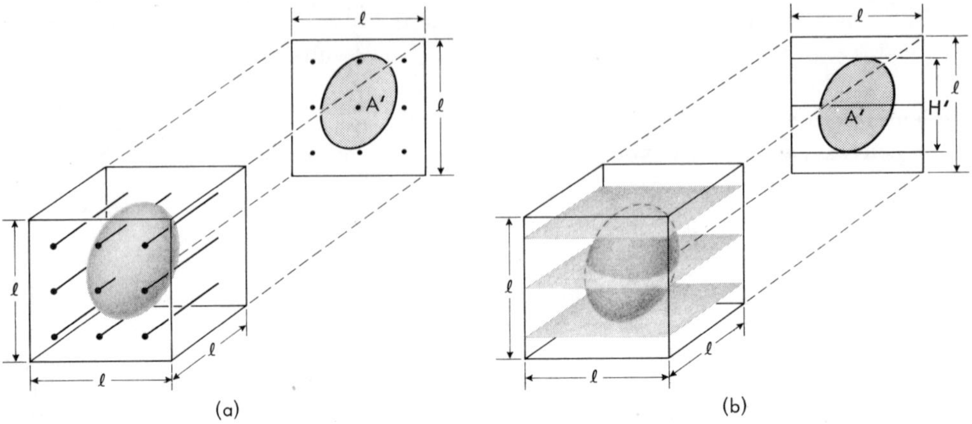

Fig. 4.1(a). Intersection of a convex particle by test lines and their projections. (**b**) Intersection of a convex particle by test planes and their projections.

We next group the system of convex particles into class intervals. For N_i particles in the ith class interval, Eq. (4.23) becomes

$$(N_L)_i = \frac{1}{V_T}(N_i A'_i),$$

where $(N_L)_i$ is now the number of intersections of particles in the ith class interval per unit length of test lines.

Summing over all particles gives

$$\sum (N_L)_i = \frac{1}{V_T} \sum N_i A'_i.$$

Dividing both sides by N, the total number of particles in the test sample, we have

$$\frac{N_L}{N} = \frac{1}{V_T}\left[\frac{N_1A'_1 + N_2A'_2 + N_3A'_3 + \cdots}{N}\right] = \frac{\bar{A}'}{V_T},$$

where $\bar{A}'$ is the mean of the projected areas of the system of randomly oriented particles. Rearranging yields

$$N_L = \frac{N}{V_T}\bar{A}'$$

or, in our notation

$$\boxed{N_L = N_V \bar{A}'} \text{ mm}^{-1}. \quad (4.24)$$

This is the basic equation relating spatial

Next, the system of convex particles are grouped into class intervals. For N_i particles in the ith class interval, Eq. (4.23a) becomes

$$(N_A)_i = \frac{1}{V_T}(N_i H'_i)$$

where $(N_A)_i$ is now the number of intersections of particles in the ith class interval per unit area of test planes.

Summing over all particles gives

$$\sum (N_A)_i = \frac{1}{V_T} \sum N_i H'_i.$$

Dividing both sides by N, the total number of particles in the test sample, we have

$$\frac{N_A}{N} = \frac{1}{V_T}\left[\frac{N_1H'_1 + N_2H'_2 + N_3H'_3 + \cdots}{N}\right] = \frac{\bar{H}'}{V_T},$$

where $\bar{H}'$ is the mean of the projected heights of the system of randomly oriented particles. Rearranging yields

$$N_A = \frac{N}{V_T}\bar{H}'$$

or, in our notation

$$\boxed{N_A = N_V \bar{H}'} \text{ mm}^{-2}. \quad (4.24a)$$

This is the basic equation [6] relating

and planar quantities when random test lines are passed through a system of randomly oriented and positioned convex particles. Since N_L, the number of intersections of particles per unit length of test line, can be measured, we need to evaluate only $\bar{A}'$ in order to obtain N_V, the number of particles per unit test volume.

We can determine $\bar{A}'$ analytically for objects with simple shapes. The mean projected area is given by

$$\bar{A}' = \int_{A'_{\min}}^{A'_{\max}} A'\mathscr{P}(A')\, dA', \tag{4.25}$$

where $\mathscr{P}(A')\, dA'$ is the probability that a given particle has a projected area between A' and $A' + dA'$ on the projection plane.

We illustrate the application of Eq. (4.25) with a thin circular disk of radius r and thickness t, where $r \gg t$. The projected area on the projection plane is

$$A' = \pi r^2 \cos \phi, \tag{4.26}$$

where ϕ is the angle between a test line and the normal to the disk. According to Eq. (2.29), the probability $\mathscr{P}(\phi)\, d\phi$ that the disk makes an angle between ϕ and $\phi + d\phi$ is $\sin \phi\, d\phi$. Then, making the indicated substitutions in

$$\mathscr{P}(A') = \mathscr{P}(\phi)\left|\frac{d\phi}{dA'}\right|, \tag{4.27}$$

we obtain

$$\mathscr{P}(A') = \sin \phi \left|\frac{1}{\pi r^2 \sin \phi}\right| = \frac{1}{\pi r^2}. \tag{4.28}$$

Using this result in Eq. (4.25) gives

$$\bar{A}' = \frac{1}{\pi r^2}\int_0^{\pi r^2} A'\, dA' = \frac{\pi r^2}{2} \tag{4.29}$$

and, when substituted in Eq. (4.24),

$$N_V = \frac{2}{\pi r^2} N_L \text{ mm}^{-3}, \tag{4.30}$$

in terms of the disk radius r. If desired, Eq. (4.30) can be expressed in other terms, such as the thickness t, mean intercept length $\bar{L}_3$, number per unit area N_A, etc. Tables 4.1 and 4.2 consist of a compilation of useful equations for simple convex particles.

spatial and planar quantities when random test planes are passed through a system of randomly oriented and positioned convex particles. Since N_A, the number of intersections of particles per unit test area, can be measured, we need to evaluate only $\bar{H}'$ in order to obtain N_V, the number of particles per unit test volume.

We can determine $\bar{H}'$ analytically [40] for objects with simple shapes. The mean projected height is given by

$$\bar{H}' = \int_{H'_{\min}}^{H'_{\max}} H'\mathscr{P}(H')\, dH' \tag{4.25a}$$

where $\mathscr{P}(H')\, dH'$ is the probability that a given particle has a projected height between H' and $H' + dH'$ on the projection plane.

A thin circular disk of radius r and thickness t, where $r \gg t$, is used to illustrate the application of Eq. (4.25a). The projected height of the disk is

$$H' = 2r \sin \phi, \tag{4.26a}$$

where ϕ is the angle between the normals to the test plane and the particle. The probability $\mathscr{P}(\phi)\, d\phi$ that the disk normal lies within the angle ϕ and $\phi + d\phi$ is $\sin \phi\, d\phi$ (see Eq. 2.29). Then, making indicated substitutions in

$$\mathscr{P}(H') = \mathscr{P}(\phi)\left|\frac{d\phi}{dH'}\right|, \tag{4.27a}$$

we obtain

$$\mathscr{P}(H') = \sin \phi \left|\frac{1}{2\, r \cos \phi}\right| = \frac{\tan \phi}{2r}. \tag{4.28a}$$

Using this result in Eq. (4.25a) gives

$$\bar{H}' = \frac{1}{2r}\int_0^{\pi/2} H' \tan \phi\, dH' = \frac{\pi r}{2} \tag{4.29a}$$

and, when substituted in Eq. 4.24a,

$$N_V = \frac{2}{\pi r} N_A \text{ mm}^{-3}, \tag{4.30a}$$

in terms of the disk radius r. As indicated for Eq. (4.30), other terms may be used here instead of r; the equations are compiled in Tables 4.1 and 4.2.

TABLE 4.1

Properties of particles with surfaces of revolution

	1.	2.	3.	4.
Particle shape \ Symbol	V Volume	S Surface	$\bar{A}'$ Mean projected area	$\bar{H}'$ Mean projected height
Sphere	$\frac{4}{3}\pi r^3$	$4\pi r^2$	πr^2	$2r$
Disk	$\pi r^2 t$	$2\pi r^2$	$\pi r^2/2$	$\pi r/2$
Cylinder	$\pi r^2 h$	$2\pi r(r + h)$	$\frac{\pi}{2}(r^2 + rh)$	$\frac{1}{2}(\pi r + h)$
Rod	$\pi r^2 l$	$2\pi rl$	$\frac{\pi}{2} rl$	$l/2$
Hemisphere	$\frac{2}{3}\pi r^3$	$3\pi r^2$	$\frac{3}{4}\pi r^2$	$\left(1 + \frac{\pi}{4}\right)r$
Prolate spheroid (worked out for $b/a = \frac{1}{2}$)	$\frac{4}{3}\pi ab^2$ $= 8.38b^3$	$2\pi\left(b^2 + ab\frac{\sin^{-1}e}{e}\right)$ $= 23.04b^2$*	$5.76b^2$	$2.76b$
Oblate spheroid (worked out for $b/a = \frac{1}{2}$)	$\frac{4}{3}\pi a^2 b$ $= 2.10a^3$	$2\pi\left(a^2 + \frac{b^2}{2e}\ln\frac{1+e}{1-e}\right)$ $= 6.58a^2$*	$1.645a^2$	$1.70a$

*$e = \sqrt{1 - (b/a)^2}$; for $b/a = \frac{1}{2}$, $e = 0.866$.

5.	6.	7.	8.		
$\bar{A}$ Mean intercept area	$\bar{L}_3$ Mean intercept length	N_V Number per unit volume	r, t, h, l, a, b True dimensions of particles	Ref.	
$\frac{2}{3}\pi r^2$	$4r/3 = 8N_L/3\pi N_A$	$\pi N_A^2/4N_L$	$r = \dfrac{2N_L}{\pi N_A} = \dfrac{3V_V}{4N_L}$	7, 26	Radius, r
$2rt$	$2t$	$2N_A^2/\pi N_L$	$r = N_L/N_A$ $t = V_V/2N_L$	7	Radius, r $r \gg t$ Thickness, t
$\dfrac{2\pi r^2 h}{\pi r + h}$	$\dfrac{2rh}{r + h}$	(if $r = h$) $0.732\dfrac{N_A^2}{N_L}$ $= \dfrac{2N_A}{\pi r + h}$	$h = V_V/\pi N_V r^2$ $r = N_L/2N_A \pm$† $\left(\dfrac{N_L^2}{4N_L^2} - \dfrac{(\pi - 1)V_V}{2\pi N_A}\right)^{1/2}$	7	$r \approx h$ Radius, r Height, h
$2\pi r^2$	$2r = 2N_L/\pi N_A$	$2N_A/l$	$r = N_L/\pi N_A$ $l = 2N_A/N_V$	7	Radius, r Length, l $l \gg r$
$\dfrac{8\pi r^2}{3(4 + \pi)}$	$8r/9$	$0.739\dfrac{N_A^2}{N_L}$	$r = 1.522\dfrac{N_L}{N_A}$	40	Radius, r
$3.025b^2$	$1.46b$	$0.756\dfrac{N_A^2}{N_L}$	$b = V_V/1.46N_L$ $= 0.479N_L/N_A$	26, 28	$2b$ $2a$
$1.235a^2$	$1.278a$	$0.569\dfrac{N_A^2}{N_L}$	$a = V_V/1.278N_L$ $= 1.034N_L/N_A$	26, 28	$2b$ $2a$

†For $r > h$, use + sign. For $r < h$ use − sign.

TABLE 4.2 *Properties of regular polyhedrons*

	1.	2.	3.	4.	5.
Symbol / Particle shape	V Volume	S Surface	$\bar{A}'$ Mean projected area	$\bar{H}'$ Mean projected height	$\bar{A}$ Mean intercept area
Cube	a^3	$6a^2$	$3a^2/2$	$3a/2$	$2a^2/3$
Rectangular parallelepiped	abc	$2(ab + bc + ca)$	$\frac{ab + bc + ca}{2}$	$\frac{a + b + c}{2}$	$\frac{2abc}{a + b + c}$
Square plate	a^2t	$2a^2 + 4at$	$a^2/2$	a	at
Square rod	a^2c	$2a^2 + 4ac$	ac	$c/2$	$2a^2$
Hexagonal prism	$\frac{3\sqrt{3}}{2}a^2c$	$3\sqrt{3}\,a^2 + 6ac$	$\frac{3\sqrt{3}\,a^2}{4} + \frac{3ac}{2}$	$\frac{3a + c}{2}$	$\frac{3\sqrt{3}\,a^2c}{3a + c}$
Tetrahedron	$0.1179a^3$	$1.732a^2$	$0.433a^2$	$0.9123a$	$0.129a^2$
Octahedron	$0.4714a^3$	$3.464a^2$	$0.866a^2$	$1.175a$	$0.4a^2$
Truncated octahedron	$11.314a^3$	$26.785a^2$	$6.696a^2$	$3.0a$	$3.77a^2$
Rhombic dodecahedron	$2a^3$	$6\sqrt{2}\,a^2$	$\frac{3\sqrt{2}}{2}a^2$	$2.0a$	a^2
Pentagonal dodecahedron	$7.663a^3$	$20.646a^2$	$5.161a^2$	$2.57a$	$2.97a^2$

6. L_3 Mean intercept length	7. N_V Number per unit volume	8. a, b, c, t True dimensions of particles	Ref.	
$2a/3$	$2N_A^2/3N_L$	$a = N_L/N_A$	24, 25 26, 30	Edges, a
$\frac{2abc}{ab + bc + ca}$	(If $c = 3a, b = 2a$) $11N_A^2/18N_L$	(If $c = 3a$, $b = 2a$) $a = 6N_L/11N_A$	26, 30	Edges, $a \neq b \neq c$
$2t$	$N_A^2/2N_L$	$a = 2N_L/N_A$	26, 30	Edge, a Thickness, t $a \gg t$
a	$4aN_A^2/cN_L$	$a = N_L/2N_A$	26, 30	Ends, a Length, c $c \gg a$
$\frac{2\sqrt{3}\,ac}{\sqrt{3}\,a + 2c}$	(If $c = a$) $0.7N_A^2/N_L$	(If $c = a$) $a = 0.715N_L/N_A$	24, 26 30	Edges, a Height, c
$0.2725a$	$0.131N_A^2/N_L$	$a = 0.421N_L/N_A$	26, 30	Edges, a
$0.545a$	$0.625N_A^2/N_L$	$a = 0.136N_L/N_A$	26, 30	Edges, a
$1.69a$	$0.744N_A^2/N_L$	$a = 0.45N_L/N_A$	22, 24 25, 26 27, 30	Edges, a
$\frac{4}{3\sqrt{2}}\,a$	$\frac{3\sqrt{2}}{8}\,N_A^2/N_L$	$a = 0.945N_L/N_A$	24, 26 30	Edges, a
$1.485a$	$0.784N_A^2/N_L$	$a = 0.497N_L/N_A$	25, 26 31	Edges, a

4.3.2 Properties of Convex Bodies of One Shape and Size

Fullman's calculations [7] for the sphere, disk, cylinder, and rod are summarized and extended in the top four lines of Table 4.1. The volume and surface area of each type of particle are given first, followed by the average projected area and height. The last four columns in Table 4.1 list quantities of interest for an aggregate of particles of uniform size and the same shape. Note that most quantities are expressed in terms of measurements that can be made on the two-dimensional section plane, for example, N_L, N_A, and V_V. However, in some cases, two-dimensional measurements alone are not sufficient to completely define the particle characteristics (see, for example, the rod).

The general equations and additional relationships used to calculate the quantities listed in Tables 4.1 and 4.2 are assembled below. There is some redundancy, but this should facilitate the selection of the desired equation. Note that projected quantities are distinguished by a prime.

Column number in Tables 4.1 and 4.2	General equations
1.	$V = \bar{A}'\bar{L}_3 = \bar{A}\bar{H}'$
2.	$S = 4\bar{A}' = 4V/\bar{L}_3$
3.	$\bar{A}' = S/4 = V/\bar{L}_3$
4.	$\bar{H}' = V/\bar{A} = \bar{A}'\,\bar{L}_3/\bar{A}$
5.	$\bar{A} = V/\bar{H}' = \bar{A}'\,\bar{L}_3/\bar{H}'$
6.	$\bar{L}_3 = 4V/S = V/\bar{A}' = \bar{A}N_A/N_L$
7.	$N_V = N_A/\bar{H}' = N_L/\bar{A}'$
7–8.	$V_V = N_V\bar{V} = N_A\,\bar{A} = N_L\bar{L}_3$

(4.31)

It is of interest to note that N_V in column 7 can be expressed as a constant times N_A^2/N_L, where the value of the constant is $\bar{A}'/\bar{H}'^2$. Also, in column 8, we see that the particle dimension can be represented by a constant times N_L/N_A, where the value of this constant is a function of the ratio $\bar{L}_3^2/\bar{A}$.

Of special interest in Table 4.2 is the expression quoted by Mack [26] and Hilliard [40] for the mean projected height of convex polyhedrons,

$$\bar{H}' = \frac{\sum l_i(\pi - \theta_i)}{4\pi}, \tag{4.32}$$

where the angle (less than π) between the faces intersecting at the ith edge is θ_i and where this edge is of length l_i. For a cube of edge length a, for example, Eq. (4.32) becomes

$$\bar{H}' = \frac{12a(\pi - \pi/2)}{4\pi} = 3a\left(1 - \frac{1}{2}\right) = \frac{3a}{2}. \tag{4.33}$$

An excellent source book for angles and dimensions of various polyhedrons is that by Cundy and Rollett [27]. We shall derive many of the relationships stated in Eq. (4.31) in Chapter 6, which deals with projected images.

The bottom two lines of Table 4.1 deal with the prolate and oblate spheroids of one arbitrarily selected eccentricity. DeHoff and Rhines [28] and Wicksell [29] have treated ellipsoidal particles in general, so their papers should be consulted if calculations for other eccentricities are needed. To determine whether the particles are prolate or oblate spheroids, see the criteria proposed in Chapter 5 (Section 5.4.2).

Results similar to those in Table 4.1 are available for simple particles with flat sides (regular polygons); these are tabulated in Table 4.2. Most entries for these particles are due to Myers [30], who calculated the basic quantity $\bar{H}'$. Note, however, that the dimensions adopted for some of the particles in Table 4.2 differ from the dimensions used by Myers. Data for the important pentagonal dodecahedron are included in the table. It has been stated [31] that the pentagonal dodecahedron, with its twelve five-sided faces, approaches the characteristics of the average metal grain and cell [4] closer than any other regular polyhedron. We shall say more on this matter in Chapter 7 when we discuss the subject of shape.

4.3.3 Relationships for Convex Bodies of Different Sizes

Tabulations given in Tables 4.1 and 4.2 pertain only to systems of particles of the same size and shape. For size distributions of particles of the same shape, mean values obtained from test sections can be calculated by extension of methods developed for monodispersed systems.

Mean values of V, S, N_V, and r for polydispersed systems of spheres or thin circular disks have been derived by Fullman [7]. He divides the particles into class intervals, each of constant size, applies the equations described above to each class interval, then sums up all values over the range of particle sizes. However, another measurement is required in order to express the average values in terms of measurable two-dimensional quantities. For a distribution of spheres, the required quantity is the mean value $\bar{m}$ of the reciprocals of the circle diameters on the test section.

Some of the equations obtained by Fullman for spheres having a distribution of sizes are reproduced below.

$$\left.\begin{aligned} \bar{V} &= \pi V_V/2\bar{m}N_A \text{ mm}^3 \\ \bar{S} &= 2\pi N_L/\bar{m}N_A \text{ mm}^2 \\ \bar{r} &= \pi/4\bar{m} \text{ mm} \\ N_V &= 2\bar{m}N_A/\pi \text{ mm}^{-3} \end{aligned}\right\} \quad (4.34)$$

The terms in these four equations are defined above. Fullman has also formulated the equations for polydispersed systems of thin disks. The derivations are rather lengthy, so we shall give only the final results here.

$$\left.\begin{aligned} \bar{r} &= \pi/4\bar{E} \\ \bar{t} &= \pi\bar{G}/4\bar{E} \\ \bar{V} &= \pi^2 V_V/8\bar{E}N_A \\ \bar{S} &= \pi^2 N_L/2\bar{E}N_A \\ N_V &= 8\bar{E}N_A/\pi^2 \end{aligned}\right\} \quad (4.35)$$

where $\bar{E}$ is the mean value of the reciprocals of the disk lengths on the section plane and $\bar{G}$ is the mean value of the ratios of the width-to-length dimensions of disk sections. The original articles by Fullman [7] should be consulted for full details.

4.3.4 Relationships Between N_V and N_A

The calculation of N_V for a particulate system from measurements made solely on the plane of polish has always been of interest to scientists. In Chapter 5 we shall discuss methods for obtaining the size distribution of spherical particles from which N_V may be obtained. In Tables 4.1 and 4.2, values of N_V for systems of convex particles of the same size and shape are available. Unfortunately, there is no simple, general equation in terms of quantities obtainable from the plane of polish alone [40].

Equation (4.24a), for convex particles of any shape, is general and exact, but $\bar{H}'$ cannot be determined directly in many cases. For a monodispersed system of spheres, however, it is clear that the diameter D is equal to H' and in a polydispersed system of spheres, $\bar{D}$ is equal to $\bar{H}'$. Thus, for spheres of equal size, we may write

$$N_V = N_A/D \text{ mm}^{-3} \tag{4.36}$$

and for a polydispersed system of spherical particles, we have

$$\boxed{N_V = N_A/\bar{D}} \text{ mm}^{-3}. \tag{4.37}$$

Equation 4.36 can be written separately for each of the k class intervals in the form $(N_A)_{i,j} = (N_V)_j D_j$, where the subscript i, j notation represents section sizes, i, originating from spheres of size j:

$$\left.\begin{aligned}
(N_A)_{i,1} &= (N_V)_1 D_1 \\
(N_A)_{i,2} &= (N_V)_2 D_2 \\
&\ \ \vdots \\
(N_A)_{i,j} &= (N_V)_j D_j \\
&\ \ \vdots \\
(N_A)_{i,k} &= (N_V)_k D_k
\end{aligned}\right\} \tag{4.38}$$

The sum of all the terms on the left, equated to the sum of all the terms on the right, gives

$$\begin{aligned}
N_A &= (N_A)_{i,1} + (N_A)_{i,2} + \cdots + (N_A)_{i,j} + \cdots + (N_A)_{i,k} \\
&= (N_V)_1 D_1 + (N_V)_2 D_2 + \cdots + (N_V)_j D_j + \cdots + (N_V)_k D_k,
\end{aligned} \tag{4.39}$$

where N_A represents the total number of all sections per unit area of the test plane. Finally, by dividing through by N_V, the number of particles per unit test volume, we obtain

$$\frac{N_A}{N_V} = \frac{(N_V)_1 D_1 + (N_V)_2 D_2 + \cdots + (N_V)_j D_j + \cdots + (N_V)_k D_k}{N_V} = \bar{D}, \tag{4.40}$$

which is Eq. (4.37).

Saltykov [6] derives four expressions for a polydispersed system of spheres that have rather interesting similarities. They are:

$$\sum D_j^3(N_V)_j = \frac{3}{2}\sum d_i^2(N_A)_i \quad \text{mm}^0, \tag{4.41}$$

$$\sum D_j^2(N_V)_j = \frac{4}{\pi}\sum d_i(N_A)_i \quad \text{mm}^{-1}, \tag{4.42}$$

$$\sum D_j(N_V)_j = \sum d_i^0(N_A)_i \quad \text{mm}^{-2}, \tag{4.43}$$

$$\sum D_j^0(N_V)_j = \frac{2}{\pi}\sum d_i^{-1}(N_A)_i \quad \text{mm}^{-3}, \tag{4.44}$$

where D_j is the particle diameter in each class interval j; $(N_V)_j$ is the number of particles of size j per unit volume; d_i is the diameter of the intercepted circles in each class interval i; and $(N_A)_i$ is the number of intercepted circles per unit area in each class interval i. Both j and i have integer values up to k, the number of the maximum class interval.

We shall not go through the derivations of these four equations, but shall note the basis of their origins. Equation (4.41) is simply an expression of the equality of volume and area ratios; Eq. (4.42) comes about through the use of the "total projection" which will be discussed in Chapter 6; Eq. (4.43) is the same as Eq. (4.39); and Eq. (4.44) is equivalent to Fullman's Eq. (4.34) for N_V.

Note that only in Eq. (4.44) can N_V for spheres be determined from parameters that are measured directly on the microstructure. We merely sum the inverse diameters of all circular sections per unit area and multiply by the coefficient 0.6366. Then, knowing N_V, we can readily obtain the mean particle diameter $\bar{D}$ from Eq. (4.37).

4.4 PARTICLES EMBEDDED IN INTERNAL BOUNDARIES

Under special circumstances it may be necessary to distinguish between particles (or other objects) located at internal surfaces and those distributed elsewhere throughout the matrix. When we are studying grain growth, for example, the particles obstructing the grain boundaries assume primary importance [44]. We may also wish to measure the grain-boundary density of voids, twin nuclei, cracks, or other features that originate at boundaries.

4.4.1 Basic Relationships

Let us designate by N_S the number of objects per unit area of the internal surface, and reserve N_A for the number of objects per unit area of the test plane. We define N_S as

$$N_S = \frac{(N_V)_b}{S_V} \quad \text{mm}^{-2}, \tag{4.45}$$

where $(N_V)_b$ refers to the number of particles at boundaries per unit test volume, and S_V is the boundary area per unit test volume.

Another quantity of interest here is N_l, the number of particles lying at the boundary traces in the section plane per unit length of boundary trace. This quantity is defined by

$$N_l = \frac{(N_A)_b}{L_A} \text{ mm}^{-1}, \tag{4.46}$$

where $(N_A)_b$ refers to the number of particles at the boundaries per unit area of the section plane and L_A is the length of boundary trace per unit area of the section plane.

The quantities S_V and L_A can be determined without special assumptions, while $(N_A)_b$ and N_l depend on simple particle counts. In order to obtain N_S, however, we must evaluate $(N_V)_b$, and this requires some assumptions about the particles.

Let us assume that all particles are spherical and have the same radius r. Moreover, let us postulate grain boundaries that are randomly oriented and positioned in space, and particles that are uniformly placed throughout space. Thus the particles at grain boundaries will also be uniformly distributed with respect to space. By considering a slice of thickness $2r$ parallel to the plane of polish, we may write

$$(N_A)_b = 2r\,(N_V)_b \text{ mm}^{-2}, \tag{4.47}$$

by analogy with Eq. (4.36). This result yields

$$N_S = \frac{(N_A)_b}{2rS_V} = \left(\frac{\pi}{4}\right)\frac{N_l}{2r} \text{ mm}^{-2}, \tag{4.48}$$

in terms of measurable quantities and r. Expressions for the radius are given in Table 4.1; however, note that N_L refers to $(N_L)_b$, the number of intersections of particles at the boundary per unit length of test line, and N_A means $(N_A)_b$.

The relationship of N_S to N_V is also desired. By setting up the ratio of $(N_V)_b$ to N_V, we can substitute in Eq. (4.45), which gives

$$N_S = 2rN_V \text{ mm}^{-2}. \tag{4.49}$$

The derivation of this equation is left as a problem at the end of the chapter. It should be emphasized that particles are frequently located preferentially at boundary surfaces. The three equations immediately above are based on the assumption of a uniform distribution of particles in space.

4.4.2 Application

The foregoing discussion has application to the problem considered by Zener and quoted by Smith [32]. Grain growth, in the presence of particles, is impeded where the migrating boundary contacts the particles. This is seen in Fig. 4.2 from Gladman [44]. When the particles exert the maximum restraining effect, the grain radius R depends on the radius r of spherical particles approximately as

$$\boxed{R = \frac{4r}{3V_V}} \text{ mm}, \tag{4.50}$$

where V_V refers to the volume fraction of all particles. The constant 4/3 is obtained by using Eq. (4.49) and correcting other errors in the previously published expression [32].

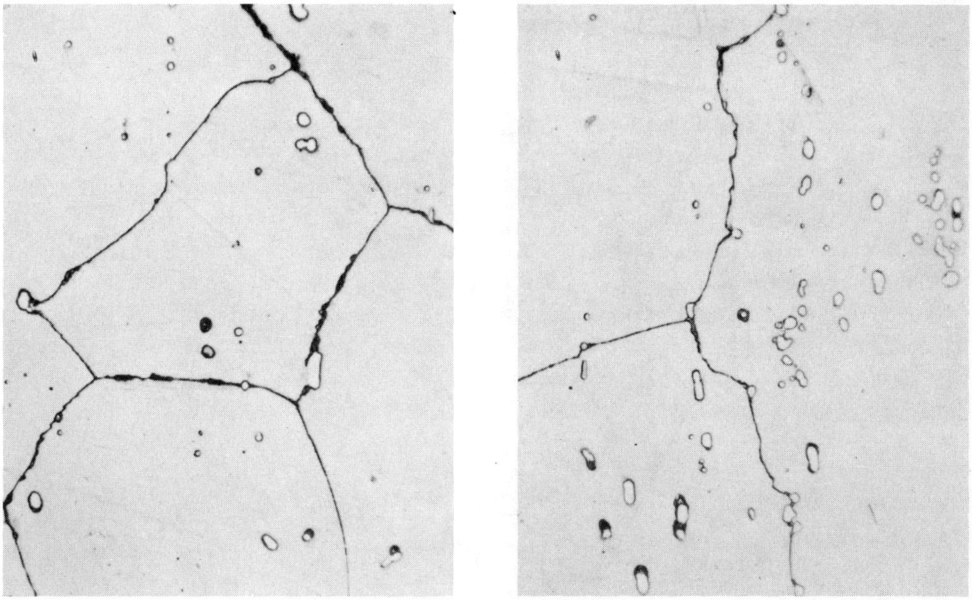

Fig. 4.2. Particles impeding grain boundaries in aluminum-treated steel [44] (750×).

Fullman's [7] expression for a system of spherical particles having the same radius

$$r = \frac{3V_V}{4N_L} \text{ mm} \tag{4.51}$$

is introduced into Eq. (4.50), giving

$$R = \frac{1}{N_L} = \sigma \text{ mm.} \tag{4.52}$$

Thus the optimum restraining effect of spherical particles should occur when the mean particle spacing, σ, equals the grain radius, R (and not the diameter, as reported previously [33]). It is noteworthy that particle size, volume fraction, or other parameters do not enter into the critical condition expressed by Eq. (4.52).

4.5 CONTIGUITY

In particulate (as well as polyphase) systems with small volume fractions of the second phase, there is negligible contact between particles if they are randomly dispersed. In other cases, there may be appreciable contact among volume elements in a microstructure (see Fig. 4.3).

The degree of contact can exert a significant influence on the properties of a structure. For example, the optimum spacing of fissionable particles in dispersion-type reactor fuels, the stress distributions in particle-strengthened alloys, and the susceptibility to crack propagation in a brittle second phase—all depend to some extent on the contacts between specific microstructural volume elements. In such systems

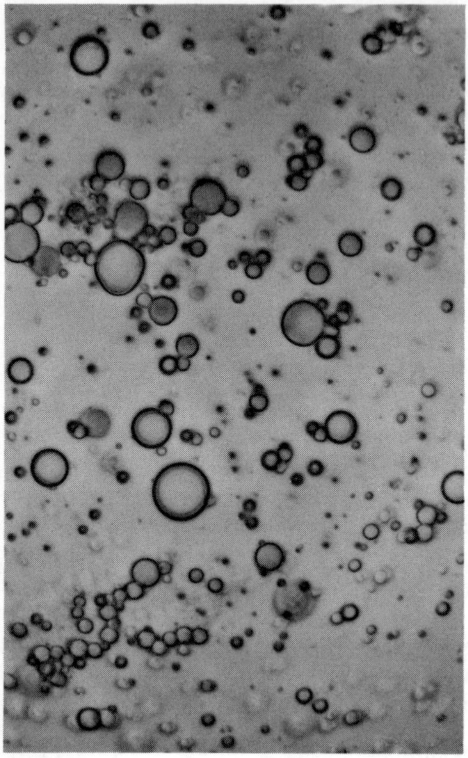

Fig. 4.3. Contiguity and continuity in an emulsion of resin in water [43] (1000×).

we may wish to inquire not only into the mean interparticle spacings, but also into the degree of contact, or the *contiguity*, of the particles.

4.5.1 Exact Relationships

Gurland [34] defines the contiguity C as the fraction of the total interface area of a phase that is shared by particles of the same phase. For a particulate structure with particles of α-phase in a homogeneous matrix of β-phase, the contiguity of α-phase is given by

$$C_{\alpha\alpha} = \frac{2(S_V)_{\alpha\alpha}}{2(S_V)_{\alpha\alpha} + (S_V)_{\alpha\beta}}, \tag{4.53}$$

where $(S_V)_{\alpha\alpha}$ and $(S_V)_{\alpha\beta}$ are the interface areas per unit volume between α-α and α-β phases, respectively. In this case, surfaces shared by two α-particles are counted twice.

By utilizing Eq. (2.31), we may express the contiguity as

$$\boxed{C_{\alpha\alpha} = \frac{4(P_L)_{\alpha\alpha}}{4(P_L)_{\alpha\alpha} + 2(P_L)_{\alpha\beta}}} \tag{4.54}$$

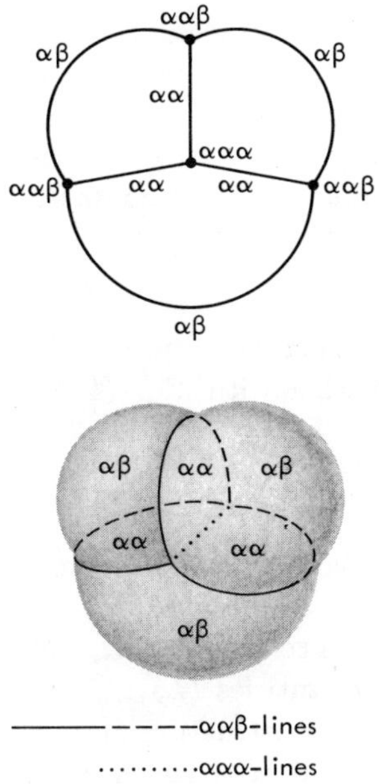

a) Two-dimensional section through α-particles in a β-matrix.

b) Three-dimensional α-particles in a β-matrix.

Fig. 4.4. Contiguous surfaces, lines, traces, and points. α = volume (α-phase); β = volume (β-phase); $\alpha\alpha$ = surface (between α-phases); $\alpha\beta$ = surface (between α- and β-phases); $\alpha\alpha\alpha$ = line (where three α-phase edges meet); $\alpha\alpha\beta$ = line (where two α-phase and one β-phase edges meet).

where $(P_L)_{\alpha\alpha}$ and $(P_L)_{\alpha\beta}$ are the number of intersections per unit length of test line with $\alpha\alpha$ and $\alpha\beta$ interface traces. Figure 4.4 illustrates the pertinent geometrical factors involved in calculating the contiguity.

Other parameters have been proposed by Cahn and Hilliard [35] for describing quantitative aspects of particle contact. One such parameter is $(L_V)_{\alpha\alpha\beta}$, the $\alpha\alpha\beta$ perimeter length per unit volume around contact areas between α-α particles; another parameter is $(L_V)_{\alpha\alpha\alpha}$, the $\alpha\alpha\alpha$ edge length per unit volume between three contiguous α-particles. These terms are depicted schematically in Fig. 4.4 for α particles in a β-matrix.

Both $(L_V)_{\alpha\alpha\beta}$ and $(L_V)_{\alpha\alpha\alpha}$ are obtained from counts on the plane of polish of the points created at the junction of three interface traces, that is, $(P_A)_{\alpha\alpha\beta}$ and $(P_A)_{\alpha\alpha\alpha}$, respectively. According to Eq. (2.62), edge lines (such as the dotted line in Fig. 4.4b) are related to points (as indicated by $\alpha\alpha\alpha$ in Fig. 4.4a) by the expression

$$(L_V)_{\alpha\alpha\alpha} = 2(P_A)_{\alpha\alpha\alpha} \text{ mm}^{-2}. \tag{4.55}$$

The total perimeter length per unit volume around $\alpha\alpha$-surfaces, $(L_V)_{\alpha\alpha}$, consists of contributions from both $\alpha\alpha\alpha$- and $\alpha\alpha\beta$-lines, which in turn involve both $\alpha\alpha\alpha$ and $\alpha\alpha\beta$ junction points, as shown in Fig. 4.4. We add up the contributions from each type of junction point to get

$$\boxed{(L_V)_{\alpha\alpha} = 6(P_A)_{\alpha\alpha\alpha} + 2(P_A)_{\alpha\alpha\beta}} \text{ mm}^{-3}, \tag{4.56}$$

where the coefficient 6 includes a factor of three because each $\alpha\alpha\alpha$-edge is shared by three $\alpha\alpha$-contacts. Each $\alpha\alpha\beta$-edge, on the other hand, adjoins only one $\alpha\alpha$-contact.

An alternative method for obtaining $(L_V)_{\alpha\alpha}$ is to count interface traces instead of triple points. Cahn and Hilliard [35] derive an expression for $(L_V)_{\alpha\alpha}$ in terms of the numbers, $(N_A)_{\alpha\alpha}$ or $(N_A)_{\alpha\beta}$, of interface traces (from $\alpha\alpha$- or $\alpha\beta$-contacts) seen in a unit area on the section plane. The problem is to relate the number of traces to the number of triple points. Since each $\alpha\alpha\alpha$-edge is the terminus of 3 $\alpha\alpha$-contacts, and each $\alpha\alpha\beta$-edge is the terminus of one $\alpha\alpha$-contact,

$$(N_A)_{\alpha\alpha} = \tfrac{3}{2}(P_A)_{\alpha\alpha\alpha} + \tfrac{1}{2}(P_A)_{\alpha\alpha\beta}\ \mathrm{mm}^{-2}; \tag{4.57}$$

the factor of $\frac{1}{2}$ in each term arises because each interface trace has two termini on the section plane. By multiplying Eq. (4.57) by four and comparing with Eq. (4.56), we see that

$$\boxed{(L_V)_{\alpha\alpha} = 4(N_A)_{\alpha\alpha}}\ \mathrm{mm}^{-2}. \tag{4.58}$$

Thus the perimeter lengths $(L_V)_{\alpha\alpha}$ can be obtained either by counting contact traces (Eq. 4.58) or by counting "triple" points (Eq. 4.56).

Cahn and Hilliard [35] generalize the treatment given above to include $\alpha\alpha$, $\alpha\beta$, $\alpha\gamma$, etc., combinations, but the procedures are the same. The quantities $(L_V)_{\alpha\alpha\alpha}$ and $(L_V)_{\alpha\alpha\beta}$ may or may not be useful, depending on the type of microstructure and on the use to be made of the measurements. By a suitable combination of parameters, we can derive others, such as the average perimeter-to-area ratio per contact, $(L_V)_{\alpha\alpha}/(S_V)_{\alpha\alpha}$.

The above expressions dealing with contiguity are perfectly general, and can be evaluated by simple measurements on the section plane. For some purposes, additional parameters may be required that are not, in general, measurable on a single section plane. One of the more important of these is the average number of contacts per unit volume, $(N_V)_{\alpha\alpha}$, between particles of the same phase.

4.5.2 Approximate Relationships

By assuming that all $\alpha\alpha$-contacts are flat circular areas of uniform size, we can obtain $(N_V)_{\alpha\alpha}$ simply from measurements on random planes. We use the expression for N_V of thin circular disks from Table 4.1, since the result is the same for two-dimensional circles. Modifying the nomenclature appropriately, we have

$$(N_V)_{\alpha\alpha} = \frac{2(N_A)^2_{\alpha\alpha}}{\pi(P_L)_{\alpha\alpha}}\ \mathrm{mm}^{-3} \tag{4.59}$$

where $(N_A)_{\alpha\alpha}$ and $(P_L)_{\alpha\alpha}$ are defined above. Note that N_L for thin disks here becomes $(P_L)_{\alpha\alpha}$ for the circular contact areas.

The additional assumption that all particles are uniform-sized spheres enables the average number of contacts per grain, $(N_c)_{\alpha\alpha}$, to be calculated readily. The latter quantity is defined by Gurland [34] as

$$(N_c)_{\alpha\alpha} = \frac{2(N_V)_{\alpha\alpha}}{(N_V)_{\alpha}}, \tag{4.60}$$

where the term in the numerator represents the number of α-grain contacts and the denominator is the total number of α-grains, both per unit volume.

$(N_V)_\alpha$ is given in Table 4.1 as N_V for uniform spheres, and $(N_V)_{\alpha\alpha}$ appears in Eq. (4.59). Cahn and Hilliard [35] treat the derivation of $(N_c)_{\alpha\alpha}$ in a more general fashion by incorporating a shape factor and the average and standard deviation of perimeter lengths. Even so, their expression cannot be evaluated by measurements on a single section plane unless a particle shape is assumed.

Another attribute of particulate systems is called [34, 36] the "continuity" of the particles, which is defined as the probability of the presence of infinitely long chains of connected particles. Considerations of the probabilities of such chain formation fall under the theory of branching chains [37], a discussion of which would lead us too far afield. For the reader interested in further details, this subject is treated in papers by Gurland [34] and Forscher [36].

4.6 STATISTICAL DIAMETERS

To round out this chapter on particles and grains, let us consider the important class of parameters known as *statistical diameters*. The latter are useful because—since they are based on a simple linear dimension—they offer a way to treat irregular particles quantitatively. Such particles do not possess a unique diameter, but do have a definite volume and surface area which vary respectively as the cube and square of a linear dimension. Thus, in order to calculate any of several mean diameters, we measure some arbitrary "diameter" in a consistent manner on a large number of particles.

4.6.1 Definitions of Mean Diameters

Before listing the usual statistical diameters [2, 3], let us first define the arithmetic, geometric, and harmonic mean diameters of a polydispersed system of N particles of the same shape. If $N_1, N_2, N_3, \ldots.$ are the number of particles having "diameters" $D_1, D_2, D_3, \ldots$, respectively, then we may define the following mean diameters.

$$\text{Arithmetic mean diameter, } \bar{D}: \quad \bar{D} = \frac{\sum (N_i D_i)}{\sum N_i} = \frac{N_1 D_1 + N_2 D_2 + \cdots + N_N D_N}{N_1 + N_2 + \cdots + N_N}. \tag{4.61}$$

The arithmetic mean diameter is simply the sum of the diameters of all particles divided by the total number of particles. (This follows directly from Eq. (4.61) for all $N_i = 1$, that is, only one particle of each size.)

$$\text{Geometric mean diameter, } D_g: \quad D_g = (D_1 D_2 D_3 \cdots D_N)^{1/N}. \tag{4.62}$$

A more convenient way of expressing D_g is in the logarithmic form, i.e.,

$$\log D_g = \frac{\sum (N_i \log D_i)}{\sum N_i} = \frac{N_1 \log D_1 + N_2 \log D_2 + \cdots + N_N \log D_N}{N_1 + N_2 + \cdots + N_N}. \tag{4.63}$$

The geometric mean diameter owes its importance in particle-size analysis to the fact that many particle-size distributions conform satisfactorily to the log-normal law (see Section 5.5). In this case, the geometric mean represents the value with the greatest frequency of particles and is, therefore, a better representative of particle size than the arithmetic mean. The geometric mean diameter is always smaller than the arithmetic mean diameter.

$$\text{Harmonic mean diameter, } D_h\text{:}\quad D_h = \left[\frac{1}{\sum N_i} \sum N_i/D_i\right]^{-1}. \tag{4.64}$$

The harmonic mean diameter is inversely related to the sum of the reciprocals of the measured diameters. This reciprocal relationship is seen more readily in the form

$$\frac{1}{D_h} = \left(\frac{N_1}{D_1} + \frac{N_2}{D_2} + \cdots + \frac{N_N}{D_N}\right)\bigg/(N_1 + N_2 + \cdots + N_N). \tag{4.65}$$

D_h is always less than D_g, therefore $\bar{D} \geq D_g \geq D_h$. The inequalities increase with the degree of dispersion of diameters in the system of particles. The harmonic mean diameter is employed frequently to correlate with specific surface areas; it also occurs as $\bar{m}$ and $\bar{E}$ in Eqs. (4.34) and (4.35), which deal with properties of polydispersed systems of spheres and disks. Note that $\bar{D}$, $\log D_g$, and $1/D_h$ represent averages that are weighted by number; that is, by N_i and $\sum N_i$.

In selecting an "average" diameter, one should give attention to its application. One should seek that average which is most likely to correlate with a given property of the material. For example, if the physical property being investigated is dependent on the particle interface area, we should use a mean diameter related to the surface area, such as the harmonic mean diameter. The arithmetic and geometric mean diameters, on the other hand, have no readily apparent physical significance.

4.6.2 Definitions of Statistical Diameters

The statistical diameters are defined so as to relate to definite geometrical properties of the particles, such as particle volume, surface area, linear dimension, etc. Moreover, the statistical diameters are weighted differently, by numbers or frequencies, by weight or volume, or by particle surface. A convenient general expression [38] for classifying the statistical diameters is

$$D_{m,n} = \left[\frac{\sum N_i D_i^m}{\sum N_i D_i^n}\right]^{1/(m-n)}, \tag{4.66}$$

where m and n are integers whose values determine the various types of averages. Thus, for $m = 1$, $n = 0$, we have $\bar{D}$, and for $m = -1$, $n = 0$, we obtain D_h. The statistical diameters given below are identified by subscripts according to the convention adopted in Eq. (4.66).

$$\text{Mean surface diameter, } D_{2,0}\text{:}\quad D_{2,0} = \left(\frac{\sum N_i D_i^2}{\sum N_i}\right)^{1/2}. \tag{4.67}$$

Here $D_{2,0}$ is that mean diameter whose corresponding mean surface area, $S = K_2(D_{2,0})^2$, divided into the total particle surface, gives the total number of particles. The shape factor K_2 cancels out in the above equation.

$$\text{Mean volume diameter, } D_{3,0}\text{:}\quad D_{3,0} = \left(\frac{\sum N_i D_i^3}{\sum N_i}\right)^{1/3} \tag{4.68}$$

The mean volume diameter may be thought of as that mean diameter whose corresponding mean volume, $V = K_3(D_{3,0})^3$, divided into the total particle volume, gives the total number of particles. The shape factor K_3 also cancels out here.

$$\text{Mean length diameter, } D_{2,1}\text{:} \quad D_{2,1} = \frac{\sum N_i D_i^2}{\sum N_i D_i}. \tag{4.69}$$

Here there is essentially a surface area term, weighted with respect to a linear dimension. Actually the ratio represents the second moment divided by the first moment of particle diameters.

$$\text{Mean volume-surface diameter, } D_{3,2}\text{:} \quad D_{3,2} = \frac{\sum N_i D_i^3}{\sum N_i D_i^2}. \tag{4.70}$$

This mean diameter is weighted with respect to surface area and is appropriate for correlations involving particle surface area. All particle-size distributions that have the same value of $D_{3,2}$ will have the same total interface area. None of the other mean diameters can provide this correlation.

$$\text{Mean weight (or volume) diameter, } D_{4,3}\text{:} \quad D_{4,3} = \frac{\sum N_i D_i^4}{\sum N_i D_i^3}. \tag{4.71}$$

$D_{4,3}$ is equally valid for weight or volume averaging because the particle density (gm/cc) cancels out. This diameter can be considered to be weighted with respect to volume.

Although it is possible to define still other mean diameters, the above list is restricted to those mean diameters found to be most useful [2, 3].

The values of these statistical diameters are systematically different for a given size distribution of particles. As a rule, it appears that the magnitudes of the mean diameters increase with the sum of the indices m and n [2]. The difference between the statistical diameters diminishes as the uniformity (i.e., of sizes) increases. When the particles have the same size, the statistical diameters all have the same value.

Because of the large number of possible mean diameters, it becomes mandatory to specify the type of mean that is used. As for the selection of a particular mean diameter, the choice should be determined by the proposed application of the data.

Methods have been described [2, 3] to determine average particle size in particulate systems in which both size *and* shape change. The subject is difficult and not too well developed, but for a discussion of the problems involved, see the paper by Green [39], as well as Chapter 7.

REFERENCES

1. S. I. Tomkeieff, "Linear Intercepts; Areas and Volumes," *Nature* **155** (1945) 24; correction, page 107
2. G. Herden, *Small Particle Statistics*, second edition, London: Butterworths (1960)
3. J. M. DallaValle, *Micromeritics, The Technology of Fine Particles*, second edition, New York: Pitman (1948)
4. C. S. Smith, "Grain Shapes and Other Metallurgical Applications of Topology," *Metal Interfaces*, ASM, Cleveland, O. (1952) 65

5. D. McLean, *Grain Boundaries in Metals*, Oxford: Clarendon Press, London (1957) 81, 92, 110
6. S. A. Saltykov, *Stereometric Metallography*, second edition, Moscow: Metallurgizdat (1958)
7. R. L. Fullman, "Measurement of Particle Sizes in Opaque Bodies," *Trans. AIME* **197** (1953) 447; 1267
8. J. Gurland, "Spatial Distribution of Discrete Particles," in *Quantitative Microscopy*, edited by R. T. DeHoff and F. N. Rhines, New York: McGraw-Hill (1968) 278
9. E. E. Underwood, discussion of "The Effect of a Dispersed Second Phase on the Recrystallization of Aluminium-Copper Alloys," *J. Inst. Met.* **92** (1963–64) 124
10. E. E. Underwood, unpublished research (1963)
11. J. Gurland, "The Measurement of Distribution, Spacing, Contact and Continuity of Particles in a Matrix," *Powder Metallurgy in the Nuclear Age*, Plansee Proceedings (1961), edited by F. Benesovsky (published in 1962) 507
12. S. Chandrasekhar, "Stochastic Problems in Physics and Astronomy," *Rev. Mod. Phys.* **15,** 1 (Jan. 1943) 86
13. M. F. Ashby and R. Ebeling, "On the Determination of the Number, Size, Spacing and Volume Fraction of Spherical Second-Phase Particles from Extraction Replicas," *Trans. AIME* **236** (1966) 1396
14. K. H. Westmacott, C. W. Fountain, and R. I. Stirton, "On the Spacing of Dispersed Obstacles," *Acta Met.* **14** (1966) 1628
15. U. F. Kocks, "On the Spacing of Dispersed Obstacles," *Acta Met.* **14** (1966) 1629
16. H. S. M. Coxeter, *Regular Polytopes*, second edition, New York: Macmillan (1963)
17. L. A. Lyusternik, *Convex Figures and Polyhedra*, New York: Dover (1963)
18. H. Hadwiger, *Altes und Neues über konvexe Körper*, Basel: Birkhäuser Verlag (1955)
19. M. G. Kendall and P. A. P. Moran, *Geometrical Probability*, No. 10 of Griffin's Statistical Monographs and Courses, London: Charles Griffin and Co. (1963)
20. A. Cauchy, "Note sur divers théorèmes relatifs à la rectification des courbes, et à la quadrature des surfaces," Paris: C. R. Acad. Sci. **13** (1841) 1060
21. C. S. Smith and L. Guttman, "Measurement of Internal Boundaries in Three-Dimensional Structures by Random Sectioning," *Trans. AIME* **197** (1953) 81
22. J. J. B. Rutherford, R. H. Aborn, and E. C. Bain, "The Relation Between the Grain Areas on a Plane Section and the Grain Size of a Metal," *Metals and Alloys* **8** (Dec. 1937) 345
23. *Metals Handbook*, Cleveland, Ohio: Amer. Soc. for Metals (1948) 405
24. H. F. Kaiser, "The Estimation of Inter-Granular Surfaces and Volumes," *Metals and Alloys* **9** (January 1938) 23
25. F. C. Hull and W. J. Houk, "Statistical Grain Structure Studies: Plane Distribution Curves of Regular Polyhedrons," *Trans. AIME* **197** (1953) 565
26. C. Mack, "On Clumps Formed when Convex Laminae or Bodies Are Placed at Random in Two or Three Dimensions," *Proc. Cambr. Phil. Soc.* **52** (1956) 246
27. H. M. Cundy and A. P. Rollett, *Mathematical Models*, second edition, London: Oxford University Press (1961)
28. R. T. DeHoff and F. N. Rhines, "Determination of Number of Particles per Unit Volume from Measurements Made on Random Plane Sections: The General Cylinder and the Ellipsoid," *Trans. AIME* **221** (1961) 975

29. S. D. Wicksell, "The Corpuscle Problem I. A Mathematical Study of a Biometric Problem," *Biometrika* **17** (1925) 84; Part II, **18** (1926) 151
30. E. J. Myers, "Sectioning of Polyhedrons," Proc. First Int. Congress for Stereology, Vienna Medical Academy, Vienna (1963) 15/1; 15/7
31. M. Paulus, "Méthode d'étude de la structure granulaire, I," *Métaux Corrosion-Industries*, **37**, 448 (Dec. 1962) 447
32. C. S. Smith, "Grains, Phases, and Interfaces: An Interpretation of Microstructure," *Trans. AIME* **175** (1948) 15
33. E. E. Underwood and G. K. Manning, "Effects of Creep Stress on Processes Occurring at Elevated Temperatures in Al-Cu Alloys: Particle Dissolution, Grain Growth, and Precipitation," *Mem. Sci. Rev. Métallurgie* **60**, 9 (1963) 648
34. J. Gurland, "The Measurement of Grain Contiguity in Two-Phase Alloys," *Trans. AIME* **212** (1958) 452
35. J. W. Cahn and J. E. Hilliard, "The Measurement of Grain Contiguity in Opaque Samples," *Trans. AIME* **215** (1959) 759
36. F. Forscher, "Analysis of Continuity of One Phase in a Powder Mixture of Two Phases," *J. Franklin Inst.* **259** (1955) 107
37. W. Feller, *Probability Theory and Its Applications*, Volume I, New York: John Wiley (1950) 272
38. J. E. Hilliard, "Grain-Size Estimation," *G.E. Res. Lab. Report* No. 62-RL-3133M, December 1962
39. H. Green, "The Effect of Nonuniformity and Particle Shape on Average Particle Size," *J. Franklin Inst.* **204** (1927) 713
40. J. E. Hilliard, "The Calculation of the Mean Caliper Diameter of a Body for Use in the Analysis of the Number of Particles per Unit Volume," *Stereology*, edited by H. Elias, *Proc. Second Int. Cong. for Stereology*, New York: Springer-Verlag (1967) 211
41. J. Gurland, "A Study of Contact and Contiguity of Dispersions in Opaque Samples," *Stereology*, edited by H. Elias, *Proc. Second Int. Cong. for Stereology*, New York: Springer-Verlag (1967) 250
42. J. E. Hilliard, "Estimating Grain Size by the Intercept Method," *Met. Prog.* (May 1964) 99
43. T. G. Rochow, "Resinography, The Microscopy of Resins and Their Plastics," Symposium on Light Microscopy, S. T. P. No. 143, Am. Soc. Testing Mat. (1952) 81
44. T. Gladman, "On the Theory of the Effect of Precipitate Particles on Grain Growth in Metals," *Proc. Roy. Soc.* **294A** (1966) 298

PROBLEMS

4.1. What is the mean free distance, λ, between graphite nodules in Fig. 4.5(a)? What is the mean intercept length, $\bar{L}_3$, of the nodules in Fig. 4.5(b)? If $\lambda + \bar{L}_3$ (times N_L) does not equal 1, to what would you ascribe the discrepancy?

4.2. Derive the relationship for α-particles in a β-matrix given by Gurland [41]

$$\bar{L}_{\alpha\alpha} = \frac{\bar{L}_\alpha}{1 - C_{\alpha\alpha}}$$

between the mean intercept length across α-phase clumps ($\bar{L}_{\alpha\alpha}$), the mean intercept

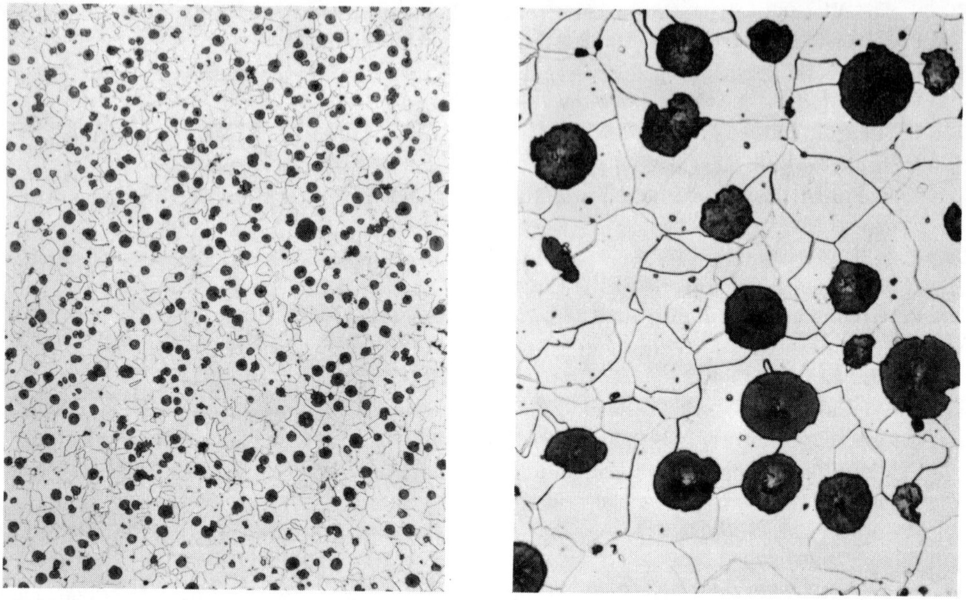

(a) As-cast structure (100X) (b) Same structure (500X)

Fig. 4.5. Ferritic cast iron with approximately spherical graphite nodules. (Courtesy of Casting Materials Co., Inc., White Plains, N.Y.)

length across particles of α-phase considered individually ($\bar{L}_\alpha$), and the contiguity ($C_{\alpha\alpha}$). [Note that Gurland uses λ_α instead of $\bar{L}_{\alpha\alpha}$.]

4.3. Using the relationships in Table 4.1 and Eqs. (4.31), derive expressions for the thickness t and mean intercept length $\bar{L}_3$ of a thin circular disk.

4.4. Compare the mean projected heights, $\overline{H}'$, for a truncated octahedron and pentagonal dodecahedron of the same volume.

4.5. Calculate the quantities in Table 4.1 (columns 1 through 6) for a double convex lens of diameter $2a$ and thickness $2t$. [*Note*: Consider the double convex lens to consist of two spherical segments of a sphere of radius r placed back to back. Then $S = 2\pi(a^2 + t^2)$ and $V = (\pi t/3)(t^2 + 3a^2)$.]

4.6. What is the value of N_S (Eq. 4.48) in Fig. 4.5(b), assuming that all particles have the same radius and lie in the grain boundaries?

4.7. Derive Eq. (4.49) under the assumptions that the particles are uniformly distributed in space, that they have the same radius r, and that all particle centers lie in the boundary.

4.8. What is the relationship between the radius r of a system of spherical particles of the same size and the surface area per unit volume, S_V?

5

PARTICLE- AND GRAIN-SIZE DISTRIBUTIONS

"It is important that the methods of particle-size analysis be thoroughly understood. . . We must realize the manifold limitations of each method in order to appreciate the true significance of the measurements made. . . . The characteristics and behavior of small particles can be understood satisfactorily only when the methods of determining size-distributions are placed on a firm basis."

J. M. DallaValle

5.1 INTRODUCTION

In this chapter we consider a more complex problem than those treated previously. We wish to determine the spatial size distribution of grains, cells, precipitate particles, air bubbles in concrete, etc. (hereafter called "particles" for short), from measurements on a random test plane. Whereas the relationships dealt with in Chapter 2 are capable of statistically exact solutions, this is not the case for size distributions of real particles with irregular shapes. Some progress has been made in the treatments of nonspherical shapes [1–11]; however, most analytical methods depend on the assumption of spherical particles [7–25].

5.2 BASIC CONCEPTS OF PARTICLE SIZE ANALYSIS

A system of particles is often represented simply by the mean particle diameter or by the number of particles per unit volume. However, in order to characterize a particulate structure more completely, a knowledge of the distribution of particle sizes is required. The distribution curve is a rather unwieldy device for most practical applications, so numerical parameters that will adequately define the distribution curve are used instead [10, 17]. Three numerical parameters commonly employed are:

1) $\bar{D}$, the mean diameter,
2) $\sigma(D)$, the standard deviation, and
3) N_V, the total number of particles per unit volume.

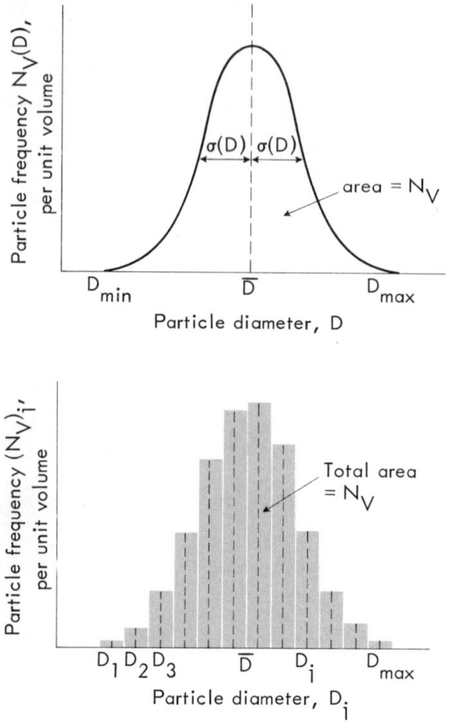

a) Continuous frequency distribution curve.

b) Discontinuous frequency distribution curve (histogram).

Fig. 5.1. Schematic representation of frequency distribution curves and important defining parameters.

The relationship of these three quantities to the frequency distribution curve* is portrayed schematically in Fig. 5.1. It is important to note that a continuous frequency distribution curve is not always used. Rather, we subdivide the diameters into discrete size groups (class intervals) and deal with an assumed constant diameter within each class interval.

5.2.1 Definition of Terms

For a polydispersed system of spheres, the arithmetic mean diameter is

$$\bar{D} = \frac{1}{N_V}[(N_V)_1 D_1 + (N_V)_2 D_2 + \cdots + (N_V)_j D_j + \cdots + (N_V)_{\max} D_{\max}]$$

$$= \frac{1}{N_V} \sum (N_V)_j D_j, \qquad (5.1)$$

where $(N_V)_1$, $(N_V)_2$, ... $(N_V)_j$, ..., $(N_V)_{\max}$ are the number of spheres in each class interval with diameters D_1, D_2, ..., D_j, ..., $D_{\max}$, respectively, and N_V is the total number of particles, all per unit volume.

---

*The typical frequency distribution curve for a particulate system has its peak nonsymmetrically nearer to the smaller diameters, and is skewed toward the larger particles. Thus the log-normal distribution is more appropriate for small particle statistics. Further details are given in Section 5.5.1.

The standard deviation $\sigma(D)$ of sphere diameters D_j may be expressed [26, 27] as the root-mean-square deviation

$$\sigma(D) = [\overline{(D_j - \bar{D})^2}]^{1/2} = \left[\frac{1}{N_V} \sum (N_V)_j(D_j - \bar{D})^2\right]^{1/2} \tag{5.2}$$

or, in an equivalent form, as

$$\sigma(D) = [\overline{D_j^2} - \bar{D}^2]^{1/2} = \left[\frac{1}{N_V} \sum (N_V)_j D_j^2 - \left(\frac{1}{N_V} \sum (N_V)_j D_j\right)^2\right]^{1/2}, \tag{5.3}$$

where $\overline{D_j^2}$ denotes the mean of the squares of the various values of D_j and $\bar{D}^2$ is the square of the mean diameter.

As pointed out previously, N_V bears a simple relationship to N_A, the total number of circular sections per unit area [12, 14–17, 29]. According to Eq. (4.36), for a monodispersed system of spherical particles of diameter D cut by random test planes, we have $N_V = N_A/D$. For a polydispersed system of spheres, Eq. (4.40) gives the overall relationship $N_V = N_A/\bar{D}$, where $\bar{D}$ is the mean diameter of all spheres [17, 33, 48].

The latter equation can be deduced readily for a system of randomly positioned spherical particles. On the average, spheres are intersected by a unit test plane only if their centers lie within a distance $\bar{D}/2$ on either side of the unit plane, i.e., within a volume $\bar{D}$. The number of spheres within the volume $\bar{D}$ must equal N_A. Therefore the number of spheres per unit volume, N_V, is equal to $N_A/\bar{D}$.

If the polydispersed system of spheres is divided into class intervals, we can write

$$(N_V)_j = \sum_i (N_A)_{i,j}/D_j \ \text{mm}^{-3}, \tag{5.4}$$

where the j's refer to sphere diameters and the i's refer to section diameters. Since all particles in each class interval are assumed to have the same size, this expression can be applied in turn to each class interval. For a particular sphere size j in Eq. (5.4), the term $\sum_i (N_A)_{i,j}$ expresses the number of sections of all sizes per unit area obtained from particles of a fixed size j. This quantity cannot be obtained directly from the plane of polish, of course, but must be calculated. Another quantity of interest is $\sum_j (N_A)_{i,j}$, which represents the number per unit area of sections of constant size i, obtained from spheres of all possible sizes. This is a directly measurable quantity.

5.2.2 Basic Relationships

Figure 5.2 illustrates a polydispersed system of spheres and the relationship between three-dimensional and two-dimensional features [28]. Note that the circular sections have a distribution of sizes, even though they are obtained from spheres of uniform diameter D_j.

The problem we wish to solve is how to work backward from the distribution of two-dimensional sections to the distribution of three-dimensional spheres. The problem is further complicated by the fact that spheres with different diameters, $j, j+1, j+2$, etc., will contribute sections of the same size i, so the source of each section must be determined.

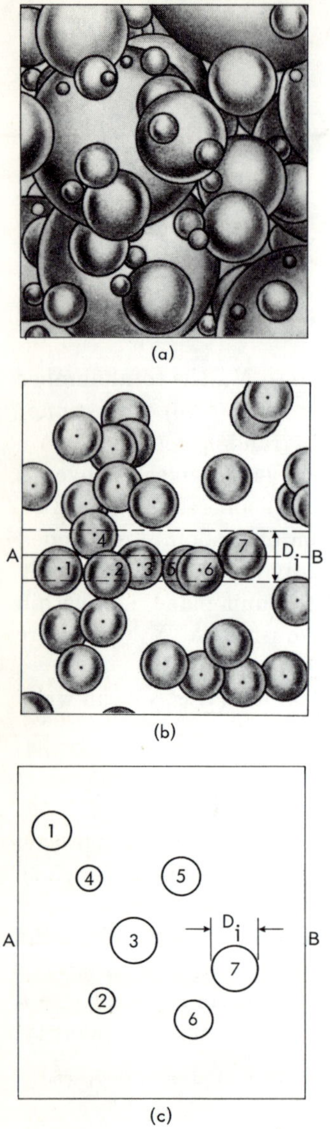

a) Polydispersed system of spheres.

b) Monodispersed system of spheres of diameter D_j.

c) Circular sections resulting from a plane intersecting the monodispersed system through AB.

Fig. 5.2. Features of a distribution of spherical particles in space [28].

First, in order to deduce how many sections of each size come from spheres in each class interval, we evaluate the probabilities of random planes intersecting a sphere to give sections of each size. If $\mathscr{P}_{i,j}$ is the probability of a test plane intersecting a sphere of diameter j to yield sections of diameter i, then

$$\mathscr{P}_{i,j} = \sum_j (N_A)_{i,j} \Big/ \sum_i (N_A)_{i,j}, \tag{5.5}$$

where the terms have been defined above. For spheres of fixed diameter j, the numerator of Eq. (5.5) represents the number of sections of *one* size i per unit area, while the denominator represents the total number of sections of *all* sizes per unit area.

this is wrong $P_{i,j} = \dfrac{(N_A)_{i,j}}{\sum_i (N_A)_{i,j}}$

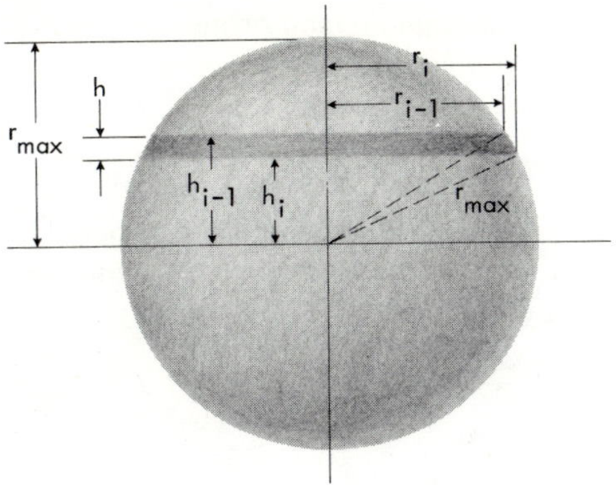

Fig. 5.3. Geometry involved in the intersection of a sphere by a plane within the slice h.

Substituting Eq. (5.5) into Eq. (5.4) gives the general expression

$$(N_V)_j = \frac{\sum_j (N_A)_{i,j}}{\mathscr{P}_{i,j}} \frac{1}{D_j} \quad \text{mm}^{-3}, \tag{5.6}$$

from which $(N_V)_j$ may be determined, provided the proper values of the probability, sphere diameter, and number of sections per unit area are known. $\sum_j (N_A)_{i,j}$ is either measured directly from the plane of polish or calculated, depending on the combination of values of i and j; $\mathscr{P}_{i,j}$ is calculated for the appropriate section radius and class interval; and D_j is either the diameter of the largest section observed on the plane of polish or some fractional value thereof, depending on the number of class intervals.

The $(N_V)_j$ may be either an unknown or a known, depending on the particular step in the sequence of calculations. When we know $(N_V)_j$, then we use Eq. (5.6) in the form

$$\sum_j (N_A)_{i,j} = \mathscr{P}_{i,j} D_j (N_V)_j \quad \text{mm}^{-2}. \tag{5.6a}$$

The probability of a random plane intersecting a sphere to give different section radii [16, 29, 30] must be obtained indirectly. The calculation is based on simple geometrical considerations. Consider a sphere of maximum radius* $r_{\max}$ cut randomly by a plane.

The probability of a plane cutting the sphere within a slice of height h is equal to

$$\mathscr{P}_{i,j} = \frac{h}{r_{\max}} = \frac{h_{i-1} - h_i}{r_{\max}}, \tag{5.7}$$

where the symbols are described by Fig. 5.3. The probability may be expressed in terms of the radii, as

$$\mathscr{P}_{i,j} = \frac{1}{r_{\max}} [\sqrt{(r_{\max})^2 - (r_{i-1})^2} - \sqrt{(r_{\max})^2 - (r_i)^2}]. \tag{5.8}$$

*The expression "maximum radius" refers here to the sphere radius of each class interval in turn. Thus when considering the fifth class interval, $r_{\max} = r_5$; when considering the fourth, $r_{\max} = r_4$, etc.

It is immaterial whether a random plane intersects one sphere many times or if many identical spheres are intersected by one plane; the probabilities are the same.

5.2.3 Illustration of the Calculation of Size Distributions of Spheres

Let us apply the principles presented above to a simple example. Take a polydispersed system of spheres that has been separated into five class intervals. The number of spheres per unit volume in each class interval is

$$(N_V)_1, \quad (N_V)_2, \quad (N_V)_3, \quad (N_V)_4, \quad \text{and} \quad (N_V)_5,$$

and their corresponding radii are

$$\tfrac{1}{5}r, \quad \tfrac{2}{5}r, \quad \tfrac{3}{5}r, \quad \tfrac{4}{5}r, \quad \text{and} \quad r,$$

or, in simpler notation,

$$r_1, \quad r_2, \quad r_3, \quad r_4, \quad \text{and} \quad r_5.$$

The total number of circular sections of a given size per unit area, regardless of origin, is

$$(N_A)_1, \quad (N_A)_2, \quad (N_A)_3, \quad (N_A)_4, \quad \text{and} \quad (N_A)_5,$$

where the section radii vary from

$$0 \text{ to } r_1, \quad r_1 \text{ to } r_2, \quad r_2 \text{ to } r_3, \quad r_3 \text{ to } r_4, \quad \text{and} \quad r_4 \text{ to } r_5,$$

respectively.

Note that the latter $(N_A)_i$ terms represent the measured quantities.

Figure 5.4 illustrates the possibilities arising from a random plane intersecting particles of all five sizes. For example, spheres of the largest class interval ($r_5 = \frac{1}{2}D_5$) yield sections of all radii when cut by a random plane. Alternatively, the observed number of sections of the smallest size per unit area, $(N_A)_1$, is the sum of contributions from spheres of all sizes. Expressed in descriptive notation, the latter statement becomes

$$(N_A)_1 = \sum_j (N_A)_{i,j} = (N_A)_{1,1} + (N_A)_{1,2} + (N_A)_{1,3} + (N_A)_{1,4} + (N_A)_{1,5}, \qquad (5.9)$$

where the i's refer to the (constant) section size and the j's to the sphere diameters. Similar equations can be written for the other section sizes $(N_A)_2$, $(N_A)_3$, etc. Unfortunately, $(N_A)_1$ in Eq. (5.9) cannot be used directly, since $(N_A)_{1,1}$ is the quantity required for calculation.

We wish to calculate from Eq. (5.6) the number of spheres per unit volume in each class interval $(N_V)_1$, $(N_V)_2$, $(N_V)_3$, . . . containing spheres of radius r_1 $(= D_1/2)$, r_2 $(=D_2/2)$, r_3 $(= D_3/2)$, . . . , respectively. In order to do so, we determine the number of sections having the same radius as their spheres $(N_A)_{1,1}$, $(N_A)_{2,2}$, $(N_A)_{3,3}$, etc. We start with the largest particles and their sections. The reason for this stems from the fact that only the largest sections of the largest particles can be unambiguously identified on the plane of polish as belonging to a particular class interval.

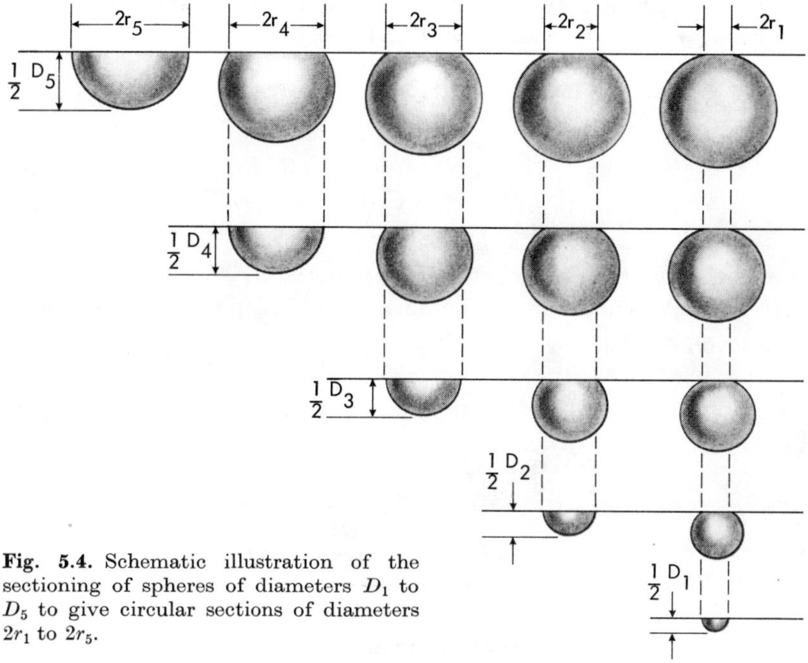

Fig. 5.4. Schematic illustration of the sectioning of spheres of diameters D_1 to D_5 to give circular sections of diameters $2r_1$ to $2r_5$.

The appropriate forms of Eq. (5.9) for our five class intervals are as follows:

$$\left.\begin{aligned}(N_A)_{5,5} &= (N_A)_5\\ (N_A)_{4,4} &= (N_A)_4 - (N_A)_{4,5}\\ (N_A)_{3,3} &= (N_A)_3 - (N_A)_{3,5} - (N_A)_{3,4}\\ (N_A)_{2,2} &= (N_A)_2 - (N_A)_{2,5} - (N_A)_{2,4} - (N_A)_{2,3}\\ (N_A)_{1,1} &= (N_A)_1 - (N_A)_{1,5} - (N_A)_{1,4} - (N_A)_{1,3} - (N_A)_{1,2}\end{aligned}\right\} \tag{5.10}$$

Since $(N_A)_5$ is a measured quantity, $(N_A)_{5,5}$ is known; $\mathscr{P}_{5,5}$ is calculated from Eq. (5.8); and D_5 is measured from the plane of polish. We can now calculate $(N_V)_5$ from Eq. (5.6) for the case of $i = j = 5$.

Knowing $(N_V)_5$, we proceed to calculate $(N_V)_4$. From Eq. (5.8) we obtain the probability of hitting a size-5 sphere to get a size-4 section. This value of $\mathscr{P}_{4,5}$ in Eq. (5.6a) gives $(N_A)_{4,5}$, which, when subtracted from the measured value $(N_A)_4$, yields $(N_A)_{4,4}$ according to Eq. (5.10). Using the appropriate probability for hitting a size-4 sphere to get a size-4 section, and knowing $(N_A)_{4,4}$, we now solve Eq. (5.6) for $(N_V)_4$.

This sequence of calculations is continued down to the smallest particle size. It is seen that each subtracted term depends on the previously calculated values of $(N_V)_5$, $(N_V)_4$, etc., and that the number of smallest particles may be inaccurate because of accumulated errors.

The application of Eqs. (5.6), (5.6a), (5.8), and (5.10) to the procedure outlined above may be clarified by an example [16] of the calculations involved.

Calculation of $(N_V)_5$: The probability, from Eq. (5.8), of a random test plane intersecting spheres of radius r_5 to form sections with radii between r_4 and r_5 is

$$\mathscr{P}_{5,5} = \frac{1}{r_{\max}} (r_5^2 - r_4^2)^{1/2} = (1 - 0.8^2)^{1/2} = 0.6$$

and according to Eq. (5.6),

$$(N_V)_5 = \frac{(N_A)_{5,5}}{0.6} \frac{1}{2r_5} = \frac{1.667(N_A)_{5,5}}{2r_5}. \tag{5.11}$$

This, then, is the equation used to calculate $(N_V)_5$, since both $(N_A)_{5,5}$ and r_5 are known. Note that 1.667 $(N_A)_{5,5}$ represents the total number of sections per unit area due to spheres of radius r_5 (compare Eqs. 4.36 and 5.11).

Calculation of $(N_V)_4$: The probability of obtaining sections with radii between r_3 and r_4 from spheres of radius r_5 is

$$\begin{aligned}\mathscr{P}_{4,5} &= \frac{1}{r_5} [(r_5^2 - r_3^2)^{1/2} - (r_5^2 - r_4^2)^{1/2}] \\ &= \sqrt{1 - 0.6^2} - \sqrt{1 - 0.8^2} = 0.2.\end{aligned}$$

Hence, using Eq. (5.6a), we get

$$(N_A)_{4,5} = 0.2(2r_5) \frac{1.667(N_A)_{5,5}}{2r_5} = 0.333(N_A)_{5,5}.$$

According to Eq. (5.10), $(N_A)_{4,4}$ is found from

$$(N_A)_{4,4} = (N_A)_4 - 0.333(N_A)_{5,5},$$

since $(N_A)_4$ and $(N_A)_{5,5}$ are known.

The probability of intersecting a sphere of radius r_4 to give sections with radii between r_3 and r_4 is

$$\mathscr{P}_{4,4} = \frac{1}{r_4} (r_4^2 - r_3^2)^{1/2} = (1 - 0.75^2)^{1/2} = 0.661.$$

From Eq. (5.6),

$$(N_V)_4 = \frac{(N_A)_4 - 0.333(N_A)_{5,5}}{0.661(2r_4)} = \frac{1.89(N_A)_4 - 0.63(N_A)_{5,5}}{2r_5}. \tag{5.12}$$

Note that the denominator can always be expressed in terms of r_5. Since both $(N_A)_4$ and $(N_A)_{5,5}$ are known, substitution in Eq. (5.12) gives us $(N_V)_4$.

By an extension of this procedure, step by step, until all class intervals are exhausted, expressions similar to Eqs. (5.11) and (5.12) are developed for $(N_V)_3$, $(N_V)_2$, and $(N_V)_1$. The coefficients in the numerators may be collected in a table of coefficients [16] for ready use. Usually, however, more than five class intervals are required for greater accuracy. Methods of calculation that are more suitable for most particulate systems will be described next.

5.3 METHODS FOR DETERMINING SIZE DISTRIBUTIONS OF SPHERES

Numerous analytical procedures are available for determining particle size distributions of spheres from section size distributions. Several of these procedures are discussed in detail [31] and their salient features compared [32] on the basis of experimental measurements on the same sample. Apparently, the first analysis was formulated by Wicksell in 1923 and published [12] in 1925, in order to solve a corpuscular problem in anatomy. Hagermann's early analysis [13] of the size distribution of particles in minerals is said [1] to be incorrect. Some later analyses of particle size distributions are given by Bach [11], Saltykov [7], and Nicholson and Merckx [9].

The importance of the log-normal distribution in small particle statistics is well known, and several analytical procedures have been based specifically on such a distribution [4, 17, 34]. The subject of grain size distributions has been treated in great detail by Schücker [25], as well as by Exner [35, 36] and others. There is considerable overlap in methodology, however, and we shall make no distinction between systems of grains (space-filling) or systems of particles (volume fractions less than one).

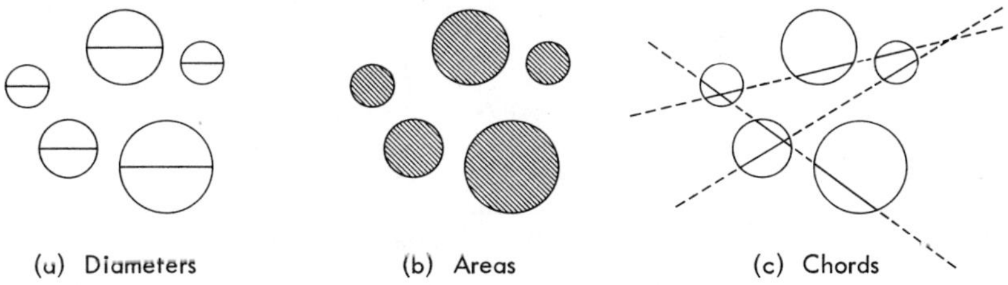

Fig. 5.5. Types of measurements required for obtaining particle size distributions [17].

5.3.1 General Characteristics of Methods

We shall discuss here several of the more representative and important methods for determining spatial size distributions from two-dimensional measurements. In general, the analyses are based on the assumptions that (a) the particles have spherical shapes and (b) the distribution of particle sizes can be represented by a discontinuous distribution. Provided that the number of class intervals is not too small, we can expect the greatest inaccuracies to arise from nonspherical particles. The assumption of sphericity for "equiaxed" grains or "spheroidal" particles is usually satisfactory in practice, however.

Analytical methods for calculating the size distribution of spheres may be classified [17] into three major categories, according to the type of planar measurement performed. Figure 5.5 illustrates the three categories: those based on measurements of (a) section diameters, (b) section areas, and (c) section chords. Table 5.1 lists representative procedures according to this classification, and compares the principal features of the methods.

TABLE 5.1

Characteristics of analytical methods for obtaining size distributions of spheres from planar measurements [32]

Method	Table of coefficients needed?	Independent calculation of each class interval?	Class interval scale required*	References
Diameters				
Wicksell	Yes	Yes	$D_{max}/15$	12
Scheil	Yes	No	$D_{max}/15$	14, 15
Schwartz	Yes	Yes	$D_{max}/10$	16
Schwartz-Saltykov	Yes	Yes	D_{max}/k ($k \leq 15$)	17
Areas				
Johnson	Yes	No	ASTM grain size	18
Johnson-Saltykov	Yes	No	Absolute scale, not dependent on D_{max}	17
Saltykov	No	Yes	Absolute scale, based on A/A_{max} ($k \leq 12$)	7
Chords				
Spektor	No	Yes	Continuous, or D_{max}/k	19
Lord and Willis	No	Yes	D_{max}/k	20
Cahn and Fullman	No	Yes	Continuous, or D_{max}/k	8
Bockstiegel	No	Yes	D_{max}/k	45

*D_{max} is the diameter of the largest grain; k is an arbitrary number of subdivisions.

One of the earliest analyses to be prompted by metallurgical problems is due to Scheil [14, 15]. His method is rather inflexible and involves lengthy calculations. It requires (a) a given number of equispaced class intervals based on D_{max}, the maximum grain diameter, (b) a recalculation of the table of coefficients if the number of class intervals is changed, and (c) sequential calculations of the number of particles in each class interval, proceeding from the largest diameter to the smallest.

The Schwartz-Saltykov [16, 17] version is a considerable improvement over Scheil's method in that only one table of coefficients is required for any number of class intervals up to 15. Moreover, the number of particles per unit volume in any particular class interval may be calculated directly. Wicksell's [12] solution for spherical particles is practically the same as the Schwartz-Saltykov solution.

Although a table of coefficients must be used in the Johnson-Saltykov (area) method [17, 18], the values of the coefficients are not dependent on the magnitude of D_{max}. The class intervals are based on an equispaced logarithmic scale in terms of absolute dimensions. The logarithmic sliding scale is entered only for those sizes that are measured experimentally.

Except for Saltykov's area method [7], the chord methods are the only ones that require neither tables of coefficients nor sequential calculations. The planar size distributions may be treated either as continuous or discontinuous functions. Although derived independently, the Spektor [19] and Lord and Willis [20] treatments are fully equivalent; Spektor's final working equation appears in slightly simpler form. On the other hand, Lord and Willis include a graphical method of solution that is unique.

From the standpoint of ease of gathering the experimental data, the chord methods are preferable. The counting and sizing of chord lengths by automatic electronic scanning devices [5] promise to be especially rapid. There seems to be a tendency with any intercept method, however, to pick up a false number of small chords, especially from particles having irregular section perimeters.

From a practical point of view, measurements of the precise magnitudes of diameters, areas, or chords are not required. If desired, transparent templates may be used to grade the various sections automatically into predetermined class intervals as the sections are measured. Graded reticules are available for standard microscopes, if one prefers to work directly from the polished specimen. Care must be taken, however, that no sections are missed in the chosen area.

Concentric circles facilitate the sorting of particle sections by areas. Linear scales, divided into the desired number of class intervals, may be used for classifying section diameters or chord lengths. Class intervals have been set to linear scales [16], to logarithmic scales [17], or to geometric scales [21]; there are special advantages to each type. One advantage of using circular templates is that irregular areas are approximated as circles, and this lessens the errors associated with complex profiles.

As the particle shape becomes more irregular, an increasing number of spurious sections are generated on the sectioning plane by the extra corners. This tends to increase the apparent number of smaller section sizes. At a still smaller scale, a major practical problem is the detection and measurement of sections in the smallest class intervals, and some of the smallest sections may be missed. Thus the experimental distribution curve may rise too steeply, and the calculated number of particles per unit volume in the first class intervals may turn out negative. One can readily see that this can happen in those methods that depend on subtraction between successive pairs of terms (see, for example, Eqs. 5.30 and 5.34).

Let us now turn to the five methods for spheres selected for a more complete exposition. Those chosen are the Schwartz-Saltykov (diameter) analysis [16, 17]; the Saltykov (area) analysis [7]; the Spektor (chord) analysis [19]; the Lord and Willis graphical (chord) analysis [20]; and the Cahn and Fullman (chord) analysis [8]. We shall discuss the treatment provided by the latter authors for lamellar structures later. Only the Schwartz-Saltykov method requires a table of coefficients. A numerical example is given in order to illustrate the use of the table. The other methods are dependent only on the final working equations, except for the aforementioned graphical solution proposed by Lord and Willis.

5.3.2 Schwartz-Saltykov (Diameter) Analysis

Since complete descriptions of this analysis are given elsewhere [17, 31], only the essentials will be reproduced here. The particle sizes may be broken down into any number of groups up to 15, depending on the accuracy required. If a photomicro-

TABLE 5.2

Nomenclature used in Schwartz-Saltykov method

Number of group	Particle diameter, mm	Number per mm^3	Section diameter, mm	Number per mm^2
1	Δ	$(N_V)_1$	0 to Δ	$(N_A)_1$
2	2Δ	$(N_V)_2$	Δ to 2Δ	$(N_A)_2$
3	3Δ	$(N_V)_3$	2Δ to 3Δ	$(N_A)_3$
.	.	.	.	.
.	.	.	.	.
.	.	.	.	.
j	$j\Delta$	$(N_V)_j$	$(j-1)\Delta$ to $j\Delta$	$(N_A)_j$
.	.	.	.	.
.	.	.	.	.
.	.	.	.	.
k	$k\Delta$	$(N_V)_k$	$(k-1)\Delta$ to $k\Delta$	$(N_A)_k$

graph is used, a magnification may be selected so that the largest diameter D_{max} can be expressed by an integer number of scalar units, preferably not less than 7 or 8.

Saltykov introduces the symbol Δ, defined as the ratio of D_{max} to the total number of groups, k. The diameter of particles in the first group equals Δ, that of the second group equals 2Δ, etc., and the diameter of the largest particles, $D_{max} = k\Delta$. Table 5.2 presents the nomenclature adopted here for designating the sizes and numbers of particles and sections.

To properly identify the number of sections of a particular class interval, one must specify both the section size and the particle size. As an example, $(N_A)_{2,3}$ indicates the number of sections per unit area of the second size group (with diameters between Δ and 2Δ) formed exclusively from particles of the third size group (with diameters equal to 3Δ).

With this notation, we write the total number of sections of the first size group per unit area as

$$(N_A)_1 = (N_A)_{1,1} + (N_A)_{1,2} + (N_A)_{1,3} + \cdots + (N_A)_{1,j} + \cdots + (N_A)_{1,k}; \quad (5.13)$$

for the second size group as

$$(N_A)_2 = (N_A)_{2,2} + (N_A)_{2,3} + (N_A)_{2,4} + \cdots + (N_A)_{2,j} + \cdots + (N_A)_{2,k}; \quad (5.14)$$

etc. The total number of particles per unit volume is

$$N_V = (N_V)_1 + (N_V)_2 + (N_V)_3 + \cdots + (N_V)_j + \cdots + (N_V)_k. \quad (5.15)$$

The probabilities that a random plane will intersect spheres of the various diameters to give particular sections are calculated as described in Section 5.2.3. Equations are then set up for the $(N_V)_j$ in terms of the $(N_A)_{i,j}$. The latter terms can be eliminated by substitution in Eqs. (5.13), (5.14), etc., leaving the $(N_V)_j$ expressed as functions of the $(N_A)_i$; that is, the observed values.

The general equation for the number of particles of any size group per unit volume is

$$(N_V)_j = \frac{1}{\Delta} \sum_{i=j}^{k} \alpha_i (N_A)_i, \tag{5.16}$$

where the coefficients α_i have been calculated according to the procedures indicated above. Expansion of Eq. (5.16) gives the working equation

$$\boxed{(N_V)_j = \frac{1}{\Delta} [\alpha_j (N_A)_j - \alpha_{j+1}(N_A)_{j+1} - \alpha_{j+2}(N_A)_{j+2} - \cdots - \alpha_k (N_A)_k],} \tag{5.17}$$

where i and j assume integer values from 1 to k, and k equals the total number of class intervals. The index j is, of course, a constant in each size group, but i varies from $i = j$ through $i + 1$, $i + 2$, etc., up to the kth class interval. Table 5.3 gives the values of the coefficients for each j, up to 15. However, one may use any lesser number of groups, k, as desired, up to $k = 15$.

To solve Eq. (5.17) for $(N_V)_4$, for example, we proceed horizontally across Table 5.3 on the $(N_V)_4$ level and extract the coefficients α_4, α_5, . . . in columns under $(N_A)_4$, $(N_A)_5$, . . . up to the maximum size group. Then we substitute experimental values for $(N_A)_4$, $(N_A)_5$, . . . in Eq. (5.17) and perform the indicated numerical operations.

It is also possible to calculate the total number of particles per mm³ directly by using the coefficients along the bottom line of the table, according to the expression

$$N_V = \frac{1}{\Delta} \sum_{i=1}^{k} \alpha_i (N_A)_i. \tag{5.18}$$

In this case, all coefficients are positive.

Experimental data supplied by Saltykov [17] can be used to illustrate the application of the Schwartz-Saltykov analysis. Table 5.4 is largely self-explanatory.

The diameter of the largest section on the photomicrograph at a magnification of 2000× is 8 mm; therefore the data are conveniently divided into 8 class intervals. The value of Δ obtained from D_m/k is 0.004 mm/8 = 0.0005 mm. Five hundred section diameters, measured within a convenient area, give the calculated numbers of sections per mm² (see column 3 of Table 5.4).

We can now solve for the values of $(N_V)_j$ in any class interval simply by substituting in Eq. (5.17). Suppose that we want to find $(N_V)_8$, the number of particles per mm³ in the largest class interval. Then j and i in Eq. (5.17) equal 8, and all higher terms drop out. We see that

$$(N_V)_8 = 0.2582 \frac{(N_A)_8}{\Delta} = 0.2582 \left(\frac{600}{0.0005}\right) = 310{,}000 \text{ mm}^{-3}.$$

The coefficient 0.2582 for the eighth group of particles is found in Table 5.3 opposite $(N_V)_8$ and $(N_A)_8$. Let us take another example of the calculation. The number of particles per mm³ in the third group (diameter = 1.5μ) is

$$(N_V)_3 = \frac{1}{0.0005} [0.4472(N_A)_3 - 0.1382\,(N_A)_4 - 0.0408\,(N_A)_5 - 0.0178\,(N_A)_6 - 0.0093(N_A)_7 - 0.0057\,(N_A)_8] = 18.78 \times 10^6 \text{ mm}^{-3}.$$

TABLE 5.3

Coefficients for Schwartz-Saltykov method of calculating particle size distributions

$j \downarrow$ \ $i \rightarrow$	Coefficients, α_i						
	$(N_A)_1$	$(N_A)_2$	$(N_A)_3$	$(N_A)_4$	$(N_A)_5$	$(N_A)_6$	$(N_A)_7$
$(N_V)_1$	+1.0000	0.1547	0.0360	0.0130	0.0061	0.0033	0.0020
$(N_V)_2$		+0.5774	0.1529	0.0420	0.0171	0.0087	0.0051
$(N_V)_3$			+0.4472	0.1382	0.0408	0.0178	0.0093
$(N_V)_4$				+0.3779	0.1260	0.0386	0.0174
$(N_V)_5$					+0.3333	0.1161	0.0366
$(N_V)_6$						+0.3015	0.1081
$(N_V)_7$							+0.2773
$(N_V)_8$							
$(N_V)_9$							
$(N_V)_{10}$							
$(N_V)_{11}$							
$(N_V)_{12}$							
$(N_V)_{13}$							
$(N_V)_{14}$							
$(N_V)_{15}$							
N_V	+1.0000	+0.4227	+0.2583	+0.1847	+0.1433	+0.1170	+0.0988

$$(N_V)_j = \frac{1}{\Delta}\left[\alpha_i(N_A)_i - \alpha_{i+1}(N_A)_{i+1} - \alpha_{i+2}(N_A)_{i+2} - \cdots - \alpha_k(N_A)_k\right]$$

TABLE 5.4

Size distribution data for spheroidized cementite in 1% carbon steel

1	2	3	4	5	6
Group no.	Range of section diameters, μ	Number of sections per mm^2, $(N_A)_i$	Diameter of particles, D_j, μ	Number of particles per mm^3, $(N_V)_j$	Distribution, %
1	0.0–0.5	4,500	0.5	1.48×10^6	3.26
2	0.5–1.0	16,500	1.0	9.34×10^6	20.54
3	1.0–1.5	26,550	1.5	18.78×10^6	41.30
4	1.5–2.0	15,600	2.0	9.93×10^6	21.86
5	2.0–2.5	5,850	2.5	2.80×10^6	6.17
6	2.5–3.0	4,350	3.0	2.35×10^6	5.18
7	3.0–3.5	1,050	3.5	0.46×10^6	1.01
8	3.5–4.0	600	4.0	0.31×10^6	0.68
Totals		$N_A = 75{,}000$		$N_V = 45.45 \times 10^6$	100.0

$(N_A)_8$	$(N_A)_9$	$(N_A)_{10}$	$(N_A)_{11}$	$(N_A)_{12}$	$(N_A)_{13}$	$(N_A)_{14}$	$(N_A)_{15}$
0.0013	0.0009	0.0006	0.0005	0.0004	0.0003	0.0002	0.0001
0.0031	0.0021	0.0015	0.0010	0.0009	0.0006	0.0006	0.0004
0.0057	0.0037	0.0026	0.0018	0.0013	0.0010	0.0007	0.0007
0.0095	0.0058	0.0038	0.0027	0.0020	0.0016	0.0012	0.0009
0.0168	0.0094	0.0059	0.0040	0.0028	0.0021	0.0016	0.0013
0.0346	0.0163	0.0091	0.0058	0.0041	0.0028	0.0022	0.0016
0.1016	0.0329	0.0155	0.0090	0.0057	0.0040	0.0029	0.0022
+0.2582	0.0961	0.0319	0.0151	0.0088	0.0056	0.0039	0.0028
	+0.2425	0.0913	0.0301	0.0146	0.0085	0.0055	0.0039
		+0.2294	0.0872	0.0290	0.0140	0.0083	0.0054
			+0.2182	0.0836	0.0280	0.0136	0.0080
				+0.2085	0.0804	0.0270	0.0132
					+0.2000	0.0776	0.0261
						+0.1925	0.0750
							+0.1857
+0.0856	+0.0753	+0.0672	+0.0610	+0.0553	+0.0511	+0.0472	+0.0441

We can determine the number of particles of all other sizes in the same way; the numbers are listed in column 5 of Table 5.4. These values, then, give us the complete spatial size distribution curve, from which we may proceed to determine $\bar{D}$, $\sigma(D)$, and N_V, if desired.

5.3.3 Saltykov (Area) Analysis

This analysis [7] attempts to improve on older methods and succeeds in several regards. It does not require sequential calculations nor tables of coefficients, as indicated in Table 5.1. It is applicable not only to polydispersed systems of spherical particles, but, in principle, to convex particles as well. The latter possibility is based on the more general form of $N_A = N_V \bar{D}$, in which $\bar{H}'$ replaces $\bar{D}$ (see Eq. 4.24a). In this analysis, all particles must have the same shape.

In order to specify the size of a section, Saltykov adopts the ratio $A/A_{\max}$ instead of the absolute area. Thus it is apparent that the correct value of $A_{\max}$ is very important. A large number of particles must be examined on the plane of polish, especially if the particles have a complicated shape.

The scale factor used to determine the class intervals for $A/A_{\max}$ is based on a logarithmic scale of diameters with the factor $10^{-0.1}$ ($= 0.7943$). Consequently,

TABLE 5.5

Group limits and corresponding frequency of sections for a spherical particle

Group number	Relative section diameters, d/d_{max}	Relative section areas, A/A_{max}	Percent of sections per unit area, N_A
1	1.0000	1.0000–0.6310	60.749
2	0.7943	0.6310–0.3981	16.833
3	0.6310	0.3981–0.2512	8.952
4	0.5012	0.2512–0.1585	5.200
5	0.3981	0.1585–0.1000	3.134
6	0.3162	0.1000–0.0631	1.926
7	0.2512	0.0631–0.0398	1.195
8	0.1995	0.0398–0.0251	0.747
9	0.1581	0.0251–0.0158	0.469
10	0.1259	0.0158–0.0100	0.294
11	0.1000	0.0100–0.0063	0.185
12	0.0794	0.0063–0.0040	0.117
⋮	⋮	⋮	⋮

for the sectional areas a logarithmic scale with the factor $(10^{-0.1})^2 = 10^{-0.2}$ $(= 0.6310)$ must be used. Table 5.5 lists the first twelve class intervals, with the intervals based on a maximum linear particle dimension of unity.

The advantage of using A/A_{max} to group the sections lies in the unique relationship of the particle shape to the N_A versus A/A_{max} frequency distribution. By intersecting a given particle with many random planes, one can obtain a distribution of sectional areas which is characteristic of that particular particle [3]. For spherical particles, the distribution of N_A can be calculated readily. The fourth column of Table 5.5 shows this for spherical particles, based on the A/A_{max} limits established in the third column.

The calculated distribution of N_A versus A/A_{max} for spheres is shown graphically in Fig. 5.6. For particles of other shapes, the calculation of the N_A distribution versus A/A_{max} would be extremely difficult, although experimental [3] or computer-aided programs [37] should help in this regard.

The computations follow along lines described previously. The procedure is similar, except that one successively subtracts those sections contributed by larger spheres. The spherical particles have diameters D_j (or D_1, D_2, D_3, . . .) corresponding to $(N_V)_j$ [or $(N_V)_1$, $(N_V)_2$, $(N_V)_3$, . . .]. The largest particle, D_1, is related to the others by $D_2 = 0.7943D_1$, $D_3 = 0.6310D_1$, etc. (see Table 5.5).

The largest section of the largest particle is $A_1 = A_{max}$, and from Table 5.5 we see that $A_2 = 0.631A_1$, $A_3 = 0.398A_1$, etc. The limits of the section areas, A_1 to A_2, A_2 to A_3, A_3 to A_4, . . . correspond to the $(N_A)_i$ [or $(N_A)_1$, $(N_A)_2$, $(N_A)_3$, . . .], where the section sizes are fixed and the particle diameters assume all possible values. The $(N_A)_i$ terms are the observed values, counted directly from the plane of polish.

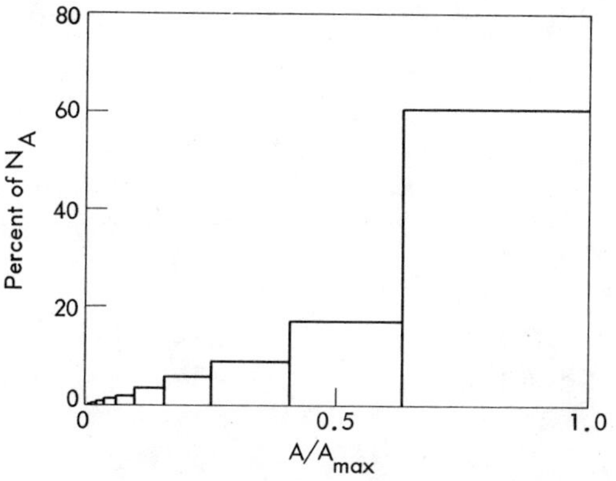

Fig. 5.6. Distribution of the number of sections versus relative area for a spherical particle [7].

The results of the successive calculations permit one to derive a general working formula which gives $(N_V)_j$ directly for any desired class interval. Note that j can have any integer value up to and including 12. For example, to calculate the value of $(N_V)_5$, one would use just the first five terms in the brackets. All terms are used only when the twelfth class interval is calculated.

$$\begin{aligned}(N_V)_j = \frac{1}{D_j}[&1.6461(N_A)_i - 0.4561(N_A)_{i-1} - 0.1162(N_A)_{i-2} - 0.0415(N_A)_{i-3} \\ &- 0.0173(N_A)_{i-4} - 0.0079(N_A)_{i-5} - 0.0038(N_A)_{i-6} - 0.0018(N_A)_{i-7} \\ &- 0.0010(N_A)_{i-8} - 0.0003(N_A)_{i-9} - 0.0002(N_A)_{i-10} - 0.0002(N_A)_{i-11}].\end{aligned} \tag{5.19}$$

Remember that for this particular derivation, the largest particle size corresponds to $j = 1$. The $i = j$, with subscripts varying through $i - 1$, $i - 2$, $i - 3$, etc., down to 1.

As an example of the calculations involved, let us determine $(N_V)_5$ from the data given in Table 5.6. The equation obtained in this case is

$$(N_V)_5 = \frac{1}{D_5}[1.6461(N_A)_5 - 0.4561(N_A)_4 - 0.1162(N_A)_3 - 0.0415(N_A)_2 - 0.0173(N_A)_1].$$

Substituting the actual data, we get

$$\begin{aligned}(N_V)_5 &= \frac{1}{0.025}[1.65(138) - 0.456(230) - 0.116(253) - 0.042(161) - 0.017(104)] \\ &= 3359 \text{ mm}^{-3}.\end{aligned}$$

The results of the complete calculation for all particle sizes are summarized in Table 5.7.

Although Eq. (5.19) has been derived for spherical particles, the same method can be used to derive similar equations for particles of other shapes. We must know the distribution of section areas $(N_A)_i$ versus A/A_{max} for the particular particle, and must replace D_j by the mean projected height, $\overline{H}_j$. The only questions to be attached to this method are whether the logarithmic scale might tend to obscure the essential features of the $(N_A)_i$ versus A/A_{max} curve (compare Fig. 5.6 and Fig. 7.15).

TABLE 5.6

Measured distribution of ferrite grain section sizes

Group number	Diameter of sections d_i, μ	Relative grain area, $A/A_{\max}$	Number of sections per mm², $(N_A)_i$
1	63.10–50.12	1.0000–0.6310	104
2	50.12–39.81	0.6310–0.3981	161
3	39.81–31.62	0.3981–0.2512	253
4	31.62–25.12	0.2512–0.1585	230
5	25.12–19.95	0.1585–0.1000	138
6	19.95–15.85	0.1000–0.0631	69
			$N_A = 955$

TABLE 5.7

Calculated distribution of ferrite grain sizes

Group number	Diameter of particles D_j, μ	Number of grains per mm³, $(N_V)_j$
1	63.10	2713
2	50.12	4341
3	39.81	8313
4	31.62	7596
5	25.12	3359
6	19.95	491
		$N_V = 26813$

Also, to fix the value of $A_{\max}$ exactly may prove troublesome. These questions aside, the method is direct, simple, and eminently useful. Of more far-reaching interest, of course, is the possibility of handling systems of convex particles of unspecified shape.

5.3.4 The Spektor (Chord) Analysis

Of the well-known analytical methods for determining particle-size distribution curves from chord-length distributions, the derivation by Spektor [19] appears to be the first. His final working equations apply to either a continuous or a discontinuous size distribution. We shall give a brief derivation of the equation for the case of the discontinuous size distribution.

The geometry involved when a sphere of diameter D is pierced by a random line is shown in Fig. 5.7. The perpendicular distance Z from the test line to the center

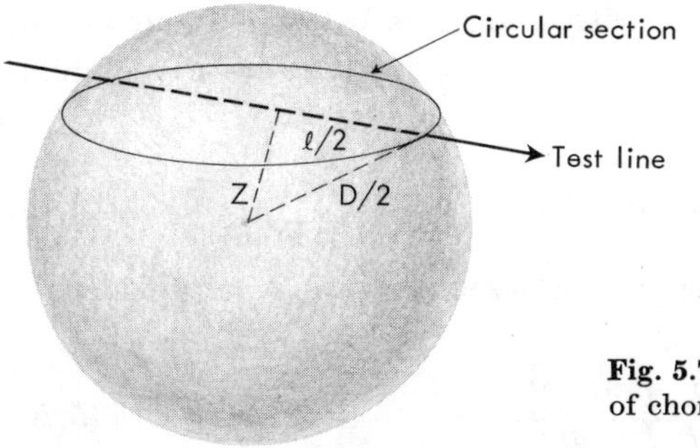

Fig. 5.7. Geometrical relationships in Spektor's method of chords [19].

of the sphere is

$$Z = [(D/2)^2 - (l/2)^2]^{1/2}, \tag{5.20}$$

where l is the chord length.

The geometrical relationships are unchanged if we pass a test line through a randomly distributed, monodispersed system of spheres of diameter D. Consider a cylindrical volume of length L and radius Z centered about the test line. The volume is $\pi Z^2 L$ and the intercepted chord lengths will vary between l and D. The number of these chords per unit length of test line is

$$(n_L)_l^D = \pi Z^2 N_V = \frac{\pi}{4}(D^2 - l^2)N_V \tag{5.21}$$

where N_V is the number of spheres per unit volume.

For a polydispersed system of spheres of diameters $D_1, D_2, \ldots, D_j, \ldots, D_{\max}$, and numbers per unit volume $(N_V)_1, (N_V)_2, \ldots, (N_V)_j, \ldots, (N_V)_{\max}$, respectively, we obtain

$$\begin{aligned}(n_L)_l^{D_{\max}} &= \sum_{D=l}^{D=D_{\max}} \frac{\pi}{4}(D_j^2 - l^2)(N_V)_j \\ &= \frac{1}{4}\sum_l^{D_{\max}} \pi D_j^2 (N_V)_j - \frac{\pi l^2}{4}\sum_l^{D_{\max}} (N_V)_j\end{aligned} \tag{5.22}$$

where $(n_L)_l^{D_{\max}}$ is the number of chords with lengths between l and $D_{\max}$ per unit length of test line. The last two terms of Eq. (5.22) are related to the total surface area and the total number of particles per unit volume and can be expressed as

$$(n_L)_l^{D_{\max}} = \frac{1}{4}(S_V)_l^{D_{\max}} - \frac{\pi l^2}{4}(N_V)_l^{D_{\max}}, \tag{5.23}$$

where $(S_V)_l^{D_{\max}}$ and $(N_V)_l^{D_{\max}}$ are, respectively, the total surface area per unit volume and the total number of particles per unit volume for particles with diameters between l and $D_{\max}$. Differentiating Eq. (5.23), assuming a continuous distribution, we obtain

$$d(n_L)_l^{D_{\max}} = \frac{1}{4}d(S_V)_l^{D_{\max}} - \frac{\pi l^2}{4}d(N_V)_l^{D_{\max}} - \frac{\pi}{2}(N_V)_l^{D_{\max}}\, l\, dl = -\frac{\pi}{2}(N_V)_l^{D_{\max}}\, l\, dl. \tag{5.24}$$

Two of the terms cancel, since, for particles with diameters between l and $l + dl$, $d(S_V)_l^{D_{max}} = \pi l^2 d(N_V)_l^{D_{max}}$. Considering that $d(n_L)_l^{D_{max}} = -d(n_L)_0^l$, further simplification is achieved, giving

$$d(n_L)_0^l = \frac{\pi}{2}(N_V)_l^{D_{max}}\, l\, dl, \tag{5.25}$$

where $(n_L)_0^l$ is the number of chords with lengths between zero and l per unit length of test line. Rearranging, we have

$$(N_V)_l^{D_{max}} = \frac{2}{\pi}\frac{1}{l}\frac{d(n_L)_0^l}{dl}, \tag{5.26}$$

where $d(n_L)_0^l/dl$ is the distribution function of chords.

For finite differences, we can rewrite Eq. (5.26) in the form

$$(N_V)_l^{D_{max}} = \frac{2}{\pi\Delta}\frac{(n_L)_{l-\Delta/2}^{l+\Delta/2}}{l}, \tag{5.27}$$

where $(n_L)_{l-\Delta/2}^{l+\Delta/2}$ is the number of chords per unit length of test line in the interval $l \pm \Delta/2$. By analogy with Eq. (5.27), we can write the difference between the equations for N_V within the limits l_1 and D_{max} and N_V within the limits l_2 and D_{max}

$$(N_V)_{l_1}^{l_2} = \frac{2}{\pi\Delta}\left[\frac{(n_L)_{l_1-\Delta/2}^{l_1+\Delta/2}}{l_1} - \frac{(n_L)_{l_2-\Delta/2}^{l_2+\Delta/2}}{l_2}\right], \tag{5.28}$$

where $(N_V)_{l_1}^{l_2}$ represents the number of particles per unit volume with diameters between l_1 and l_2.

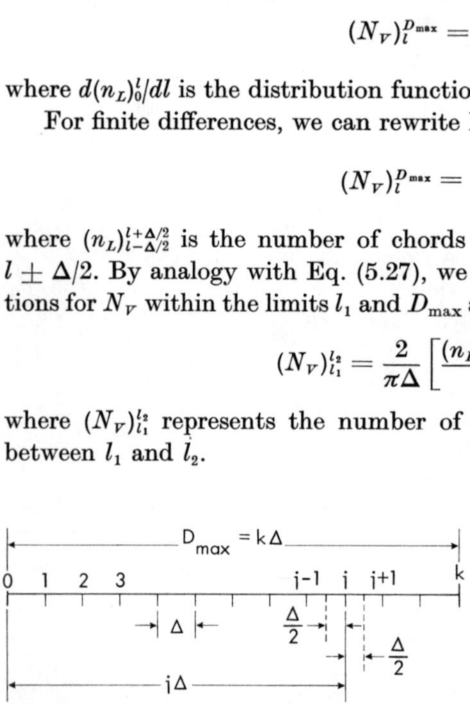

Fig. 5.8. Subdivision of chord lengths and particle diameters into class intervals [19].

Let the entire range of chord lengths be broken down into k class intervals of length Δ, as shown in Fig. 5.8. Using this new nomenclature to describe Eq. (5.28), we have

$$(N_V)_{j\Delta} = \frac{2}{\pi\Delta}\left[\frac{(n_L)_j - (n_L)_{j-1}}{\Delta(j - \frac{1}{2})} - \frac{(n_L)_{j+1} - (n_L)_j}{\Delta(j + \frac{1}{2})}\right], \tag{5.29}$$

where $(N_V)_{j\Delta}$ is the number of particles of mean diameter $j\Delta$ per unit volume in the interval $j\Delta \pm \Delta/2$, and the $(n_L)_j$ terms are the number of chords per unit length of test line. Rearranging and simplifying, we obtain the working formula

$$\boxed{(N_V)_j = \frac{4}{\pi\Delta^2}\left[\frac{(n_L)_j}{2j - 1} - \frac{(n_L)_{j+1}}{2j + 1}\right]} \tag{5.30}$$

how does (5.29) give (5.30)?

TABLE 5.8

Analysis of size distribution of ferrite grains in low carbon steel by Spektor method

1	2	3	4	5	6
Group no., j	Range of chord lengths, μ	Number of chords per mm, $(n_L)_j$	Diameter of particles, D_j, mm	Number of particles per mm^3, $(N_V)_j$	Distribution, %
1	0.0– 2.5	6	0.0025	0.68×10^5	5.5
2	2.5– 5.0	17	0.0050	1.77×10^5	14.5
3	5.0– 7.5	24	0.0075	4.26×10^5	34.9
4	7.5–10.0	19	0.0100	2.59×10^5	21.2
5	10.0–12.5	13	0.0125	1.46×10^5	11.9
6	12.5–15.0	8	0.0150	0.70×10^5	5.7
7	15.0–17.5	5	0.0175	0.24×10^5	2.0
8	17.5–20.0	4	0.0200	0.19×10^5	1.5
9	20.0–22.5	3	0.0225	0.14×10^5	1.1
10	22.5–25.0	2	0.0250	0.21×10^5	1.7
Totals	0.0–25.0	101	0.0025–0.0250	12.24×10^5	100.0

where the $(N_V)_j$ and $(n_L)_j$ are the number of particles per unit volume and the number of chords per unit length of test line, respectively, for the indicated j-size. We can see from Eq. (5.30) that no coefficients need be calculated for the $(n_L)_j$ terms. We merely select a convenient Δ value, then substitute it and the experimentally determined values of $(n_L)_j$ into Eq. (5.30). Table 5.8 gives a numerical example [19] which illustrates the method.

A value of $k = 10$ was selected for the data shown in Table 5.8. Then, $\Delta = 0.0025$ mm and the constant term preceding the brackets in Eq. (5.30) is 2.04×10^5 mm^{-2}. The number of particles per unit volume in the sixth group, for example, is

$$(N_V)_6 = 2.04 \times 10^5 \left[\frac{8}{12 - 1} - \frac{5}{12 + 1}\right] = 0.70 \times 10^5 \text{ mm}^{-3}.$$

The other values of $(N_V)_j$ are calculated from Eq. (5.30) in a similar manner.

Another advantage of the Spektor analysis is that the total grain boundary surface area per unit volume, S_V, can be obtained from the same experimental data. Since $P_L = N_L$ for space-filling grains, and here $N_L = n_L = 101$ total chords per mm,

$$S_V = 2(n_L) = 202 \text{ mm}^2/\text{mm}^3.$$

5.3.5 Lord and Willis Graphical (Chord) Analysis

In an effort to obtain the volume fraction and size distribution of spherical air bubbles in frost-resistant concrete, Lord and Willis [20] were led to the same type of analysis as that proposed by Spektor [19]. They described both an analytical solu-

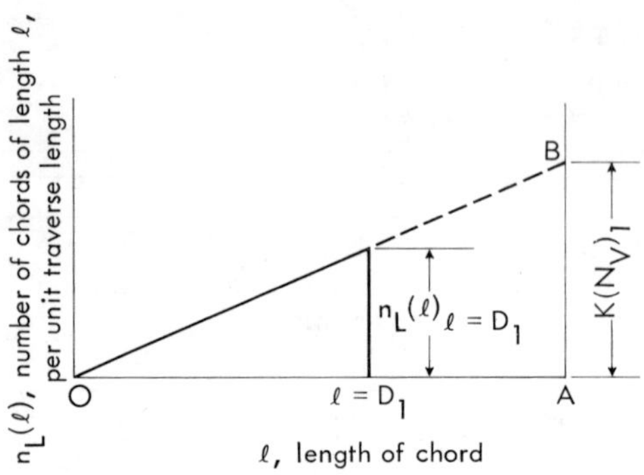

Fig. 5.9. Frequency distribution curve of chord lengths resulting from penetration of spheres of diameter D_1 by a random test line [20].

tion and a graphical procedure for converting a distribution of chord intercepts into a distribution of sphere diameters. Since their graphical solution is unique, it is reviewed here for use in those cases where an analytical solution is not desired.

The main steps in the Lord and Willis graphical solution may be listed as follows:

1) They show, for uniform spheres of diameter D_1, that the graph of the frequency distribution $n_L(l)$ of chord lengths l is a straight line from $l = 0$, $n_L(l) = 0$ to a maximum at $l = D_1$ (Fig. 5.9). Thus a monodispersed system of spheres is

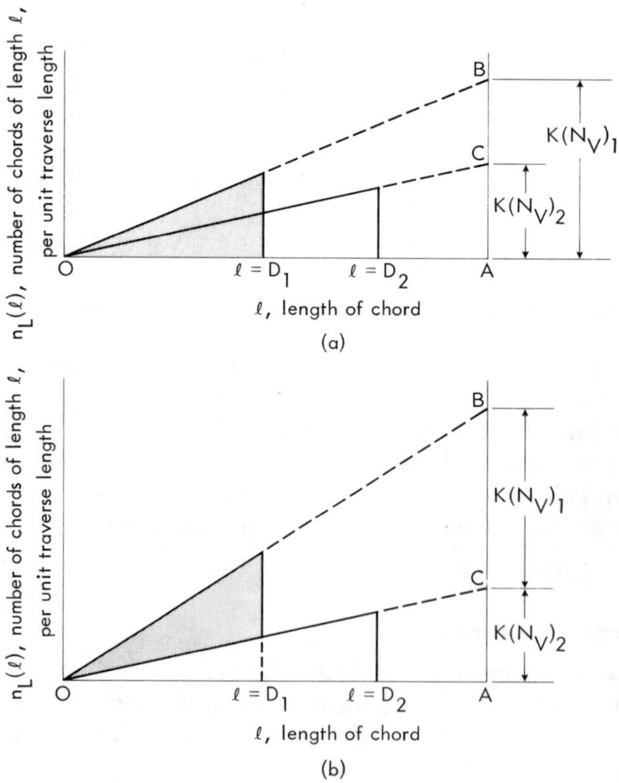

Fig. 5.10. Frequency distribution curves of chord lengths resulting from penetration of spheres of diameter D_1 and D_2 by a random test line [20].

represented by a triangle, the area of which equals the total number of chords per unit length of test line.

2) For two monodispersed systems of spheres, there will be two triangles; for three systems, three triangles; etc. These triangles can be recombined, preserving the individual triangle areas, in such a way that the resulting figure represents the total system (Fig. 5.10).
3) A frequency distribution curve of chord lengths resulting from penetration of a polydispersed system by a test line is constructed in the same manner. The number of chords from each class interval contributes a triangular area, the dimensions of which are dependent on the diameter and number of spheres in the class interval (Fig. 5.11).
4) After they have completed the frequency distribution curve, Lord and Willis list the steps for obtaining $(N_V)_j$ graphically for each class interval.

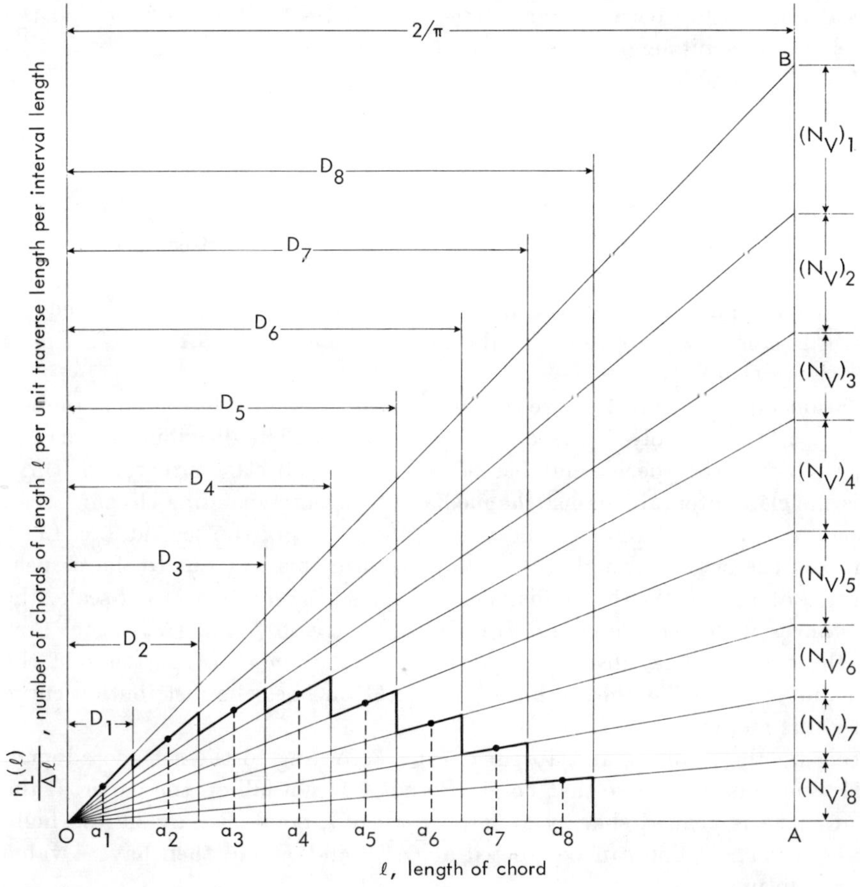

Fig. 5.11. Distribution curve of chord lengths constructed by graphical method [20].

These four steps are the basis of the graphical method. A more detailed description of the derivation of the method is given below, largely in terms of the diagrams in Figs. 5.9, 5.10, and 5.11.

The frequency distribution curve of chord lengths resulting from the penetration of a monodispersed system of spheres of diameter D_1 by a random test line is depicted in Fig. 5.9. Since the triangular area in this figure equals the total number of chords, we can write (using Eq. 5.21 with $l = 0$)

$$\tfrac{1}{2} D_1 n_L(l)_{l=D_1} = \frac{\pi}{4} D_1^2 (N_V)_1 \tag{5.31}$$

or

$$n_L(l)_{l=D_1} = \frac{\pi}{2} D_1 (N_V)_1, \tag{5.32}$$

which reveals that the number of chords of length D_1 per unit length of test line is proportional to the number of spheres per unit volume. This result is used to develop the graphical relationship between the number of spheres per unit volume and the number of chords per unit length.

From Eq. (5.32) and similar triangles in Fig. 5.9, we obtain

$$\frac{n_L(l)_{l=D_1}}{D_1} = \frac{\overline{AB}}{\overline{OA}} = \frac{(N_V)_1}{2/\pi}. \tag{5.33}$$

Therefore, when $\overline{OA}$ equals $2/\pi$, $\overline{AB}$ equals $(N_V)_1$. The coefficient of $(N_V)_1$ in Fig. 5.9, K, must equal unity under these circumstances, and the vertical distance $\overline{AB}$ will equal $(N_V)_1$ directly.

By similar reasoning, we see in Fig. 5.10(a) that when $\overline{OA}$ equals $2/\pi$, $\overline{AB}$ equals $(N_V)_1$ and $\overline{AC}$ equals $(N_V)_2$. For the equivalent arrangement in Fig. 5.10(b), $\overline{CB}$ equals $(N_V)_1$ and $\overline{AC}$ equals $(N_V)_2$.

The relationships developed above can be extended to include the construction shown in Fig. 5.11 for a polydispersed system of spheres. The distribution curve is built up by plotting the specific number of chords in each class interval, $n_L(l)/\Delta l$, where Δl is the class interval, against the median of the corresponding class interval. For example, a_1 is the median of the first class interval, a_2 of the second, etc. Lines are drawn from the origin through all points plotted in this way out to the perpendicular erected along $\overline{AB}$. With the distance $\overline{OA}$ set off $2/\pi$ units on the l-scale, the vertical distances indicated along $\overline{AB}$ are equal directly to $(N_V)_1$, $(N_V)_2$, etc. Furthermore, the total vertical distance equals the total number of particles in the system per unit volume. The heavy line in Fig. 5.11 outlines the distribution curve obtained by this method.

In setting up the ordinate and abscissa scales according to the scheme adopted in Fig. 5.11, one must take care that points B and A do not fall off the paper. If the distribution curve is peaked sharply at low values of l, point B may lie too high. In that case, a vertical line can be erected at $\overline{OA}/2$ and K will then have a value of $\tfrac{1}{2}$, rather than unity.

In the event that calculation rather than the graphical solution of the $(N_V)_j$ is preferred, Lord and Willis give the working equation

$$(N_V)_j = \frac{2}{\pi}\left[\frac{(n_L(l)/\Delta l)_j}{a_j} - \frac{(n_L(l)/\Delta l)_{j+1}}{a_{j+1}}\right], \tag{5.34}$$

where $(N_V)_j$ is the number of particles per unit volume in the class interval j; $(n_L(l)/\Delta l)_j$ is the number of chords of length l per unit length of test line, divided by the interval length Δl; and the a_j are the medians of the class intervals (l_0, l_1), (l_1, l_2), ...

Lord and Willis provide numerical data to illustrate their method of analysis. If modified suitably, the data given in Table 5.8 may also be used to illustrate the graphical analysis or the numerical analysis (using Eq. 5.34).

5.3.6 Cahn and Fullman (Chord) Analysis

Cahn and Fullman [8] have provided a chord-intercept analysis for both spheroidal and lamellar structures. The method is flexible, calculation is simple, and, as in the other chord analyses, a minimum of experimental labor is involved. Their analysis for spheres is similar in principle to that of Lord and Willis [20]; they also use much the same procedure for lamellar structures. Since the analysis for the latter will be described later in some detail, we shall here present only the final working equation for spheres. However, see the derivation of this equation by Bockstiegel [45] and his improvements thereto.

If we assume a continuous distribution, the number of spheres of diameter D per unit volume, $N_V(D)$, is given as

$$N_V(D)_{D=l} = \frac{2}{\pi}\left[\frac{n_L(l)}{l^2} - \frac{dn_L(l)}{l\,dl}\right] \tag{5.35}$$

where $n_L(l)$ is the number of chords of length l per unit length of test line.

This simple result requires only the experimental frequency curve of $n_L(l)$ versus l for the evaluation of $N_V(D)_{D=l}$. The slope $dn_L(l)/dl$ is measured at as many values of l as desired and corresponding values of $n_L(l)$ are taken from the curve. Measurement of the slopes for Eq. (5.35) may introduce some error unless the experimental chord distribution curve is well defined. Aaron, *et al.*, [32] have compared the relative accuracy of three chord distribution analyses on the same sample and find that the results are much the same. Figure 5.12, which summarizes their results with eight types of grain size analyses, shows that all methods are approximately equal in precision. See Exner [47], however, and his results from a similar comparison.

5.4 SPATIAL SIZE DISTRIBUTION OF NONSPHERICAL PARTICLES

It is understandable that the mathematical complexities increase considerably in the analysis of size distributions of nonspherical particles. As a consequence of these increased difficulties, there are few treatments available. One of the earliest papers, dealing with the distribution of ellipsoidal particles, was written by Wicksell [1] in 1926. Scheil and Wurst [2] gave an analysis in 1936, which was based on the aver-

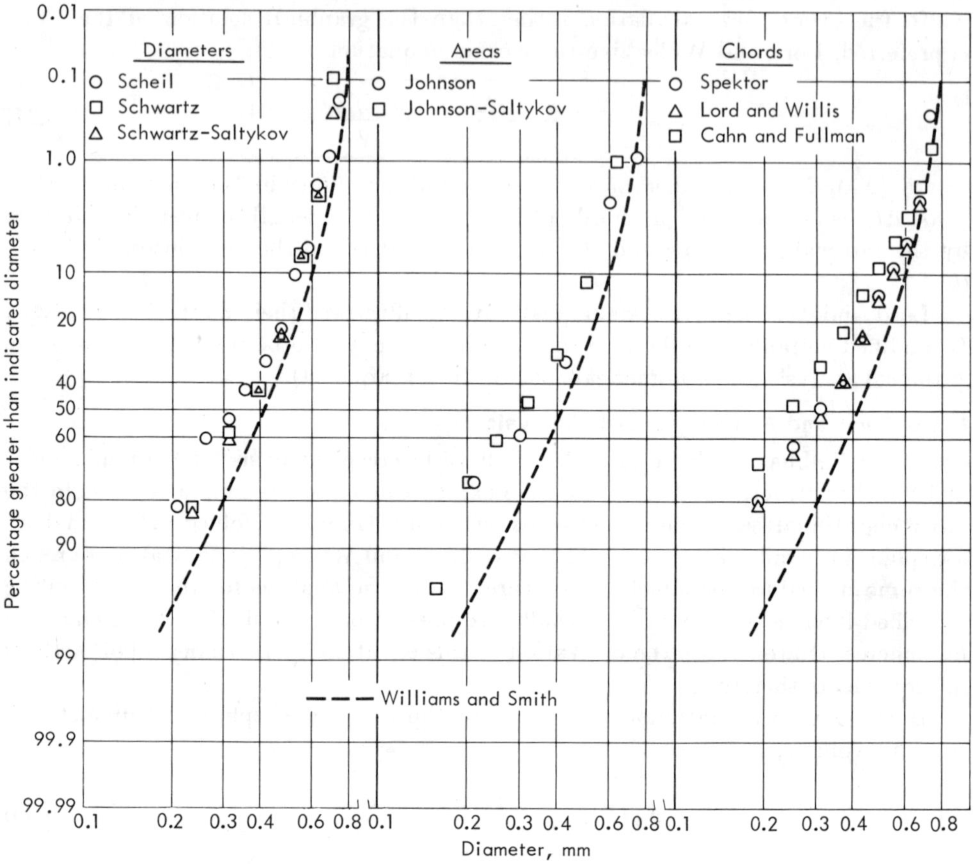

Fig. 5.12. Summation curves of grain sizes by three methods [32].

age statistical shape of grains in a sample of Armco iron. In 1956, Cahn and Fullman [8] published a method for obtaining the spacing distribution of lamellar structures by lineal analysis. Recently, analyses for ellipsoids have been published by DeHoff [6] and for pentagonal dodecahedrons by Paulus [4] and Schücker [25].

5.4.1 Wicksell's Analysis for Ellipsoids

Wicksell's paper on the size distribution of ellipsoidal particles [1] complements his earlier paper on the size distribution of spherical particles [12]. Although the length of the derivation precludes a detailed discussion here, we give a brief résumé to indicate the line of attack.

There are two equations that are fundamental to the problem of ellipsoidal particles. The first is

$$\phi(x) = x\int_x^R f(r)\,\frac{dr}{r(r^2 - x^2)^{1/2}}, \tag{5.36}$$

where $\phi(x)$ is the relative frequency function of x for particles cut by the section plane, x is the "diameter" of the section ellipse, $f(r)$ is the relative frequency function of r for particles cut by the section plane, r is the "diameter" of the central ellipse, and R is the upper limit to the "diameter." The second equation is

$$f(r)h_0 = \iiint hP(\rho)\,\Pi\,(\alpha_2, \alpha_3)\,d\rho\,d\alpha_2\,d\alpha_3, \tag{5.37}$$

where h_0 is the mean value of the "heights," h, of the particles, $P(\rho)$ is the relative frequency function of ρ for all particles, ρ is the "diameter" of a particle, $\Pi\,(\alpha_2, \alpha_3)$ is the correlation function of α_2 and α_3, defining the way in which the principal axes of the particles are directed, and α_2 and α_3 are direction cosines of the normal to the section plane with reference to the small axes of the particle.

Note that $\phi(x)$ is the observed, apparent distribution of "diameters" and $P(\rho)$ is the distribution of actual "diameters." The problem is to express $P(\rho)$ in terms of $\phi(x)$; viz., we solve Eq. (5.36) for $f(r)$, then, knowing $f(r)$, we determine $P(\rho)$ by means of the integral equation, Eq. (5.37).

The following cases are considered:

1) Particles are prolate or oblate ellipsoids of revolution and of constant form, i.e., the eccentricities do not vary. The two principal eccentricities e_2 and e_3 are either equal, or $e_3 = 0$.
2) The eccentricities vary, but with the stipulation that the forms of the particles are independent of ρ. Two cases of varying eccentricity are distinguished: (a) a spherical type, and (b) an eccentric type.

Wicksell prepared tables of coefficients for solving the above cases but we shall not present them here because they are not very useful unless there is a full explanation.

The problem treated by Wicksell is one of considerable mathematical complexity. His analysis appears to be quite rigorous for the cases considered, and when approximations are employed, the expected limits of error are stated clearly. It is unfortu nate that his comprehensive and precise paper on the size distribution of ellipsoidal particles is not known more widely.

5.4.2 DeHoff's Analysis for Ellipsoids

From 1925 to 1962, no further attempts seem to have been made to re-derive Wicksell's solution for the size distribution of ellipsoidal particles. Nor indeed did scientists—other than in the literature of anatomists—seem aware of Wicksell's results. In 1962, DeHoff [6] modified the Schwartz-Saltykov [17] analysis for spherical particles so as to obtain the size distribution of ellipsoidal particles of the same axial ratio.

The nomenclature adopted by DeHoff is basically that given previously in this chapter in Section 5.3.2, *et seq.* He considers systems of particles consisting of either prolate or oblate spheroids; however, the particle shapes must be constant within each size distribution, i.e., the axial ratios of the generating ellipses, q, must all have the same value. In a system of prolate particles, the measured "diameters" are the minor axes of the ellipses on the plane of polish; for oblate particles, the "diameters" of interest are the major axes of the ellipses.

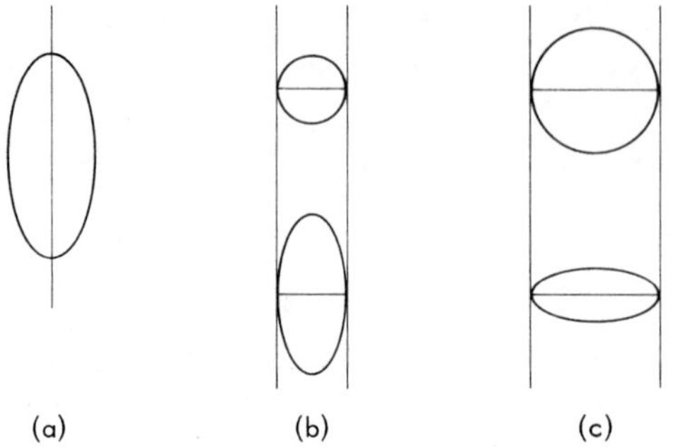

Fig. 5.13. Scheme for identifying systems of prolate or oblate spheroids on the plane of polish. (a) Most unequiaxed ellipse, gives q. (b) Prolate particles. (c) Oblate particles.

By suitable modification of the original equations for spherical particles, he obtains an expression for the number of ellipsoidal particles in each class interval per unit volume of the system:

$$(N_V)_j = \frac{1}{k(q)\Delta} \sum_{i=j}^{k} \alpha(i, j)(N_A)_i, \tag{5.38}$$

where $k(q)$ represents $k_p(q)$ or $k_o(q)$, the shape factors for prolate or oblate spheroids, respectively; Δ is the size increment; the coefficients $\alpha(i, j)$ are the same as those given in Table 5.3; the $(N_A)_i$ terms are the number of ellipses per unit area of the test plane in each class interval; and q is defined as the ratio of the minor to the major axis. Except that two additional terms (the axial ratio and the shape factor) must be obtained, the procedure followed in evaluating Eq. (5.38) is the same as that described for Eq. (5.16).

The value of q for a system of ellipsoidal particles of the same axial ratio is determined by the axial ratio of the most unequiaxed ellipse observed on the test plane (see Fig. 5.13a). Such an ellipse is formed when a particle is intersected parallel to its axis of rotation. In order to select the correct shape factor, one can perform a simple test. The largest circular section and the longest elliptical section are found on the plane of polish. If the particles are prolate, the diameter of the largest circular section will equal the *minor* axis of the longest elliptical section (see Fig. 5.13b). If, on the other hand, the particles are oblate, the circle diameter will equal the *major* axis of the ellipse (see Fig. 5.13c). When one knows q, one can select the proper value of $k_p(q)$ or $k_o(q)$ from the appropriate curve in Fig. 5.14.

The applicability of this analysis to real systems is limited because of the requirement of ellipsoids of constant axial ratio. If the actual particulate system does not conform closely enough to this requirement, then Wicksell's more general treatment may be more appropriate.

5.4.3 Scheil and Wurst's Analysis for a Statistical Grain Shape

The approach adopted by Scheil and Wurst [2] had an objective different from that of Wicksell or DeHoff. Scheil and Wurst wished to determine the distribution of grain sizes based on actual grain shapes, rather than on grains having (assumed) spherical or ellipsoidal shapes. By successive sectioning of a sample of Armco iron, Scheil

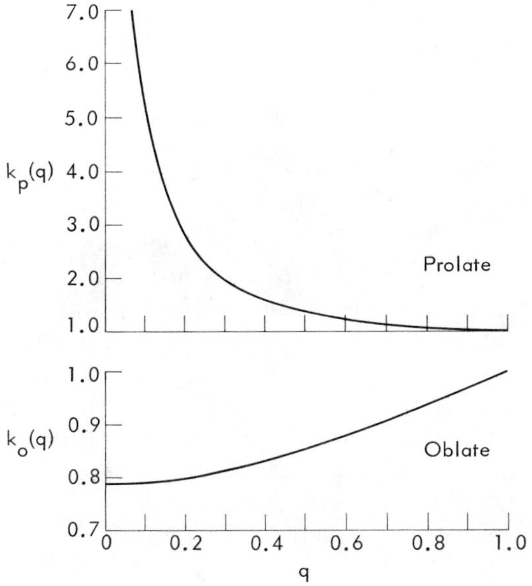

Fig. 5.14. Shape factors $k_p(q)$ and $k_o(q)$, for prolate and oblate spheroids, respectively, plotted versus q, the axial ratio of the generating ellipse [6].

and Wurst obtained average section radii throughout 68 grains, from which they calculated a statistical grain shape. (See Section 7.6.2 for further experimental details.)

They provided a table of coefficients based on 14 class intervals for this particular statistical grain shape. The derivation of the spatial size distribution from the planar size distribution is similar to that described previously by Scheil [15, 31] for spherical particles. Consequently, one must use a rather lengthy and laborious calculation method with their table for the statistical grain shape.

This type of analysis should be repeated under more modern conditions. For example, purer material should be used in order to attain more equiaxed grain shapes. Both experimental [3] and analytical [38] methods for obtaining area and size distribution curves should be refined. And finally, more flexible analytical procedures, such as are described in this chapter, should be incorporated in the method.

5.4.4 Paulus' Analysis for the Pentagonal Dodecahedron

An improved analysis conducted along lines suggested in the preceding section is given by Paulus [4]. His interest lies in the magnetic and electrical properties of ferrites as influenced by their grain-size distributions. On the basis of a comparison of average characteristics of real grains and cells, Paulus selects the pentagonal dodecahedron as the prototype shape for ferrite grains. Using Hull and Houk's [3] frequency distribution curve of diameters for the pentagonal dodecahedron, he calculates a table of coefficients in much the same way as Scheil and Wurst [2]. Paulus offers both analytical and graphical solutions. The analytical method is based on Scheil's original analysis [14], except that class intervals are spaced according to a geometrical progression. With the help of the latter simplification, plus the assumption of a log-normal distribution, Paulus prepares graphical solutions for several special cases.

The limitations inherent in a Scheil-type analysis are circumvented to some extent in Paulus' graphical solution. The idea of basing the analysis on a pentagonal dodecahedron shape and a log-normal distribution curve is good. However, when the graphical solution cannot be used, a simpler analytical method is desirable.

5.4.5 Cahn and Fullman's Analysis for Lamellar Structures

We turn now to the important analysis of lamellar structures by Cahn and Fullman [8], which is an extension of their procedure for spheres. The method applies to structures such as pearlite, in which hard lamellae are embedded regularly in a soft matrix. There are no restrictions on lamellae spacing, curvature, or thickness. Thus, instead of a calculated or analytical distribution function of lamellae spacings, this method is based on a measured lamellae spacing distribution curve.

Two experimental parameters enter Cahn and Fullman's analysis: the true spacing σ_t and the random intercept spacing σ_r. The true spacing is the perpendicular distance across two adjacent platelets, while the random spacing is the center-to-center distance between hard lamellae along random test lines on the plane of polish. These spacings, as well as others encountered in lamellar structures, were classified and defined in Section 3.6.

Consider a randomly oriented test line across the microstructure which intersects the lamellae at all possible angles. The lamellae may or may not have variable spacings and may or may not be parallel to one another. Bearing in mind the equivalence of volume and lineal fractions, if $V_V(\sigma_t)\, d\sigma_t$ is the volume fraction occupied by lamellae with true spacings σ_t to $\sigma_t + d\,\sigma_t$, then $V_V(\sigma_t)\, d\sigma_t$ is also the expected fraction of length of test line intercepted by these lamellae. The probability of randomly oriented lines in space lying within an angle θ to $\theta + d\theta$ is $\sin\theta\, d\theta$ (see Eq. 2.29). Thus the expected fraction of the test line occupied by lamellae oriented randomly in space, with true spacings between σ_t to $\sigma_t + d\sigma_t$ and with normals to the test line between θ to $\theta + d\theta$, is $\sin\theta\, d\theta\, V_V(\sigma_t)\, d\sigma_t$.

The expected fraction of test line that intersects lamellae with spacings σ_t to $\sigma_t + d\sigma_t$ and gives random intercept spacings of σ_r to $\sigma_r + d\sigma_r$ is

$$\sin\theta\, V_V(\sigma_t) \left| \left(\frac{\partial\theta}{\partial\sigma_r}\right)_{\sigma_t} \right| d\sigma_t\, d\sigma_r. \tag{5.39}$$

Since $\sigma_r = \sigma_t/\cos\theta$, then $|(\partial\theta/\partial\sigma_r)_{\sigma_t}| = \sigma_t/\sigma_r^2 \sin\theta$, and Eq. (5.39) becomes

$$\sigma_t V_V(\sigma_t) d\sigma_t d\sigma_r/\sigma^2. \tag{5.40}$$

Therefore the expected lineal fraction distribution function $g(\sigma_r)$ of lamellae with true spacings σ_t to $\sigma_t + d\sigma_t$ and with random spacings σ_r to $\sigma_r + d\sigma_r$ is

$$g(\sigma_r) d\sigma_r = \frac{d\sigma_r}{\sigma_r^2} \int_0^{\sigma_r} \sigma_t V_V(\sigma_t) d\sigma_t \tag{5.41}$$

or

$$\sigma_r^2 g(\sigma_r) = \int_0^{\sigma_r} \sigma_t V_V(\sigma_t) d\sigma_t. \tag{5.42}$$

When $\sigma_r = \sigma_t$

$$2\sigma_r g(\sigma_r) + \sigma_r^2 \frac{dg(\sigma_r)}{d\sigma_r} = \sigma_r V_V(\sigma_t)_{\sigma_t=\sigma_r} \tag{5.43}$$

or

$$V_V(\sigma_t)_{\sigma_t=\sigma_r} = 2g(\sigma_r) + \frac{dg(\sigma_r)}{d\ln\sigma_r}. \tag{5.44}$$

Thus the volume fraction distribution function of the true spacings of lamellae can be calculated from two simple terms. We need know only the lineal fraction distribution function $g(\sigma_r)$ of random spacings. Evaluation of Eq. (5.44) proceeds in much the same manner as described for Eq. (5.35).

It may be preferable to deal with a *number* distribution function of random spacings rather than with a *lineal fraction* distribution function. The two types of functions are related by

$$n_L(\sigma_r)\,d\sigma_r = \frac{g(\sigma_r)}{\sigma_r}\,d\sigma_r, \tag{5.45}$$

where $n_L(\sigma_r)\,d\sigma_r$ is the number of random spacings per unit length of test line with random spacings between σ_r to $\sigma_r + d\sigma_r$. Substituting $g(\sigma_r)$ from Eq. (5.45) into Eq. (5.44) gives

$$V_V(\sigma_t)_{\sigma_t=\sigma_r} = 3\sigma_r n_L(\sigma_r) + \sigma_r^2\frac{dn_L(\sigma_r)}{d\sigma_r} \tag{5.46}$$

which expresses the volume fraction distribution function of the true spacings of lamellae in terms of the number distribution function of random spacings.

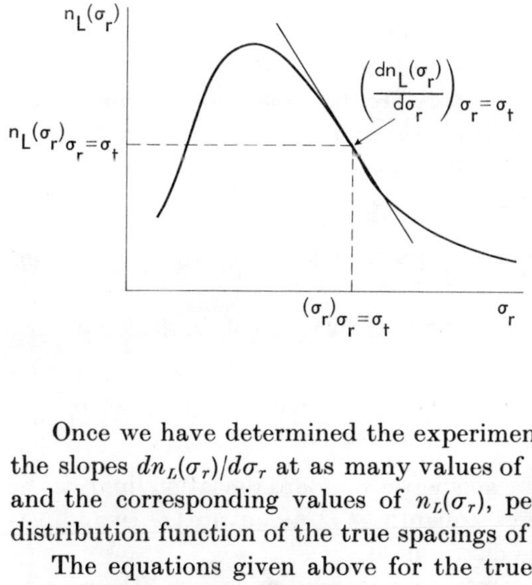

Fig. 5.15. Schematic representation of measurements required for Eq. (5.46).

Once we have determined the experimental curve of $n_L(\sigma_r)$ versus σ_r, we measure the slopes $dn_L(\sigma_r)/d\sigma_r$ at as many values of σ_r as we wish (see Fig. 5.15). The slopes, and the corresponding values of $n_L(\sigma_r)$, permit calculation of the volume fraction distribution function of the true spacings of lamellae.

The equations given above for the true-spacing distribution are quite powerful and general. They are particularly useful because of the ease with which intercept spacings can be measured. The above analysis need not be restricted to the pearlites encountered in steels. Banded rocks, randomly twinned structures, or the remarkably uniform eutectic lamellae, for example, are also amenable to this type of analysis.

5.5 SPECIAL METHODS FOR REPRESENTING PARTICLE SIZE DISTRIBUTIONS

The methods described above involve the counting and sizing of particle sections, and from these data one can obtain the complete distribution curve. By means of special assumptions, much of this laborious procedure can be side-stepped. Moreover, it is seldom that all the information provided by a distribution curve is required or utilized. Frequently one needs only certain basic parameters of the distribution, such as those listed in Section 5.2.

Several special methods for representing particle size distributions have been developed. Examples of these include the procedures by DeHoff [34] and Ashby and Ebeling [49], which are based on the assumption of a log-normal distribution function, and by Bach [11], who utilizes a Pearson distribution of type III. Hilliard [10, 46] and Bach [11] describe methods for characterizing size distributions of particles by the moments of the distribution. Nicholson and Merckx [9] provide computer programs for analyzing experimental data obtained from particle sections.

We shall not go into further detail concerning these various methods of analysis, except for the paper by DeHoff [34]. Before we talk about his paper, however, let us briefly discuss the general attributes of the log-normal distribution.

5.5.1 Log-Normal Distribution of Particle Sizes

Because of the overwhelming occurrence of log-normal distributions in small particle systems, it is well to state some basic relationships and definitions for such distributions [26, 39, 40, 41].

If a distribution of particle diameters obeys the log-normal law, a plot of the frequencies against the logarithms of the diameters will result in a symmetrical, normal curve. Plotting the diameters directly will give a skew distribution curve (see Fig 5.16). The log transformation may, therefore, be looked on as a means for transforming a skew distribution into a symmetrical normal one [17].

Let $D_1, D_2, D_3, \ldots, D_{\max}$ be a set of random particle diameters, and let $z_1, z_2, z_3, \ldots, z_{\max}$ be the transformation $z = \ln D$. Then, if

$$y = \frac{1}{\sigma_z\sqrt{2\pi}} \exp\left[-\frac{1}{2}\left(\frac{z-\bar{z}}{\sigma_z}\right)^2\right], \tag{5.47}$$

we say that the variable D is *log-normally distributed*. In terms of D, the log-normal distribution is given by

$$y = \frac{1}{\ln \sigma_g \sqrt{2\pi}} \exp\left[-\frac{1}{2}\left(\frac{\ln D - \ln D_g}{\ln \sigma_g}\right)^2\right], \tag{5.48}$$

where D_g is the geometric mean and σ_g the geometric standard deviation ($\ln \sigma_g$ is the standard deviation of the $\ln D$ plot). The two quantities D_g and σ_g are the two adjustable parameters that uniquely define this distribution.

The log-normal distribution possesses moments of any order. The ith moment about the origin is denoted by $\mu_i'(D)$, and, in general, can be expressed as

$$\mu_i'(D) = \exp\left(i \ln D_g + \tfrac{1}{2}i^2 \ln^2 \sigma_g\right), \tag{5.49}$$

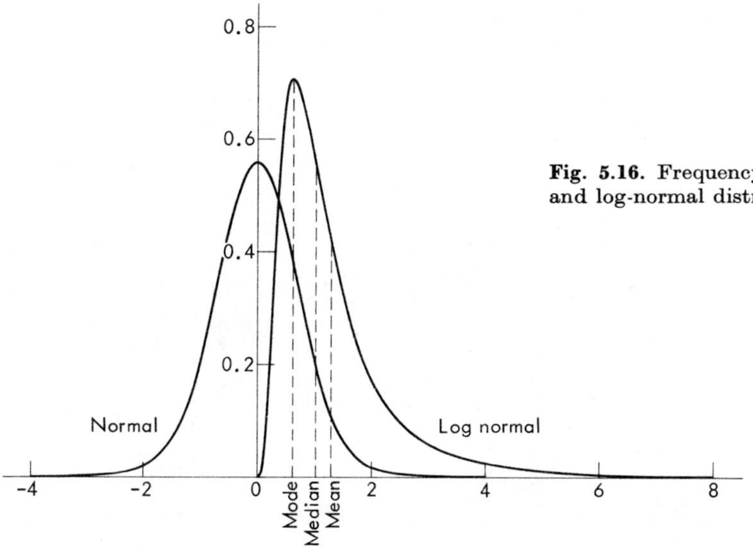

Fig. 5.16. Frequency curves of a normal and log-normal distribution [40].

where $i = 1, 2, 3, \ldots$. The first three moments of the distribution are:

$$\mu_1'(D) = \frac{1}{N_V} \sum (N_V)_j D_j = \exp(\ln D_g + 0.5 \ln^2 \sigma_g), \tag{5.50a}$$

$$\mu_2'(D) = \frac{1}{N_V} \sum (N_V)_j D_j^2 = \exp(2 \ln D_g + 2 \ln^2 \sigma_g), \tag{5.50b}$$

$$\mu_3'(D) = \frac{1}{N_V} \sum (N_V)_j D_j^3 = \exp(3 \ln D_g + 4.5 \ln^2 \sigma_g), \tag{5.50c}$$

where the first moment is related to the average size; the second moment is a measure of the deviation from the average; and the third moment indicates the degree of skewness of the distribution.

We shall not develop this subject further here; however, there are numerous texts and papers available in the literature that go into any desired degree of detail. For a practical discussion, see the report by King [42]; see also DallaValle [41] for a wider presentation. For helpful remarks on normalizing distributions other than log-normal ones, see the publication by Natrella [43].

5.5.2 Procedure Based on Log-normal Distributions

DeHoff [34] shows how N_V and the first two moments of the size distribution are related to three simple types of counting measurements. It is assumed that the distribution can be expressed by a two-parameter function and that the particles all have the same (known) shape.

The defining equations for the three counting measurements are

$$N_A = N_V \bar{D}, \tag{5.51a}$$

$$P_L = \tfrac{1}{2} S_V = \tfrac{1}{2} N_V \bar{S}, \tag{5.51b}$$

$$P_P = V_V = N_V \bar{V}, \tag{5.51c}$$

which have appeared previously in Eq. (5.4); Eq. (2.31) and Table 1.1; and Eqs. (2.23) and (4.31), respectively. The procedure is to establish relationships among these counting measurements and the two parameters, D_g and σ_g.

To accomplish this, we modify the three equations given by DeHoff [44]:

$$\bar{D} = \alpha\bar{L} = \alpha\mu_1'(D), \tag{5.52a}$$

$$\bar{S} = \beta\overline{L^2} = \beta\mu_2'(D), \tag{5.52b}$$

$$\bar{V} = \gamma\overline{L^3} = \gamma\mu_3'(D), \tag{5.52c}$$

where L is some linear measure of the "size" of the particle, for example, the diameter D for spheres; α, β, and γ are shape (or form) factors (see Section 7.6.1); and $\mu_i'(D)$ are the moments of the distribution.

Substitution of Eqs. (5.52a, b, c) in Eqs. (5.51a, b, c), respectively, gives

$$N_A = N_V\alpha\mu_1'(D) \tag{5.53a}$$

$$P_L = \tfrac{1}{2}N_V\beta\mu_2'(D) \tag{5.53b}$$

$$P_P = N_V\gamma\mu_3'(D), \tag{5.53c}$$

or, according to Eqs. (5.50a, b, c),

$$N_A = N_V\, \alpha \exp(\ln D_g + 0.5 \ln^2 \sigma_g), \tag{5.54a}$$

$$P_L = \tfrac{1}{2}N_V\beta \exp(2 \ln D_g + 2 \ln^2 \sigma_g), \tag{5.54b}$$

$$P_P = N_V\, \gamma \exp(3 \ln D_g + 4.5 \ln^2 \sigma_g), \tag{5.54c}$$

where now the two parameters of the log-normal distribution are given in terms of the particle shape factors, N_V and the three counting measurements. We can eliminate N_V from these equations and obtain it independently from Eqs. (551a, b, c) and (5.52a, b, c).

By appropriate manipulations, we express the two parameters and N_V in convenient form:

$$\ln D_g^2 = \ln\left(\frac{256\alpha^5\gamma^3}{\beta^8}\right) + \ln\left(\frac{P_L^8}{N_A^5P_P^3}\right), \tag{5.55a}$$

$$\ln^2 \sigma_g = \ln\left(\frac{\beta^2}{4\alpha\gamma}\right) + \ln\left(\frac{P_PN_A}{P_L^2}\right), \tag{5.55b}$$

$$\ln N_V = \ln\left(\frac{\beta^3}{8\alpha^3\gamma}\right) + \ln\left(\frac{P_PN_A^3}{P_L^3}\right). \tag{5.55c}$$

Representing the three shape-factor terms by

$$K_D = \frac{256\alpha^5\gamma^3}{\beta^8}, \tag{5.56a}$$

$$K_\sigma = \frac{\beta^2}{4\alpha\gamma}, \tag{5.56b}$$

$$K_N = \frac{\beta^3}{8\alpha^3\gamma}, \tag{5.56c}$$

we write in final form

$$\ln D_g^2 = \ln\left(\frac{P_L^8}{N_A^5 P_P^3}\right) + \ln K_D, \tag{5.57a}$$

$$\ln^2 \sigma_g = \ln\left(\frac{P_P N_A}{P_L^2}\right) + \ln K_\sigma, \tag{5.57b}$$

$$\ln N_V = \ln\left(\frac{P_P N_A^3}{P_L^3}\right) + \ln K_N. \tag{5.57c}$$

For spheres of diameter $D = L$, $K_D = 32/27\ \pi^5$, $K_\sigma = 3\pi/2$, and $K_N = 3\pi^2/4$. For other simple particle shapes, see the master curves provided by DeHoff [44].

The three parameters D_g, σ_g, and N_V can also be obtained without knowledge or specification of the distribution curve. Hilliard [10, 46] has devised an analysis which gives the moments of any order, as well as N_V, for size distributions of convex particles of the same shape or different shapes, if known. It is not necessary to assume any particular form of distribution, although the method is applicable to log-normal distributions. His method requires measurement of chord lengths in certain cases, which may be somewhat more laborious than the three counting measurements, unless instrumented. However, confidence limits on the estimated properties are easily calculated from the data, and N_V can be estimated without determining the complete distribution of intercept lengths.

REFERENCES

1. S. D. Wicksell, "The Corpuscle Problem II: Case of Ellipsoid Corpuscles," *Biometrika* **18** (1926) 151
2. E. Scheil and H. Wurst, "Statistische Gefügeuntersuchungen II: Messung der raümlichen Kristallgrösse," *Z. Metallk.* **28**, 11 (1936) 340
3. F. C. Hull and W. J. Houk, "Statistical Grain Structure Studies; Plane Distribution Curves of Regular Polyhedrons," *Trans. AIME* **197** (1953) 565
4. M. Paulus, "Méthode d'étude de la structure granulaire. Parts I and II," *Métaux, Corrosion-Industries* **37**, 448 (Dec., 1962) 447; *ibid.*, **38**, 449 (Jan., 1963)
5. R. J. Allio and C. H. Randall, "The Quantitative Analysis of Microstructure with Densitometer Data," *Trans. AIME* **224** (1962) 221
6. R. T. DeHoff, "The Determination of the Size Distribution of Ellipsoidal Particles from Measurements Made on Random Plane Sections," *Trans. AIME* **224** (1962) 474
7. S. A. Saltykov, "The Determination of the Size Distribution of Particles in an Opaque Material from a Measurement of the Size Distribution of Their Sections," *Stereology*, edited by H. Elias, *Proc. Second Int. Cong. for Stereology*, New York: Springer-Verlag (1967) 163
8. J. W. Cahn and R. L. Fullman, "On the Use of Lineal Analysis for Obtaining Particle Size Distribution Functions in Opaque Samples," *Trans. AIME* **206** (1956) 610

9. W. L. Nicholson and K. R. Merckx, "Unfolding Particle Size Distributions," Pacific Northwest Laboratory Report BNWL-210, March 1967; see also *ibid.*, *Stereology*, edited by H. Elias, *Proc. Second Int. Cong. for Stereology*, New York: Springer-Verlag (1967) 191
10. J. E. Hilliard, "The Direct Determination of the Number of Convex Particles per Unit Volume and the Moments of Their Size Distribution by an Intercept Analysis on a Section," *Stereology*, edited by H. Elias, *Proc. Second Int. Cong. for Stereology*, New York: Springer-Verlag (1967) 195
11. G. Bach, "Size Distribution of Particles Derived from the Size Distribution of Their Sections," *Stereology*, edited by H. Elias, *Proc. Second Int. Cong. for Stereology*, New York: Springer-Verlag (1967) 174
12. S. D. Wicksell, "The Corpuscle Problem I: A Mathematical Study of a Biometric Problem," *Biometrika* **17** (1925) 84
13. T. H. Hagermann, *Geol. Fören. i Stockholm Förhand.*, **46** (1924) 325
14. E. Scheil, "Die Berechnung der Anzahl und Grössenverteilung kugelförmiger Kristalle in undurchsichtigen Körpern mit Hilfe der durch einen ebenen Schnitt erhaltenen Schnittkreise," *Z. anorg. u. allg. Chem.* **201** (1931) 259
15. E. Scheil, "Statistiche Gefügeuntersuchungen, I," *Z. Metallk.* **27** (1935) 199
16. H. A. Schwartz, "The Metallographic Determination of the Size Distribution of Temper Carbon Nodules," *Metals and Alloys*, **5**, 6 (June 1934) 139
17. S. A. Saltykov, *Stereometric Metallography*, second edition, Moscow: *Metallurgizdat* (1958)
18. W. A. Johnson, "Estimation of Spatial Grain Size," *Met. Prog.* **49** (Jan., 1946) 87
19. A. G. Spektor, "Distribution Analysis of Spherical Particles in Non-Transparent Structures," *Zavod. Lab.* **16,** 2 (1950) 173
20. G. W. Lord and T. F. Willis, "Calculation of Air Bubble Size Distribution from Results of a Rosiwal Traverse of Aerated Concrete," *ASTM Bulletin*, **177** (Oct., 1951) 56
21. E. D. Hyam and J. Nutting, "The Tempering of Plain Carbon Steel," *J. Iron Steel Inst.* **184** (October 1956) 148; see also discussion, *ibid.* **186** (1957) 351
22. F. Lenz, "Zur Grössenverteilung von Kugelschnitten," *Z. Wissen. Mikr.* **63,** 1 (1956) 50
23. G. Bach, "Über die Grössenverteilung von Kugelschnitten in durchsichtigen Schnitten endlicher Dicke," *Z. Wissen. Mikr.* **64** (1959) 265
24. S. A. Tyler and L. Sanathanan, "On the Estimation of Size Distribution from Plane Sections," personal communication, Argonne National Laboratory (1965)
25. F. Schücker, "Grain Size," *Acta Polytech. Scand.*, Chemistry Including Metallurgy Series, No. 54, Stockholm (1966)
26. G. Herdan, *Small Particle Statistics*, second edition, London: Butterworths (1960)
27. M. R. Spiegel, *Theory and Problems of Statistics*, New York: Schaum (1961) 71
28. E. P. Papadakis, "Grain-Size Distribution in Metals and its Influence on Ultrasonic Attenuation Measurements," *J. Acoustical Soc. of Amer.* **33,** 11 (Nov., 1961) 1616
29. R. L. Fullman, "Measurement of Particle Sizes in Opaque Bodies," *Trans. AIME* **197** (1953) 447
30. J.J.B. Rutherford, R. H. Aborn, and E. C. Bain, "The Relation Between the Grain Areas on a Plane Section and the Grain Size of a Metal," *Metals and Alloys* **8** (Dec., 1937) 345

31. E. E. Underwood, Chapter 6, "Particle Size Distribution," in *Quantitative Microscopy*, edited by R. T. DeHoff and F. N. Rhines, New York: McGraw-Hill (1968) 149

32. H. B. Aaron, R. D. Smith, and E. E. Underwood, "Spatial Grain-Size Distribution from Two-Dimensional Measurements," paper 16 in *Proc. First Int. Cong. for Stereology*, Vienna, Austria, April 18–20, 1963

33. R. T. DeHoff and F. N. Rhines, "Determination of the Number of Particles per Unit Volume from Measurements Made on Random Plane Sections: The General Cylinder and the Ellipsoid," *Trans. AIME* **221** (1961) 975

34. R. T. DeHoff, "The Estimation of Particle Size Distributions from Simple Counting Measurements Made on Random Plane Sections," *Trans. AIME* **233** (1965) 25

35. H. E. Exner, "Die mikroskopische Bestimmung von räumlichen Korngrössenverteilungen in undurchsichtigen Stoffen," *Praktische Metallographie* **3**, 8 (1966) 334

36. H. E. Exner, "Analyse der Grössenverteilung von Körnern, Poren und Pulverteilchen," *Z. Metallk.* **57** (1966) 755

37. E. J. Myers and M. J. Sinnott, "Quantitative Metallography for Particles Having Polyhedral Shapes," Chapter 3 in *Computer Applications in Metallurgical Engineering*, Amer. Soc. for Metals, Metals Park, Ohio (1964) 17

38. J. L. Meijering, "Interface Area, Edge Length, and Number of Vertices in Crystal Aggregates with Random Nucleation," *Phillips Res. Rep.* **8** (1953) 270

39. A. Goldman, H. D. Lewis, and R. H. Moore, "On the Proper Use of Transformations of Log Normal Functions in Small Particle Statistics," Los Alamos Scientific Lab., LA-3262, 9 July 1965

40. J. Aitchison and J.A.C. Brown, *The Lognormal Distribution*, Cambridge, England: Cambridge University Press (1963)

41. J. M. DallaValle, *Micromeritics*, second edition, New York: Pitman (1948) Chapter 5

42. J. R. King, "Graphical Data Analysis with Probability Papers," published by TEAM, Lowell, Mass., 1965

43. M. G. Natrella, "The Use of Transformations," Chapter 20 in *Experimental Statistics*, NBS Handbook 91, Aug. 1, 1963, Washington, D. C.: U. S. Govt. Printing Office

44. R. T. DeHoff, "The Determination of the Geometric Properties of Aggregates of Constant-Size Particles from Counting Measurements Made on Random Plane Sections," *Trans. AIME* **230** (1964) 764

45. G. Bockstiegel, "Eine einfache Formel zur Berechnung räumlicher Grössenverteilungen aus durch Linearanalyse erhaltenen Daten," *Z. Metallk.* **57** (1966) 647

46. J. E. Hilliard, "Direct Determination of the Moments of the Size Distribution of Particles in an Opaque Sample," *Trans. AIME* **242** (1968) 1373

47. H. E. Exner, "Modellversuche zu den Umrechnungsverfahren von linearen und ebenen in räumliche Grössenverteilungen," *Metall* **21** (1967) 431

48. J. E. Hilliard, "The Calculation of the Mean Caliper Diameter of a Body for Use in the Analysis of the Number of Particles per Unit Volume," *Stereology*, edited by H. Elias, *Proc. Second Int. Cong. for Stereology*, New York: Springer-Verlag (1967) 211

49. M. F. Ashby and R. Ebeling, "On the Determination of the Number, Size, Spacing, and Volume Fraction of Spherical Second-Phase Particles from Extraction Replicas," Trans. *AIME* **236** (1966) 1396

PROBLEMS

5.1. Using data given in Table 5.4 (columns 4 and 5), calculate (a) $\bar{D}$ and $\sigma(D)$; (b) D_g and σ_g.

5.2. Plot the following size data in class intervals of (a) 10, (b) 25, (c) 50, and (d) 100.

Size	No.	Size	No.	Size	No.	Size	No.	Size	No.	Size	No.	Size	No.
43	1	81	4	95	1	113	1	136	2	153	1	168	1
49	1	82	2	101	2	114	2	138	1	154	1	176	1
55	1	84	1	102	2	118	1	139	1	156	1	179	1
57	1	85	3	103	1	119	1	142	2	157	1	181	1
65	1	88	1	104	2	121	1	143	1	158	3	191	1
69	1	89	2	105	1	122	1	147	2	159	1	203	1
73	1	90	1	106	1	123	1	149	1	163	1	206	1
75	2	92	1	107	3	128	1	150	1	164	1		
76	1	93	1	109	2	131	1	151	1	165	1		
79	1	94	1	112	3	133	2	152	1	166	1		

What would you say about the reliability of the histogram with class intervals of 10? of 100?

5.3. Calculate the arithmetic mean, m, and standard deviation, s, of the following four frequency distributions. Based on the values of m, s, and total number of cases, what would you conclude about these four distributions?

Class limits	Frequencies			
	f_1	f_2	f_3	f_4
20– 29	1			
30– 39	4	2		
40– 49	6	5	12	
50– 59	8	10	12	34
60– 69	10	16	12	12
70– 79	16	17	12	6
80– 89	18	18	12	4
90– 99	16	12	12	6
100–109	10	10	12	12
110–119	8	7	12	34
120–129	6	5	12	
130–139	4	3		
140–149	1	1		
150–159		1		
160–169		1		

5.4. Calculate the moment coefficient of skewness, $\alpha_3 = \mu_3'/s^3$ and the moment coefficient of kurtosis, $\alpha_4 = \mu_4'/s^4$ for the four distributions in Problem 5.3. What conclusions can you draw based on these additional data?

[*Note:* In symmetrical distributions, $\alpha_3 = 0$. For normal distributions, $\alpha_4 = 3$, and $\alpha_4 < 1.75$ signifies that the distribution is U-shaped.]

5.5. Using the data of Table 5.4, calculate $(N_V)_2$ by the diameter analysis.

5.6. Using the data of Table 5.6, calculate $(N_V)_4$ by the area analysis.

5.7. Using the data of Table 5.8, calculate $(N_V)_7$ by the chord analysis.

5.8. Derive the relationship for spheres

$$S_V = \pi N_V(D_g^2 + \sigma_g^2).$$

[*Hint*: Use Eqs. (5.3) and $S_V = N_V\bar{S}$ from Table 1.1.]

For a chord analysis, the mean intercept length would be more useful than S_V. Express the equation in terms of $\bar{L}_3$ for (a) space-filling grains, (b) discrete particles.

6

PROJECTED IMAGES

"The transmission electron-microscope technique is undoubtedly a powerful method for the quantitative study of dislocations and stacking faults, and a great deal can be learned from simple geometrical considerations alone. . . . The more detailed and quantitative investigations of dislocations by means of the new experimental techniques will become of increasing importance."
Archibald Howie

6.1 INTRODUCTION

Geometrical relationships between structural features and their projected images have only recently been subjected to serious mathematical scrutiny. Outstanding in this regard are the treatments accorded projected images of particulate systems and the linear features in crystals known as dislocations. Although we have drawn freely from the current literature on dislocations, no loss in generality is incurred thereby, because the methods developed for dislocations may also be applied to other linear systems, regardless of their source.

The geometry of projections entails problems that do not arise in sections cut through opaque materials. At the present time, rigorous mathematical treatments and exact solutions are few, and even in the simpler cases approximate solutions may be the only ones available. However, the accelerated rate at which these problems are being attacked promises early rectification of the situation.

Projected images of microstructural features are usually obtained from thin sheets, called variously (thin) foils, films, slices or sections,* depending on the particular background of the scientist. Thin foils that vary in thickness from a few hundred to a few thousand angstroms create both geometrical and experimental difficulties. Surface effects tend to distort the true bulk configurations [1, 2]; overlap of particle

*In order to avoid the semantic confusion that results when a word like "section" is used as a verb meaning to cut, and also as a noun describing a two-dimensional plane or a three-dimensional sheet, we employ the term "thin foil" in a general sense to mean any thin sheet of finite thickness (be it of tissue, rock, or metal) from which a projected image is formed.

and surface images becomes serious as the volume fraction and foil thickness increase [3, 4]; and determination of true foil thickness [5–10] poses problems in some materials. Moreover, one must ensure that the sample population is statistically adequate and that the degree of randomness is properly assessed. It is most important that thin foils be prepared without altering the true structural arrangement in the original bulk specimen. Systematic investigations are currently underway to determine the existence and extent of any such alterations in thin metal foils [1, 2, 11, 12]. It is well known that diffraction effects may prevent the complete image from registering [12]. But this is a problem that does not fall within the scope of the present book and will not be treated here. We shall confine our attention to the projected image and assume that it is an accurate representation of the true structure.

Books that touch upon the special problems of thin foil preparation and analysis include those for petrologists [13], metallurgists [14, 59], anatomists [15], and others [41, 47]. However, quantitative treatments of projection geometry have received scant attention. In this chapter we discuss the geometry of projections from a general standpoint, in terms of dimensions rather than specific structures. Of course we also consider the physical nature of the structure, if additional insight may be gained thereby.

Unless otherwise noted, by "projection" we refer to a *parallel** projection, although central projections and hemispherical projections (encountered in astronomy and field emission microscopy) have merit in special situations [43, 46]. Stereoscopic projections are of interest in connection with overlap of projected images, and are discussed in Chapter 7. Even though complete mathematical derivations will not be presented in every case, references to the original sources should prove sufficient for the reader interested in further details.

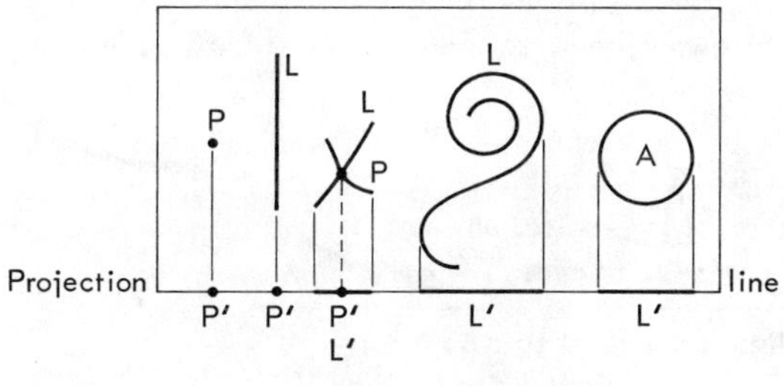

Fig. 6.1. Projection of figures in a plane onto a projection line.

6.2 DIMENSIONAL EQUIVALENTS

Before we undertake an analysis of the equations that apply to projected images, it might be well to examine the equivalence of the dimensions of real objects and the (apparent) dimensions of their projected images. Figure 6.1 gives examples of features in a plane, their projections onto a line, and the equivalent dimensions involved.

*Two figures in distinct planes are said to be derived from each other by *parallel* projection if corresponding points can be joined by parallel lines [42].

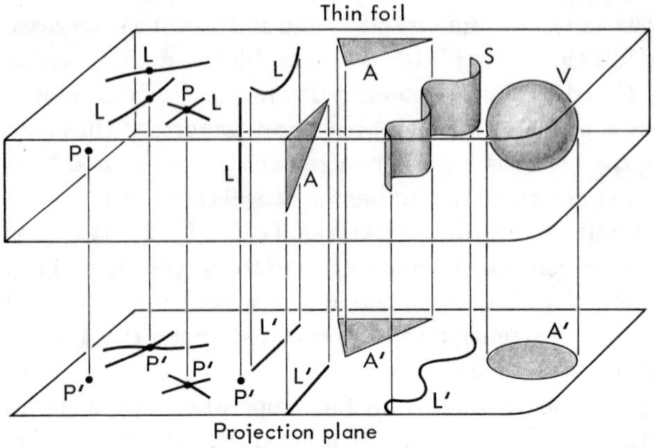

Fig. 6.2. Projection of structural features in a foil onto a projection plane.

Note that projected quantities are identified by a prime to distinguish them from their real counterparts. It is evident that points and lines may project to zero dimensions, and that lines and areas may project to one dimension. The subject of projections from a plane onto a line is not treated at length in this chapter; however, the principle is invoked later in connection with the subject of total projection of lines in space (see Section 6.6.2).

Figure 6.2 shows a set of simple objects in space and their projected images on a plane. The dimensional equivalents are indicated by primes on the corresponding symbols. Points, intersections, and vertical straight lines project as points; lineal elements and vertical surfaces may project as lines; and volume elements always project as areas. Simple examples such as these allow us to recapitulate the results in tabular form. Table 6.1 summarizes the dimensional equivalents that are possible between real features in the foil and their images on the projection plane. This table shows us that the dimensional equivalents are characterized by indeterminacies and redundancies.

TABLE 6.1

Dimensional equivalents of spatial objects and their projected images

Actual dimensions	Projected dimensions
Features in a plane projected to a line	
0	0
1	0 or 1
2	1
Features in space projected to a plane	
0	0
1	0 or 1
2	1 or 2
3	2

In addition to the structural elements, there are also the test quantities to consider, such as test points, test lines, and test areas. A test point (of a grid) placed in the projection plane corresponds to a vertical line of length t (the foil thickness) in the foil. If there are P'_T of these test points, then $L_T = P'_T t$, where L_T is the total test line length in the foil. Similarly, a test line of length L'_T on the projection plane corresponds to a vertical area in the foil, such that $A_T = L'_T t$. For test area A' on the projection plane, we find that $V_T = A'_T t$. There are no ambiguities in these relationships, since they refer only to test quantities that originate in the projection plane and not to the structural features listed in Table 6.1.

The customary combined notation, such as L'_A, consists of a structural quantity (L') and a test quantity (A'_T). Therefore the dimensions of the combined term may also differ in the foil and in the projection plane. The role of dimensional equivalents in the combined symbol is amplified in the next section.

6.3 CORRESPONDING QUANTITIES

It is apparent from Fig. 6.2 that structural elements in the foil do not always have the same dimensions as their images on the projection plane. Since the test quantities also have equivalent dimensions, it follows that the combined symbol will usually have dimensions that are different from the corresponding quantity in the foil.

In addition to the dimensional problems, there are also ambiguities that may or may not arise in the interpretation of measurements and images in the projection plane. An example of a unique measurement is a P'_L count across a system of projected lines, since the corresponding quantity can only be a P_A count in space. On the other hand, the measurement of A'_A from a set of areas in the projection plane is ambiguous, since the images could arise, either exclusively or in part, from flat surfaces (A_V), curved surfaces (S_V), or volume elements (V_V) in the foil.

In order to clarify this situation, we show simple examples of corresponding quantities schematically in Fig. 6.3, which also illustrates measurements and structures, lists the dimensions of corresponding quantities, and gives examples of equations linking the two terms. Figure 6.3 does not consider overlap, nor does it include all possible alternative structures. Equations for the actual structures encountered in real materials are treated later in this chapter.

Considerations such as these must be borne in mind when we are dealing with projected images. It is also helpful if we know, at least partially, the true structure in the foil, in order to reduce at least some of the ambiguities in the projected quantities.

6.4 EXACT EXPRESSIONS FOR PROJECTED IMAGES

Although exact relationships have not yet been developed for every case, it is possible under certain conditions to deduce the true magnitude of quantities in the foil from measurements on the projection plane. Of course, if certain features of the actual structure are known, the analysis of the projected image can be simplified considerably.

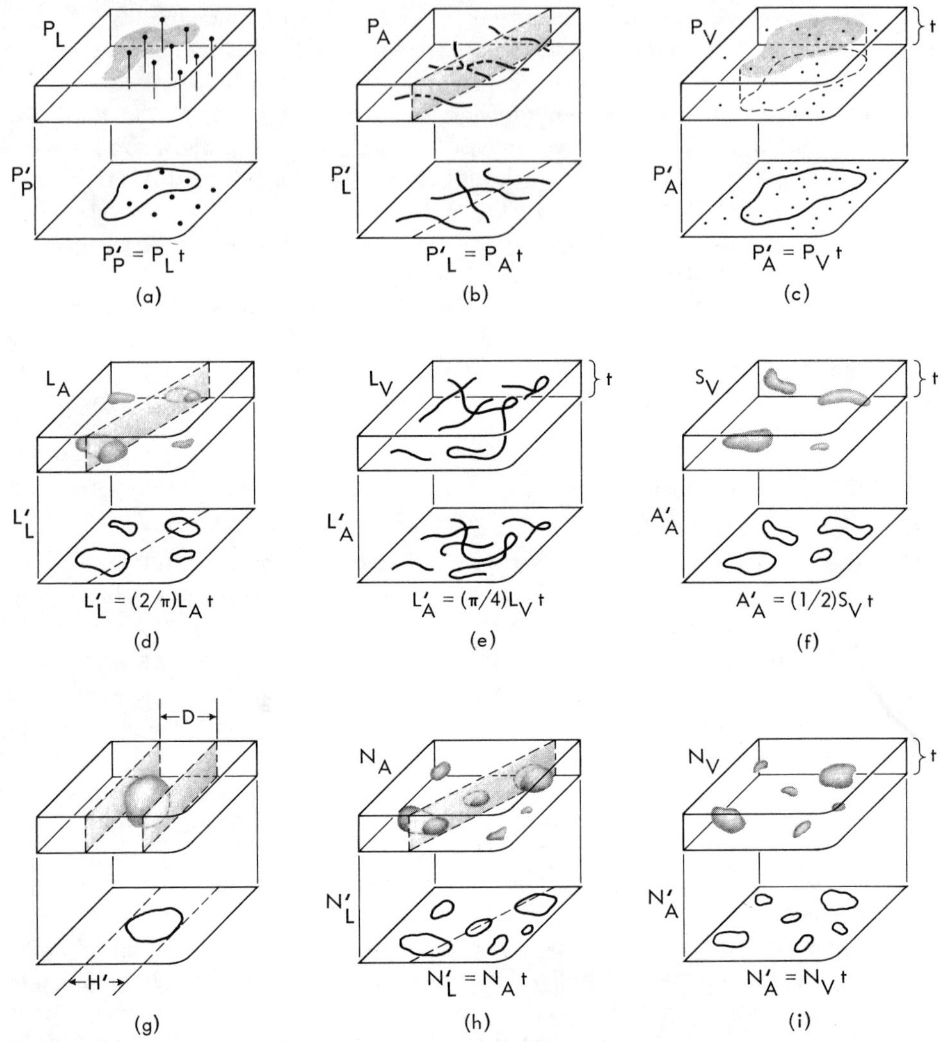

Fig. 6.3. Measurements on the projection plane and their relationships to microstructural features in thin foils.

In order to develop exact expressions for the simpler cases, one may prefer to deal with an isolated object not contained within a foil, rather than with a system of similar objects within a foil. The advantage of this procedure is that one can obtain the mean projected value by random rotation of a single object and thereby avoid certain complicating factors, such as surface or particle overlap, truncation of objects at the foil surface, and any nonrandomness that occurs within a foil. We shall introduce corrections for these factors later, when we get down to a more detailed consideration of systems of surfaces or particles.

TABLE 6.2

Interpretation of projected images

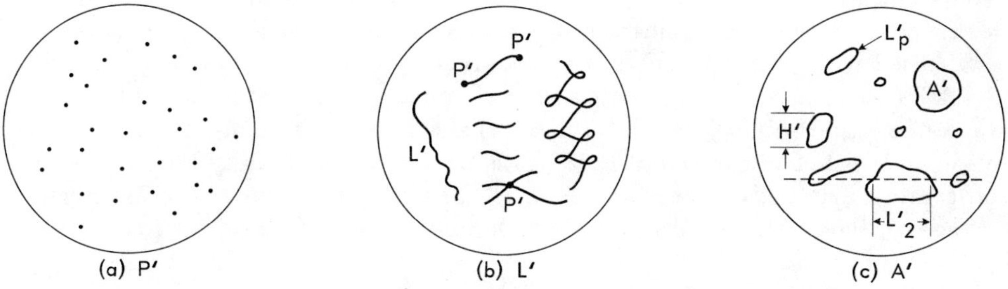

Type of feature on projection plane	Measured quantity on projection plane	Spatial features that could give indicated projection	Spatial symbol
(a) Points, P'	P'_A	Points in foil	P_V
	P'_A	Vertical line segments	L_V
(b) Lines, L'	L'_A	Lines in foil	L_V
	L'_A	Vertical planes	S_V
	Associated quantities		
	P'_A	Terminal points within foil	P_V
	P'_A	Points of intersection at foil surface	P_A
	P'_A	Intersection of lines within foil	P_V
(c) Areas, A'	A'_A	Flat surfaces	A_V
	A'_A	Curved surfaces	S_V
	A'_A	Volume elements	V_V
	Associated quantities		
	H'	Tangent diameter (or height) (see Fig. 6.5)	H, D
	L'_2	Intercept diameter (see Fig. 6.5)	—
	L'_p	Perimeter length*	—

*The subscript lower-case p serves to specify *perimeter* lengths.

A convenient way to proceed is to consider systematically the points, lines, and areas that appear on the projection plane. Table 6.2 summarizes possibilities arising from these three cases, and lists the spatial features that could give rise to the projected images.

6.4.1 Points

Although the projection of true points represents a rather trivial case, there is an ambiguity in the corresponding quantities. Projected points may arise from actual points in space and/or from vertical straight-line segments. The sources of the projected points cannot be distinguished, nor can the true length of linear features be established without supplementary information.

Two exact equations that relate points of intersection to test lines and test planes appear in Fig. 6.3(a) and (b). In part (a), there are P'_T test points in the projection plane, of which a certain number, P', falls in the projected area. For simple open surfaces without folds or overlap,* $P = P'$, where P is the number of intersections of the test lines with the surface in the foil. Since $L_T = P'_T t$, then

$$P/L_T = (P'/P'_T)/t, \tag{6.1}$$

or

$$P_L = P'_P/t \text{ mm}^{-1}, \tag{6.2}$$

for a system of simple open surfaces in a foil.

In part (b) of Fig. 6.3, there are P' intersections of a test line on the projection plane with images of lines, and this is equivalent to the P intersections of the lines in the foil by a test plane. The number of intersections are the same, that is, $P = P'$. Upon dividing through by $A_T = L'_T t$, we obtain

$$P_A = P'_L/t \text{ mm}^{-2} \tag{6.3}$$

for a system of lines in a foil.

For points, or for small volume elements that approximate points, the number in space equals the number on the projection plane, that is, $P = P'$, and since $V_T = A'_T t$, we may write

$$P_V = P'_A/t \text{ mm}^{-3}, \tag{6.4}$$

where P_V is the true point density, P'_A is the number of points per unit area of the projection plane, and t is the foil thickness (see Fig. 6.3c). In the next section we shall discuss points of emergence or intersection generated by lines.

6.4.2 Lines

Systems of projected lines are relatively straightforward to interpret, primarily because the probability of overlap is negligible for true lines. Linear elements that are of finite width or are closely interwoven into tangles (where individual lines cannot be distinguished), may cause practical difficulties.

One of the fundamental relationships for projected lines relates the mean projected length of a straight, randomly oriented line segment to the true length. Longer lines, curved or otherwise, can be considered to consist of small linear segments that are

---

*If there are reentrant folds or overlap of surfaces, $P' < P$. We must then invoke the "total" projection, which allows the exact expression to be written (see Section 6.7.4).

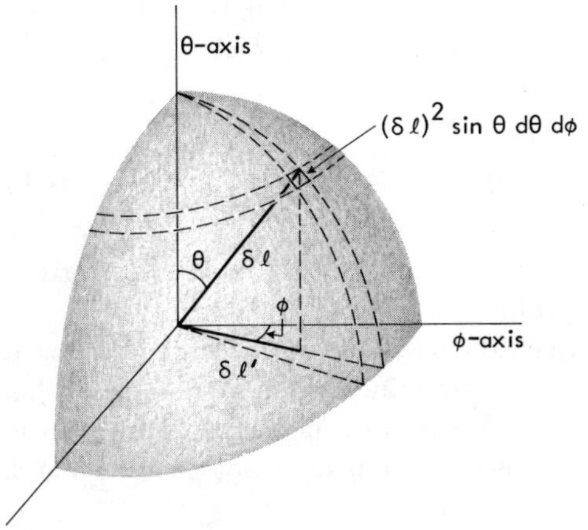

Fig. 6.4. Mean projected length of a randomly oriented linear segment.

essentially straight, so the relationship is general. Consider an elementary line segment of length δl, with one end fixed, which is free to rotate in any direction described by the angles ϕ and θ. Figure 6.4 shows one octant of the spherical surface (of area $\pi\delta l^2/2$) generated by a rotating line segment of length δl. The average projected line length is given by

$$\overline{\delta l'} = \frac{\int_0^{\pi/2}\int_0^{\pi/2} \delta l \sin\theta\, \delta l^2 \sin\theta\, d\theta\, d\phi}{\pi\, \delta l^2/2}$$

$$= \frac{2\delta l}{\pi}\int_0^{\pi/2}\int_0^{\pi/2} \sin^2\theta\, d\theta\, d\phi = \delta l \int_0^{\pi/2} \sin^2\theta\, d\theta = \frac{\pi}{4}\,\delta l.$$

Since any curved line of length L can be represented by a sequence of straight, elementary linear segments, then $L = \sum \delta l$ and

$$\boxed{L = \left(\frac{4}{\pi}\right)\bar{L}'}\ \text{mm}, \tag{6.5}$$

where $\bar{L}'$ is the mean line length on the projection plane. This, then, is the important equation relating the length of randomly oriented lines in space to the mean length of their projected images [40].

For a random system of projected lines within an area A'_T, Eq. (6.5) becomes

$$L_V = \left(\frac{4}{\pi}\right) L'_A/t\ \text{mm}^{-2}, \tag{6.5a}$$

as seen in Fig. 6.3e. Substituting $L'_A = (\pi/2)P'_L$ (see Eq. 2.49), we obtain

$$L_V = 2P'_L/t\ \text{mm}^{-2}. \tag{6.5b}$$

There are ambiguities even with a simple system of projected lines, since, as noted previously, vertical surfaces in the foil can also leave linear traces in the projected plane. This situation should not occur too frequently in random configurations, but, as always, a knowledge of the actual microstructure helps one to avoid errors.

We shall also require the general relationship between the length of a straight line segment L, randomly oriented in a plane, and its mean projected length $\bar{L}'$, in a projection line. The result [19, 21],

$$\boxed{L = \left(\frac{\pi}{2}\right)\bar{L}'} \text{ mm,} \tag{6.6}$$

is obtained by a procedure similar to that employed in deriving Eq. (6.5). The derivation of Eq. (6.6) is left for an exercise in the problem section.

It should be remarked that a randomly oriented *curved* line in a plane will obscure itself partially when randomly projected to a projection line. What we require in this case is the "total" projection (see Section 6.6.1) of lines in a plane. This problem is not encountered with a curved line in space, since overlap does not occur when a line is projected to a plane.

Figure 6.3(d) portrays a simple system of open surfaces in a foil projected to a plane. According to Eq. (6.6), the length L of the surface traces in the vertical test plane in the foil is related to the length L' of the intercepts made by the test line in the projection plane. Dividing $L = (\pi/2)\bar{L}'$ by $A_T = L'_T t$ gives

$$L_A = \left(\frac{\pi}{2}\right) L'_L/t \text{ mm/mm}^2, \tag{6.6a}$$

which relates the length of surface traces per unit area of a vertical test plane in the foil to a simple L'_L measurement in the projection plane. Note that, in the projection plane, $L'_L = A'_A = P'_P$.

We can readily obtain the mean length $\bar{L}$ of random linear segments in space from the projection plane. The mean projected length $\bar{L}'$ is given by

$$\bar{L}' = L'_A/N'_{A'}, \tag{6.7}$$

where L'_A is the total projected length of linear segments per unit area and N'_A is the number of projected linear segments per unit area. According to Eq. (6.5), we may write

$$\bar{L} = (4/\pi)L'_A/N'_A, \tag{6.8}$$

which yields

$$\bar{L} = 2P'_L/N'_A \text{ mm.} \tag{6.8a}$$

Systems of lines in a foil may also produce points on the projection plane. Line segments with ends that terminate within the foil, or lines that intersect at the foil surfaces, lead to points with special properties. Appropriate equations for these two cases are

$$P_V = P'_A/t \text{ mm}^{-3} \tag{6.4a}$$

for terminal points within the foil, and

$$P_A = P'_A/2 \text{ mm}^{-2} \tag{6.9}$$

for points of intersection at foil surfaces.

Equation (6.4a) has the same form as Eq. (6.4) in Section 6.4.1; however, in this section Eq. (6.4a) refers to linear segments with all termini lying within the foil, so the number of points in P_V and P'_A must always be even.

Equation (6.9) is based on the equality of P and P'. Since there are two foil surfaces involved but only one projection area, $A_T = 2A'_T$. Dividing through gives $P_A = P'_A/2$, which is strictly valid only when the ends of each linear segment terminate at the upper and lower foil surfaces. The lines need not be randomly oriented.

The total length of random lines per unit volume is given by Eq. (2.62), $L_V = 2P_A$. Substituting Eq. (6.9) gives

$$\boxed{L_V = P'_A} \text{ mm}^{-2}, \tag{6.10}$$

provided that all linear segments intersect both sides of the foil. Note that L_V is given by Eq. (6.5b) as well. However, the measurements in that case refer, in reality, to test planes perpendicular to the foil surfaces.

Points are also created by line images that intersect on the projection plane. The interpretation of points on the projection plane due to intersecting lines must be made with caution, since the real lines may or may not touch in space (see Fig. 6.2).

Several studies on the structure of paper [48, 49] have considered the material as a linear system in a plane. The linear fiber elements are assumed to lie randomly in a plane, with thick paper consisting of several of these planes. Employing a statistical approach, these researchers have related such parameters as the fiber contacts, fiber lengths, and attributes of the polygons formed by the fibers to important paper properties. These are well-done and important articles, although the generality would be enhanced by treating paper as a three-dimensional foil [50] with the linear structure projected to a projection plane.

6.4.3 Areas

As shown in Table 6.2, projected areas can result from the projection of surfaces or volumes in the foil. In general, these features cannot be related quantitatively to their projected images because of overlapping folds in the surfaces or concavity in the volume elements.

Exact expressions exist for simpler cases, however, and these are derived for surfaces without reentrant folds and for convex volume elements. The first derivation was attributed to Cauchy [16] in 1832, and is essentially the same for both cases. We follow the derivation suggested by Rudman [17], since it avoids certain inconsistencies [18].

The surface S of a convex particle is divided into elements $\boldsymbol{ds}$ [38, 45]. The projection dA' of each element on an arbitrary projection plane specified by its unit normal $\boldsymbol{n}$ is then

$$dA' = \boldsymbol{n} \cdot \boldsymbol{ds} = ds\,(\cos\theta),$$

where θ is the angle between normals to the elements and the projection plane. The average projected area $\overline{dA'}$ of an element ds is then

$$\overline{dA'} = \frac{ds}{4\pi} \int_0^{\pi/2} 2\pi \cos\theta \sin\theta \, d\theta = ds/4, \tag{6.11}$$

from which follows the Cauchy result

$$\boxed{S = 4\bar{A}'} \ \text{mm}^2 \tag{6.12}$$

for a convex body. In Eq. (6.11), although the average has been taken over all orientations 4π, the integration limits $0 \leq \theta \leq \pi/2$ extend only over the upper hemisphere. This is done because the lower elements of a convex body are obscured between $\pi/2 \leq \theta \leq \pi$, and hence make no contribution to the apparent projected area.

If the body is not convex (that is, if it contains reentrant surfaces), then it is possible for an element to be obscured even though $\theta < \pi/2$. Therefore, in general,

$$S \geq 4\bar{A}' \ \text{mm}^2 \tag{6.12a}$$

where the equality holds in the limit of a convex body. Kendall and Moran [19] give an alternative derivation of Cauchy's formula.

For a system of convex particles, referred to a unit area on the projection plane, we have (according to Eq. 6.12)

$$\boxed{S_V = 4A'_A/t} \ \text{mm}^{-1}. \tag{6.12b}$$

As a direct consequence of the Cauchy derivation, we may write

$$S \geq 2\bar{A}' \ \text{mm}^2 \tag{6.13}$$

for a simple open surface with random curvature, in which the equality applies to a surface without reentrant curvature or overlap.

If we divide through by $V_T = A'_T t$, then Eq. (6.13) becomes

$$S_V \geq 2A'_A/t \ \text{mm}^{-1}, \tag{6.13a}$$

where the equality holds provided that losses of projected area by overlap or reentrant surfaces are negligible (see Fig. 6.3f); otherwise the inequality becomes operative.

Quantities that can be obtained from a set of discrete areas (or closed perimeters) on the projection plane are H', the tangent diameter (or height) [25]; L'_2, the intercept "diameter" (or length); and L'_p, the perimeter length. These three quantities are important in the equations relating spatial, projected, and section quantities, and will be used later in the discussion of convex particles.

The tangent diameter H' of a particle image is the perpendicular distance between parallel lines, tangent to the perimeter at opposite sides of the image (also called Feret's diameter [44] in particle technology). The mean tangent diameter $\bar{H}'$ for all orientations of the image is also identical with the mean distance between parallel tangent planes for all orientations of the particle (see Fig. 6.3g). The intercept diameter, L'_2, is the length of a straight test line intercepted by the image of a particle. The subscript 2 refers to the dimension of the figure from which the intercept diameter is obtained. Later we shall introduce L_3 for three-dimensional figures.

Values of $\bar{H}'$ and $\bar{L}'_2$ may be obtained from the projected images, as indicated in Fig. 6.5, taken from a discussion by Smith [20]. The two measurements differ in that $\bar{L}'_2$ is determined from just those images intersected by the random test lines, while $\bar{H}'$ is measured from many directions on all images within a selected area. Note that $\bar{H}'$ is obtained by projecting from two to one dimension.

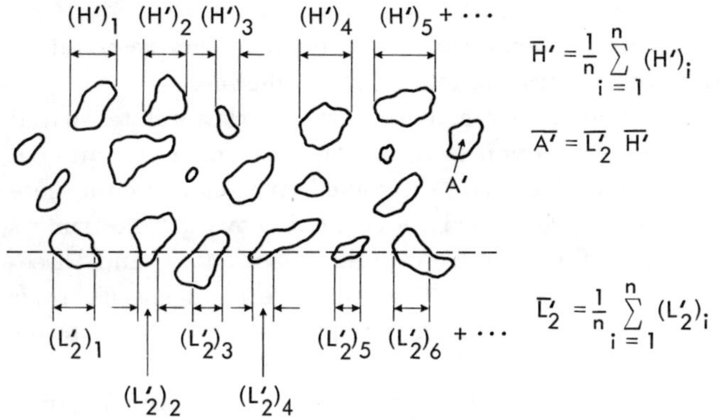

Fig. 6.5. Measurement of mean tangent height $\bar{H}'$ and mean intercept length $\bar{L}'_2$ on the projection plane [20].

The last quantity listed in Table 6.2 is L'_p, the perimeter length of an isolated surface or closed loop in the projection plane. The mean value without image overlap is defined for many loops as

$$\bar{L}'_p = L'_A/N'_A \text{ mm}, \tag{6.14}$$

where L'_A is the total (perimeter) length per unit area of the projection plane and N'_A is the number of closed loops per unit area of the projection plane. Note the similarity of Eq. (6.14) to Eq. (6.7).

6.4.4 Convex Particles

Exact relationships for convex particles in the absence of overlap and truncation appear in Fig. 6.3(h) and (i). In part (h), N, the number of particles in the foil cut by the (vertical) test plane, equals N', the number of particle images in the projection plane cut by the test line. Dividing $N = N'$ by $A_T = L'_T t$ gives

$$N_A = N'_L/t \text{ mm}^{-2}, \tag{6.15}$$

which relates the areal density in the foil to the lineal density in the projection plane, for a system of particles in a foil in the absence of truncation and overlap.

In part (i) of Fig. 6.3, the total number of particles in the foil, N, equals the total number of particle images in the projection plane, N'. Dividing $N = N'$ by $V_T = A'_T t$ gives

$$N_V = N'_A / t \text{ mm}^{-3} \tag{6.16}$$

for a system of particles in a foil projected to the projection plane.

A useful result ensues if we eliminate t from the above two equations. We see directly that

$$N_V = \left(\frac{N'_A}{N'_L}\right) N_A \text{ mm}^{-3}, \tag{6.17}$$

where the number of particles per unit volume can be obtained by simple counting measurements, provided overlap and truncation effects are negligible.

We turn now to the exact expressions for convex particles that relate spatial quantities to those measured on a test section or on a projection plane. Figure 6.6 illustrates the quantities, symbols, and relationships involved. We shall discuss these factors first under the simple conditions mentioned above, i.e., no overlap, no truncation, etc., then introduce corrections for such factors. Since both two- and three-dimensional terms are employed in the following discussion, we shall specify their origins by the subscript 2 or 3, respectively. Note the complete interchange of terms among the equations listed under "Projection," "Spatial," and "Section."

The equations listed under "Projection" and "Section" in Fig. 6.6 are similar in form. In order to emphasize the similarities in their derivations, we employ parallel columns when feasible.

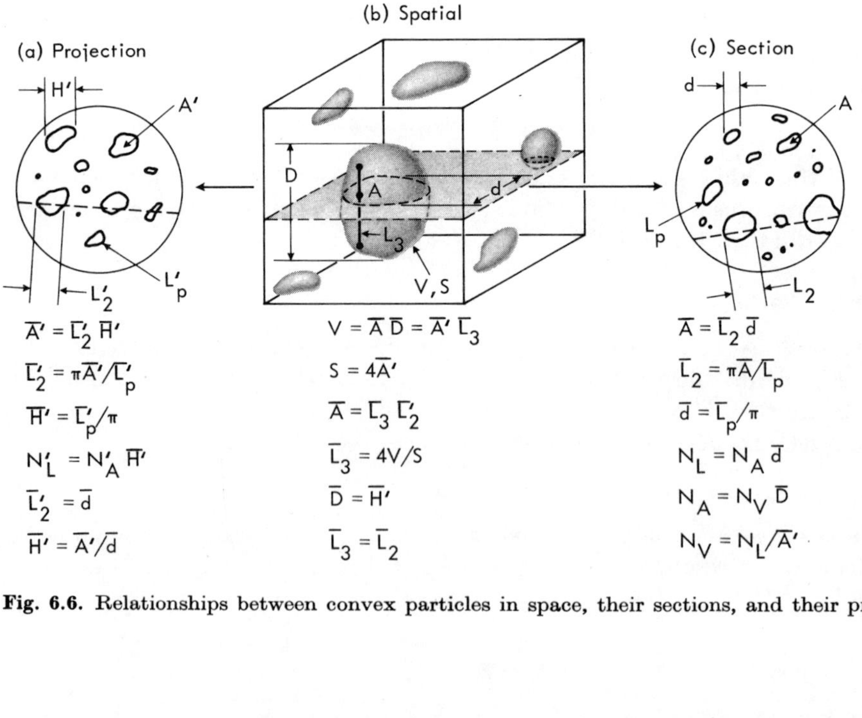

Fig. 6.6. Relationships between convex particles in space, their sections, and their projections.

Convex Particles Projected to a Plane (See Fig. 6.6a.)

The mean intercept length of a convex figure of projected area A' and projected perimeter length L'_p is given by

$$\overline{L'_2} = \pi A'/L'_p \text{ mm} \tag{6.18}$$

from Eq. (2.54).

The mean tangent height (or "diameter") of a convex figure and its projected perimeter length are related by

$$\overline{H'} = L'_p/\pi \text{ mm} \tag{6.19}$$

according to Eq. (6.6). The factor 2 does not appear here because both upper and lower halves of the perimeter project on the same line (of length H').

By combining Eqs. (6.18) and (6.19), we obtain

$$A' = \overline{L'_2}\overline{H'} \text{ mm}^2 \tag{6.20}$$

which relates the projected area of a convex figure to the product of its mean intercept length and mean tangent height (and not as given in Reference 20).

Another important relationship concerns two-dimensional convex figures randomly positioned on the projection plane. A test line of unit length is laid down randomly over the figures. On the average, only those figures will be intercepted by the test line whose centers lie within a distance $\overline{H'}/2$ on either side of the test line; i.e., within an area equal to $\overline{H'}$. Therefore, the number of figures with centers lying within the area $\overline{H'}$ is equal to N'_L, and

$$N'_A = N'_L/\overline{H'} \text{ mm}^{-2}. \tag{6.21}$$

This equation is a two-dimensional analog to the three-dimensional form expressed by Eq. (4.40).

Convex Particles Cut by Test Section (See Fig. 6.6c.)

The mean intercept length of a convex figure of intercepted area A and perimeter length L_p is given by

$$\overline{L_2} = \pi A/L_p \text{ mm} \tag{6.18a}$$

from Eq. (2.54).

The appropriate form of Eq. (6.6) for the tangent diameter of a convex figure is

$$\bar{d} = L_p/\pi \text{ mm}. \tag{6.19a}$$

The factor 2 does not appear for the reason noted under Eq. (6.19).

The area of a convex figure in the test section is related to its mean intercept length $\overline{L_2}$ and mean tangent diameter $\bar{d}$ according to

$$A = \overline{L_2}\bar{d} \text{ mm}^2 \tag{6.20a}$$

where the product of $\overline{L_2}$ and $\bar{d}$ can be thought of as a width times a height.

If a test line of unit length is laid randomly over a system of convex figures, we measure N_L, N_A is counted, and $\bar{d}$ can be calculated from the relationship

$$N_A = N_L/\bar{d} \text{ mm}^{-2}. \tag{6.21a}$$

Just as for convex projected images, only those figures will be intercepted by a unit test line whose centers lie, on the average, within a distance $\bar{d}/2$ on either side of the test line, or within an area equal to $\bar{d}$. Thus the number of figures with centers lying within the area $\bar{d}$ is N_L, or $N_A = N_L/\bar{d}$. Equation (6.21a) should be compared with Eq. (4.40).

Let us turn now to other equations for convex particles. Since the spatial quantities are generally related to the two-dimensional section quantities, the new elements introduced here are the projected quantities.

The volume, surface area, and mean projected area of a convex particle are related to the mean linear intercept, $\bar{L}_3$, by means of the equations

$$\bar{L}_3 = 4V/S = V/\bar{A}' \text{ mm}^3/\text{mm}^2, \tag{6.22}$$

which follow from Eqs. (2.36) and (6.12). Moreover, the volume,

$$V = \bar{A}'\bar{L}_3 \text{ mm}^3, \tag{6.23}$$

becomes

$$V = \bar{H}'\bar{L}_2'\bar{L}_3 \text{ mm}^3 \tag{6.24}$$

upon substitution of Eq. (6.20). The Cauchy equation $S = 4\bar{A}'$ becomes

$$S = 4\bar{H}'\bar{L}_2' \text{ mm}^2, \tag{6.25}$$

also upon substituting Eq. (6.20).

If the mean intercept area $\bar{A}$ of convex particles is introduced into these equations, other useful expressions result. We solve the identity $V_V = N_V\bar{V} = N_A\bar{A}$ for $\bar{V}$, getting

$$\bar{V} = \left(\frac{N_A}{N_V}\right)\bar{A}, \tag{6.26}$$

where N_A/N_V is equal to $\bar{D}$, as well as to $\bar{H}'$. The equality of the latter two terms is evident from Fig. 6.3g. Thus we obtain

$$\bar{V} = \bar{H}'\bar{A} \text{ mm}^3 \tag{6.27}$$

which should be compared with Eq. (6.23).

Noting that $\bar{L}_3$ is identical with $\bar{L}_2$, we obtain the symmetrical expression from Eqs. (6.23) and (6.27),

$$\bar{A}'/\bar{H}' = \bar{A}/\bar{L}_2 \text{ mm}^2/\text{mm}, \tag{6.28}$$

from which, by using Eqs. (6.20) and (6.20a), we can relate the mean intercept length on the projection plane to the mean tangent diameter on the test section

$$\bar{L}_2' = \bar{d} \text{ mm}. \tag{6.29}$$

By using Eq. (6.29), we can now write the mean intercept area from Eq. (6.20a) as

$$\bar{A} = \bar{L}_2\bar{L}_2' \text{ mm}^2. \tag{6.30}$$

Other combinations of these terms are possible, and may be required from time to time. The most important equations have been given under Eq. (4.31) and are collected in Fig. 6.6 for ready reference. It is worth while to rewrite these equations for the case of spherical particles of diameter D. This task is given as a problem at the end of this chapter.

Given the particle surface and volume, then we can calculate both $\bar{A}'$ and $\bar{L}_3$ by means of Cauchy's and Tomkeieff's equations, respectively (Eqs. 6.12 and 2.36).

However, no similar independent calculation of $\bar{A}$ and $\bar{D}$ appears possible at the present time. At least one additional quantity must be obtained, such as the mean tangent diameter from the test section, or the mean intercept length from the projection plane. Of course, $\bar{H}'$ can be calculated [22, 23, 27] for a regular polyhedron, without recourse to the extra measurement on the section or projection plane (see Section 4.3.2). However, the expressions can become quite complex, and a computer may be required for the solution.

The equations summarized in Fig. 6.6 are quite useful, and are exact when applied to isolated particles. When the equations are applied to systems of particles, however, errors are introduced because of particle overlap and particle truncation by the foil surfaces. For small volume fractions of particles and for relatively thin foils, the above equations should not err too greatly compared with the experimental errors of measurement and the commonly made assumption of convex particles of identical size and shape.

Let us now proceed to a discussion of projected oriented systems, and to a collation of equations for normal and total projections. After this, we shall apply correction terms for particle overlap and truncation at the foil surfaces to the basic equations of Fig. 6.6.

6.5 PROJECTION OF ORIENTED STRUCTURES

Previous equations in this chapter referred primarily to random structures and to random measurements made on isolated objects. We must also consider oriented structures (see Chapter 3) and the special methods required to evaluate their properties when random projections of the structure cannot be achieved from the thin foils. It is important to realize that the special equations for oriented structures may differ greatly from those for random structures.

Two types of oriented structures are encountered frequently: systems of lines in the foil and systems of surfaces in the foil. Either system may be completely or partially oriented, there may be one or more orientation axes or planes, and the total structure may contain both lines and surfaces. Inasmuch as the number of possible combinations of structural elements is prohibitively large, we shall not attempt to analyze each case. Rather we shall simply mention a few guiding principles to show how the results of Chapter 3 can be adapted to the particular case at hand.

Since we are dealing here with parallel projections, we should recall their general properties to facilitate the interpretation of oriented structures. If the real figure lies in a plane parallel to the projection plane, the figure and its image are exactly alike (congruent). Otherwise they may have different shapes. However, straight lines remain straight, tangents to curves remain tangent, parallel lines remain parallel, bisected segments remain bisected, and equal areas remain equal. In other words, the properties of straightness, tangency, parallelism, bisection, and equality of area are *invariant* under parallel projection [42].

If one is interested only in the degree and type of linear orientation as it occurs on the projection plane, the analysis is the same as given previously for lines in a plane (see Section 3.2). On the other hand, the problem may be to deduce the actual linear structure in the foil in terms of the projected oriented structure. Section 3.3 supplies most of the basic information required in such a case.

Additional information may be needed to determine the orientation direction and/or orientation planes in the three-dimensional foil space. Supplementary pictures could be taken at other angles, or the known crystallography of the structural features might allow calculation of the orientation direction and/or planes.

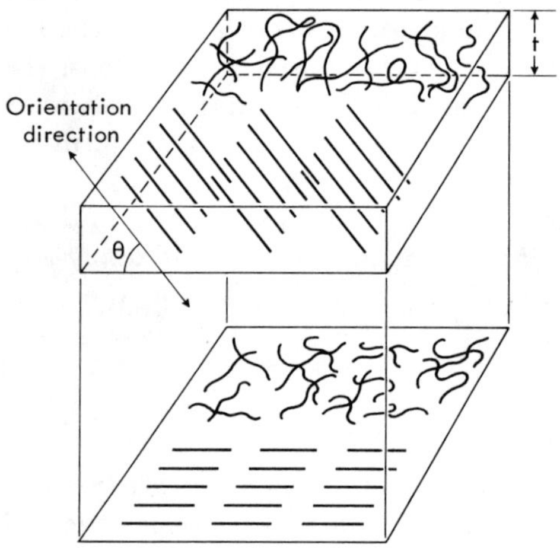

Fig. 6.7. Projection of a partially oriented system of lines.

Consider a system of parallel lines in the foil that lie within a plane not parallel to the projection plane. Once we know the plane and the direction of the oriented lines, we can readily calculate the true length by simple trigonometry from the projected images. The same is true of oriented surfaces in the foil. Partial obscuration of surfaces by overlap creates special problems for systems of surfaces, and these will be treated later in Section 6.7.2.

As an example of how a partially oriented system of lines in a thin foil may be analyzed, consider the structure sketched in Fig. 6.7. For simplicity, the oriented lines are taken to occupy the front portion of the foil, in the indicated manner, and the randomly oriented lines occupy the back portion of the foil. We assume that the foil thickness, the orientation direction, and the plane of the oriented lines have been determined independently. Moreover, all lines and line segments project clearly, and have features similar to those classified in Table 6.2(b). Two methods for obtaining L_V will be discussed for the structure shown in Fig. 6.7.

First method (counting intersections): We measure P'_L by random traverses, taking care to count the intersections with random (isometric) lines $(P'_L)_{\text{is}}$ separately from those with oriented lines $(P'_L)_{\text{or}}$. The equation for random lines is $(L'_A)_{\text{is}} = (\pi/2)(P'_L)_{\text{is}}$ and for oriented lines is $(L'_A)_{\text{or}} = (\pi/2)(P'_L)_{\text{or}}$ (using random traverses).

Alternatively, if the intersections with oriented lines only are made by traverses perpendicular to the orientation direction $(P'_L)_{\perp\perp}$, then the appropriate equation is $(L'_A)_{\text{or}} = (P'_L)_{\perp\perp}$.

Knowing $(L'_A)_{\text{is}}$ and $(L'_A)_{\text{or}}$, we can write, for random lines (Eq. 6.5b)

$$(L_V)_{\text{is}} = 2(P'_L)_{\text{is}}/t$$

and for the oriented lines

$$(L_V)_{\text{or}} = (L'_A)_{\text{or}}/t \cos\theta = \left(\frac{\pi}{2}\right)(P'_L)_{\text{or}}/t \cos\theta = (P'_L)_{\perp\perp}/t \cos\theta,$$

depending on the type of measurement employed.

Then, from the general equation $L_V = (L_V)_{\text{is}} + (L_V)_{\text{or}}$, we write for random traverses

$$L_V = 2(P'_L)_{\text{is}}/t + (\pi/2)(P'_L)_{\text{or}}/t \cos\theta.$$

Second method (counting points): We assume that the random and the oriented lines intersect both top and bottom surfaces of the foil. The points of emergence, $(P'_A)_{\text{is}}$ and $(P'_A)_{\text{or}}$, are counted separately. For random lines, $2(P_A)_{\text{is}} = (P'_A)_{\text{is}}$, and by direct substitution (Eq. 6.10)

$$(L_V)_{\text{is}} = (P'_A)_{\text{is}}.$$

For the oriented lines,

$$2(P_A)_{\text{or}} = (P'_A)_{\text{or}},$$

and from the geometry of the dislocation lines and the foil surface, as in Fig. 6.7,

$$(P_A)_{\perp\perp} = (P_A)_{\text{or}}/\sin\theta,$$

where $(P_A)_{\perp\perp}$ is the number of intersections per unit area on a plane normal to the direction of the dislocation array, and $(P_A)_{\text{or}}$ represents the point density in the surface of the foil.

Therefore, in Eq. (3.12),

$$(L_V)_{\text{or}} = (P_A)_{\perp\perp},$$

we substitute from above, giving

$$(L_V)_{\text{or}} = (P'_A)_{\text{or}}/2\sin\theta.$$

Since $L_V = (L_V)_{\text{is}} + (L_V)_{\text{or}}$, we have

$$L_V = (P'_A)_{\text{is}} + (P'_A)_{\text{or}}/2\sin\theta.$$

Comparing the final equations obtained by the two methods, we see that L_V depends in both cases in a nonlinear manner on the angular relationships among the orientation plane, orientation directions, and projection plane. Moreover, both equations for L_V depend on measurements of random and oriented components, and on the ability to distinguish which is which on the projection plane. Also, the foil thickness enters one equation and not the other, which may introduce a significant experimental error into the results. Because the determination of t is difficult, P'_A counts are preferred to P'_L intersection counts. However, the former method cannot be used unless the actual structure permits. Two methods of measuring dislocation density were adopted by Ham and Sharp [1], but unless one knows the degree of orientation in their foils, their results are difficult to assess.

Complete classification and analysis of oriented and partially oriented structures are beyond the scope of this text. Suffice it to say that if the basic methods for handling such structures are properly applied, the majority of actual structures encountered in practice can be treated correctly as the need arises.

6.6 TOTAL PROJECTIONS OF RANDOM STRUCTURES

Total projections are apt to be required whenever overlap occurs. In general, there are three types of systems to be investigated: lines in a plane, lines in space, or surfaces in space. The total projection is simply the value of the projection of *all* elements of the structure, whether obscured or not. For a sphere, the total projection of the surface is equal to twice the area of the equatorial section, since the upper and lower hemispheres each contribute a projected area of πr^2.

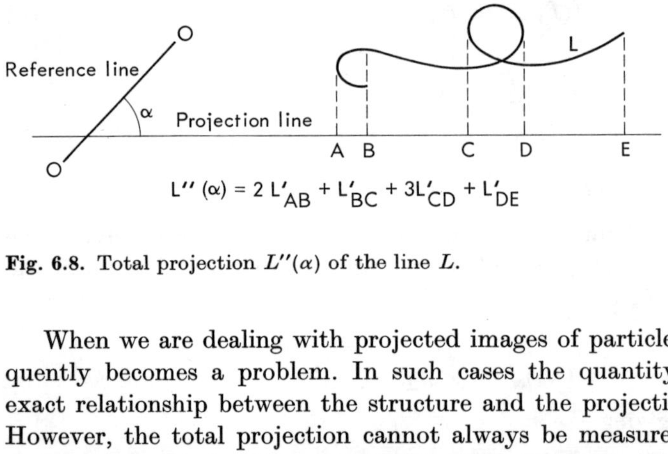

Fig. 6.8. Total projection $L''(\alpha)$ of the line L.

When we are dealing with projected images of particles or surfaces, overlap frequently becomes a problem. In such cases the quantity required for writing an exact relationship between the structure and the projection is the total projection. However, the total projection cannot always be measured. Instead, we may have to employ approximations or simplifications based on the actual projected images.

The determination of total projection of lines in a plane or in space is also of interest. In the former case, the lines are projected to a line; in the latter, we first project the lines to a plane, then project the projected lines in a plane to a projection line. We shall first derive exact relationships for the three types of systems mentioned above; applications will be made in subsequent sections.

6.6.1 Total Projection of Lines in a Plane

The concept of total projection of a line in a plane is demonstrated in Fig. 6.8, from Kendall and Moran [19]. A curved line in a plane is projected onto a projection line having directions α with respect to a fixed reference line. The total projection for the angle shown is

$$L''(\alpha) = 2L'_{AB} + L'_{BC} + 3L'_{CD} + L'_{DE} \text{ mm}, \tag{6.31}$$

where the total projection is indicated by a double prime. The value of $L''(\alpha)$, averaged over all directions, is given by

$$\boxed{\overline{L''} = \left(\frac{2}{\pi}\right) L} \text{ mm}, \tag{6.32}$$

which is analogous to Eq. (6.6) for the mean projected length of a straight line segment in a plane. Equations (6.31) and (6.32) have been used by Steinhaus [24] to obtain the length of a curve observed under the microscope. When we are analyzing the projected images of surfaces, however, the total projection is usually the quantity sought, rather than the quantity measured.

We also wish to determine the total projected line length per unit length of projection line, L''_L. Take the system of randomly oriented lines in a plane, shown in Fig. 6.9(a). Following Saltykov [25], a grid of parallel test lines, a distance Δ apart, is erected within the rectangular area of height h, perpendicular to an arbitrary projection line in the plane.

The system of lines is cut into segments whose projected lengths equal Δ or less. The number of segments per strip with projected length of at least $\Delta/2$ are counted as equal to Δ, while the shorter ones are ignored. If the number of segments per strip is m_1, m_2, m_3, etc., then the length of projected segments from each strip is Δm_1, Δm_2, Δm_3, etc., and the total projected length on the projection line is

$$L'' = \Delta(m_1 + m_2 + m_3 + \cdots). \tag{6.33}$$

For Δ sufficiently small, the strips approximate test lines and the total number of segments counted becomes the total number of point intersections, P. We divide through by L', the length on the projection line, which gives

$$\frac{L''}{L'} = \frac{m_1 + m_2 + m_3 + \cdots}{L'/\Delta} = \frac{P}{N}, \tag{6.34}$$

in which N is the number of test lines of total length $L_T = Nh$. Substituting for N yields

$$\frac{L''}{L'} = \frac{P}{L_T} h,$$

or, in our notation,

$$L''_L = (P_L)_\perp h, \tag{6.35}$$

where $(P_L)_\perp$ is the number of intersections per unit length of test lines perpendicular to the projection line. It is apparent that the ratio L''_L can be greater than unity.

Since the subscript L in L''_L represents the length of the projection line L' and h represents the height of the test area, the test area equals $L'h$, and the spatial equivalent of Eq. (6.35) can be written

$$\boxed{(L_A)_{\text{TP}} = (P_L)_\perp} \text{ mm}^{-1}, \tag{6.36}$$

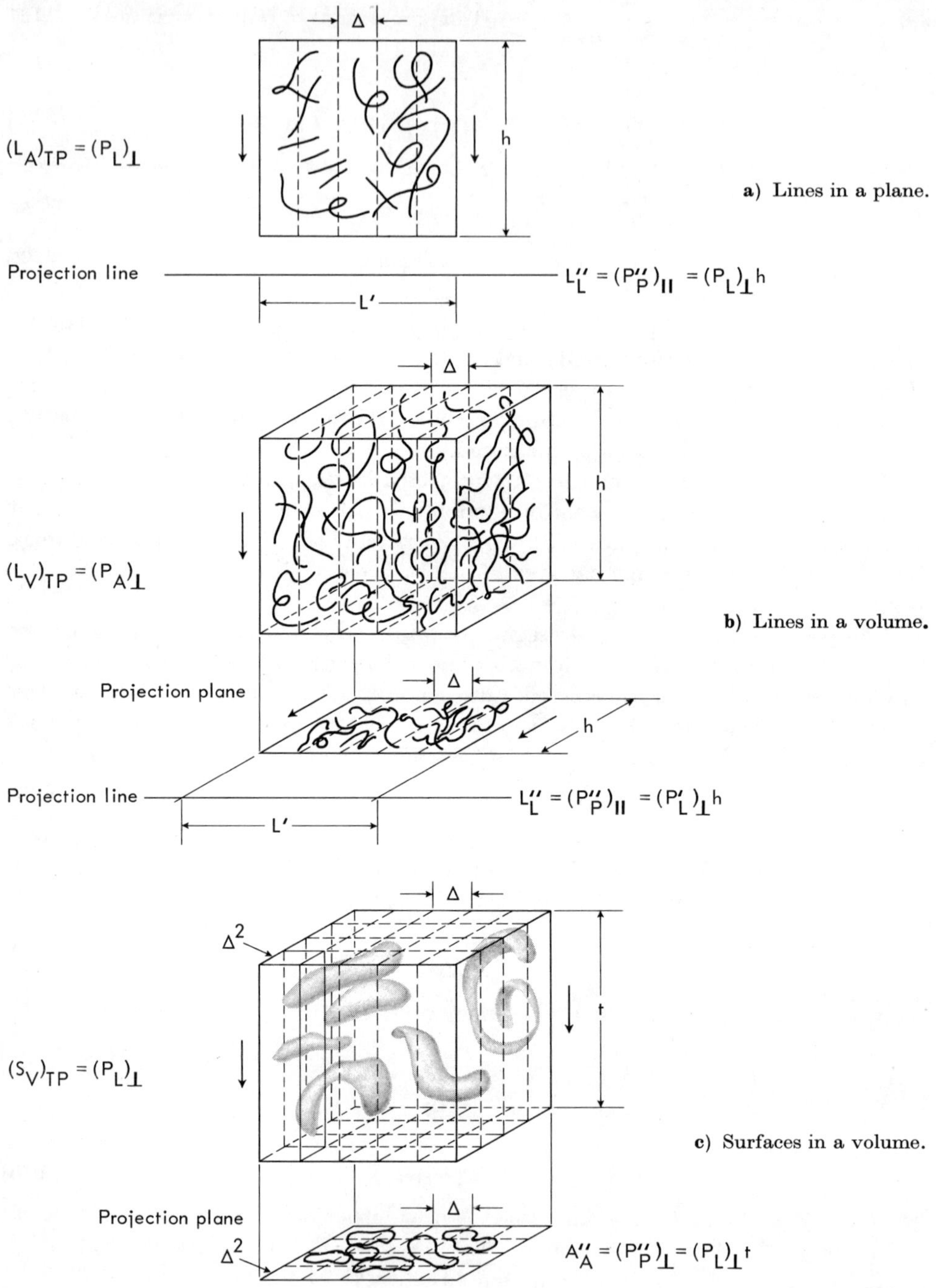

Fig. 6.9. Total projection of random lines and surfaces.

where $(L_A)_{TP}$ represents the total projection of a system of lines in a plane, projected to a line perpendicular to the test lines. Note that in $(L_A)_{TP}$ the total projected length is referred to the test area, while in L_L'' the total projected line length is referred to the projection line.

Since the number P of intersections counted in the test area equals the total number P'' of points projected to the projection line, $P = P''$. Moreover, on the projection line, N becomes P_T', and $L_T = P_T'h$. Dividing the first equality by the second yields

$$\frac{P}{L_T} = \frac{P''}{P_T'}\frac{1}{h}$$

or

$$(P_L)_\perp = (P_P'')_{||}/h \text{ mm}^{-1}, \qquad (6.37)$$

where $(P_P'')_{||}$ is the number of points (intersections) projected to the projection line, per point of the linear array of test points along (or parallel to) the projection line. Note that this ratio can be greater than unity.

Substitution of Eq. (6.37) in Eq. (6.35) yields another form of the basic equation

$$L_L'' = (P_P'')_{||}, \qquad (6.38)$$

which applies to the total projection of a system of random lines in a plane projected to a line.

Equations (6.35), (6.36), and (6.38) are entered in Fig. 6.9(a) to emphasize their interrelationships. Equation (6.36) refers to the system of random lines in a plane and the total projected length on an areal basis. Equations (6.35) and (6.38), however, refer to the projection line itself. Comparison of Eq. (6.36), $(L_A)_{TP} = (P_L)_\perp$ with Eq. (3.4), $(L_A)_{or} = (P_L)_\perp$, for a system of parallel lines in a plane, reveals interesting similarities. The latter equation represents a special case of the general equation (Eq. 6.36) for total projection, inasmuch as only for oriented lines (lying parallel to the projection line) does the total projection equal the total line length.

6.6.2 Total Projection of Lines in Space

A natural choice for the system of lines in a plane considered above is to project to a line. For a system of random lines in a volume, however, it is convenient to use a double projection, whereby the lines in space are first projected to a plane, then projected from the projection plane to a projection line.

Figure 6.9(b) presents these steps for obtaining the total projection of lines in space. We consider a rectangular prism of dimensions $h \times h \times L'$ containing a system of randomly oriented and positioned lines. A set of parallel test planes, a distance Δ apart, is placed parallel to one of the prism faces, creating slices of width Δ which contain spatially random segments of the system of lines. The test planes delineate a set of parallel test lines, a distance Δ apart, on a projection plane perpendicular to the projection direction. The resulting strips contain randomly oriented segments of projected lines on the projection plane.

It is apparent that the number of linear segments counted is the same, whether they be random segments in the test prism, random segments in the projection plane, or projected lengths parallel to the projection line. Therefore we may proceed directly to the projection plane and follow the same analysis given above for Eq. (6.35). With appropriate modifications, we get

$$L''_L = (P'_L)_\perp h = (P''_P)_{||} \tag{6.39}$$

for the total projection of lines in a volume, projected to a plane, and in turn projected to a line in the plane. As in Eq. (6.38), $(P''_P)_{||}$ is the projected point density referred to a linear array of test points along the projection line; $(P'_L)_\perp$ is the number of intersections (in the projection plane) per unit length of test lines perpendicular to the projection line.

To obtain the equivalent spatial equation, we note that for N test planes cutting the test prism, there are N test lines in the projection plane. Moreover, the number P of intersections with the system of lines by test planes in the test volume is equal to the number P' of intersections with the projected lines by test lines in the projection plane.

Furthermore the test volume $V_T = L'h^2$, the test section area $A_T = Nh^2$, and the test line length $L'_T = Nh$. If Eq. (6.39) is rewritten in the expanded form

$$\frac{L''}{L'} = \left(\frac{P'}{L'_T}\right) h,$$

then divided through by h^2,

$$\frac{L''}{L'h^2} = \frac{P'}{L'_T h} = \frac{P}{A_T},$$

we obtain the spatial equivalent of Eq. (6.39),

$$\boxed{(L_V)_{\text{TP}} = (P_A)_\perp} \ \text{mm}^{-2}, \tag{6.40}$$

where $(P_A)_\perp$ is the number of intersections with the system of lines per unit area of test planes perpendicular to the projection line. In this form, Eq. (6.40) may be compared with Eq. (3.12), $(L_V)_{\text{or}} = (P_A)_{\perp\!\perp}$, which is a special case of Eq. (6.40). Only in the special case of parallel lines can the total projection equal the total line length.

6.6.3 Total Projection of Surfaces in Space

We now consider the total projection of surfaces in space [25]. Take a system of randomly oriented and positioned surfaces, of any size or shape, within a cube of edge length t, such as is shown in Fig. 6.9(c). Identical prisms with square bases equal to Δ^2 are erected perpendicular to the cube base, which is parallel to the projection plane. The prisms carve out a number of elementary segments from the surfaces of the system. Those elementary segments whose projected areas are greater than $\frac{1}{2}\Delta^2$ are counted as Δ^2, while those with smaller projected areas are neglected.

The number of projected areas counted in each prism is designated n_1, n_2, n_3, etc., so that total projected area on the projection plane will be

$$A'' = \Delta^2(n_1 + n_2 + n_3 + \cdots). \tag{6.41}$$

As Δ approaches zero, the prisms become test lines, and the total number of surface segments can be equated to the total number P of intersections with test lines within the unit volume. Dividing through by A', the area on the projection plane, gives

$$\frac{A''}{A'} = \frac{n_1 + n_2 + n_3 + \cdots}{A'/\Delta^2} = \frac{P}{N},$$

where N is the number of test lines of total length $L_T = Nt$. Substituting for N gives $A''_A = (P/L_T)t$, or, in our notation,

$$A''_A = (P_L)_\perp t \tag{6.42}$$

where $(P_L)_\perp$ is the number of intersections with the surfaces per unit length of test lines perpendicular to the projection plane.

By analogy to the procedure followed in obtaining Eq. (6.37), we see that $(P_L)_\perp = (P''_P)_\perp/t$, which gives

$$A''_A = (P''_P)_\perp, \tag{6.43}$$

where the upper P in P''_P refers to the number of points in the projection plane generated by intersections made with test lines erected perpendicular to the projection plane, and the subscript P refers to the number of points in the two-dimensional array of test points in the projection plane. Thus the total projection of a system of surfaces is given by an ordinary point count in the projection plane, provided that all projected surfaces are observable separately. Overlap of projected areas would interfere with a direct measurement of P''_P, so special methods must be adopted for an evaluation of A''_A. These methods will be described later.

The spatial equivalent of Eq. (6.42) is obtained by dividing through by t, giving

$$\frac{A''}{A't} = (P_L)_\perp.$$

Since the test volume $V_T = A't$, we have finally

$$\boxed{(S_V)_{\text{TP}} = (P_L)_\perp} \ \text{mm}^{-1}, \tag{6.44}$$

which shows that the total projection of any system of surfaces is related directly to the number of intersections with the surfaces by test lines perpendicular to the projection plane.

A special case of Eq. (6.44) is Eq. (3.19), $(S_V)_{\text{or}} = (P_L)_{\perp\perp}$, which pertains to a system of parallel planes in a volume. Evidently the total projection can equal the total surface area only in the case of flat surfaces parallel to the projection plane.

We shall point out one final set of equations before we leave this section. Consider the three basic equations, $L_A = (\pi/2)P_L$, $L_V = 2P_A$, and $S_V = 2P_L$. Upon substituting Eqs. (6.36), (6.40), and (6.44), respectively, we obtain

$$L_A = \frac{\pi}{2}(L_A)_{\text{TP}} \ \text{mm/mm}^2, \tag{6.45}$$

$$L_V = 2(L_V)_{\text{TP}} \ \text{mm/mm}^3, \tag{6.46}$$

$$S_V = 2(S_V)_{\text{TP}} \ \text{mm}^2\text{/mm}^3. \tag{6.47}$$

TABLE 6.3

Comparison of equations related to total projections

Total projections*	Oriented systems†	Random structures
$(L_A)_{TP} = (P_L)_\perp$ (6.36)	$(L_A)_{or} = (P_L)_{\perp\!\!\perp}$ (3.4)	$L_A = (\pi/2)(L_A)_{TP}$ (6.45)
$(L_V)_{TP} = (P_A)_\perp$ (6.40)	$(L_V)_{or} = (P_A)_{\perp\!\!\perp}$ (3.12)	$L_V = 2(L_V)_{TP}$ (6.46)
$(S_V)_{TP} = (P_L)_\perp$ (6.44)	$(S_V)_{or} = (P_L)_{\perp\!\!\perp}$ (3.19)	$S_V = 2(S_V)_{TP}$ (6.47)

*Measurements made perpendicular ($\perp$) to the projection plane (or line).
†Measurements made perpendicular ($\perp\!\!\perp$) to an oriented (or) array of parallel lines or surfaces.

If a "random" structure is postulated, P_L or P_A will be the same in any direction, and the subscript $\perp$ can be ignored when substituting into the basic equations. What these last three equations represent is a means for determining the total projection of random structures independently of the foil and projection plane. Although a bulk specimen is not always available in addition to a thin foil, Eqs. (6.45), (6.46), and (6.47) can be quite useful in situations requiring the total projection. Note that none of the terms are affected by overlap in the structure.

A convenient summary and comparison of equations involving total projections are made in Table 6.3. Note the similarity in form between the equations for Total Projections and those for Oriented Systems.

To round out this discussion of total projection, it may be helpful to mention that Eqs. (6.36), (6.40), and (6.44) can also be obtained from Hilliard's [26] treatment of anisotropic structures. His three expressions, quoted in Chapter 3 as Eqs. (3.31), (3.34), and (3.37), yield the total projection of $L_A(\theta)$, $L_V(\phi, \theta)$, and $S_V(\psi, \omega)$ in terms of the expected values of $P_L(\omega)$, $P_A(\psi, \omega)$, and $P_L(\phi, \theta)$, respectively. Note that Hilliard's equations provide for angular variations due to anisotropy, whereas for the total projection of random structures we are only concerned with *one* direction, i.e., perpendicular to the projection line or plane.

6.7 APPROXIMATE EXPRESSIONS FOR PROJECTED IMAGES

In the preceding discussion, we have given exact equations for simple objects under simplified conditions, e.g., when the type of figure is known, or when there is no image overlap, truncation by foil surfaces, or reentrant surfaces. We now consider the more difficult problems encountered with actual structures. The approach adopted here is to develop the four basic equations (Section 2.2) and related expressions in terms of projected quantities.

It is possible to proceed in at least two different ways. One way is to start with a known spatial structure and deduce the proper equations linking that structure to its projection on the projection plane. A less direct way is to work with the projected image and the measurements that can be made unambiguously on the projection plane, then relate these results to the actual spatial structures. The latter procedure, although necessary under most circumstances, is unattractive here because we wish to develop the subject as fully as possible. Consequently, both approaches will be utilized whenever feasible.

6.7.1 Volume Fraction

The basic equations for volume fraction given in Eqs. (2.1) relate both spatial quantities and two-dimensional measurements made on section planes. Similar rigorous expressions should be possible between the volume fraction of random volume elements (say, particles) and measurements made on the projection plane. Although a general solution is not available at the present time, a start can be made on the problem.

No overlap or truncation. According to Eq. (6.23), the volume fraction of a system of randomly distributed convex particles having negligible overlap and truncation can be expressed by

$$\boxed{V_V = A'_A \bar{L}_3 / t,} \tag{6.48}$$

where t is the foil thickness. Since $\bar{L}_3$ cannot be obtained from the projection plane alone, an independent measurement on the test section is required or the shape of the particles must be known [39]. Relationships given in Fig. 6.6 indicate that $\bar{L}_3 = \frac{2}{3}\bar{D}$ for spherical particles and that $\bar{D} = \bar{H}'$. Thus Eq. (6.48) may be modified to read

$$V_V = \tfrac{2}{3} A'_A \bar{H}' / t, \tag{6.48a}$$

and the volume fraction is expressed in terms of quantities measurable from the projection plane except for the foil thickness. There is still the question of truncation of particles that lie near the foil surface, however, and we shall discuss this next.

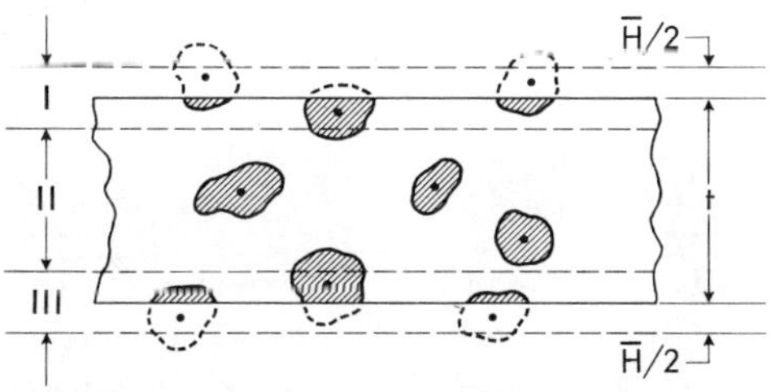

Fig. 6.10. Truncation of "convex" particles in a thin foil. (The widths of zones I or III equal the mean particle height.)

Truncation without overlap. Cahn and Nutting [3] derive an expression for the volume fraction of spherical particles that includes a correction for truncation by foil surfaces. Insofar as appropriate, we use their approach to obtain the volume fraction of a system of convex particles from projected images.

The three zones identified in Fig. 6.10 separate those particles intersected by the foil surfaces (Zones I and III) from the interior particles (in Zone II) which are not cut by foil surfaces.

With respect to the particles in Zone I, there are two types, i.e., those with centers outside the foil and those with centers inside the foil, and the expected number of both types will be the same. The area fraction of all particles intersected by the upper foil surface is equal to A_A. One-half of the area fraction is contributed by particles with centers lying above the foil surface and one-half the area fraction is contributed by particles with centers below the foil surface.

Considering only those particles with centers lying above the foil surface, we know that they will make a contribution to the projected area equal to one-half the area fraction. Similarly, those particles in Zone III with centers outside the foil will also contribute $\frac{1}{2}A'_A$ to the projected area fraction. Therefore the sum of the contributions to the projected area fraction by all particles with centers outside the foil is

$$(A'_A)_{\text{out}} = A_A = V_V. \tag{6.49}$$

The remaining particles throughout the foil have their centers inside the foil, and, regardless of truncation, will project an area fraction according to the modified Cauchy relationship (see Eq. 6.12b):

$$(A'_A)_{\text{in}} = S_V t/4 \tag{6.50}$$

Even though Eq. (6.50) considers only those particles with centers *inside* the foil, S_V refers to the surface area per unit volume of *all* surfaces within the foil. This apparent anomaly is based on the observations that truncation of particles with centers inside the foil does not affect the value of $(A'_A)_{\text{in}}$, whereas the Cauchy relationship *is* affected by truncation. It is readily apparent that the surface area lost through truncation is equal to the surface area donated by particles with centers outside the foil. Thus, on the average, the surface area per inside particle is unchanged by truncation, and the Cauchy relationship remains valid.

Summing both contributions to the observed projected area fraction, we state the exact expression (in the absence of overlap) as

$$A'_A = (A'_A)_{\text{out}} + (A'_A)_{\text{in}}, \tag{6.51}$$

where the subscripts "out" and "in" refer to the location of the particle centers. Substituting and rearranging gives

$$\boxed{V_V = A'_A - \tfrac{1}{4}S_V t}\,, \tag{6.52}$$

which expresses the volume fraction of randomly positioned convex particles for the case of truncation at the foil surfaces. Moreover, it is assumed that overlap is negligible, i.e., that V_V is small or that t is not large with respect to $\bar{H}$, so that the measured projected area fraction can be used without corrections.

In practice, the determination of the two spatial terms S_V and t creates difficulties, since usually they cannot be measured from the projection plane alone [4]. These terms can be obtained either by independent measurements, or approximated by appropriate simplifying assumptions.

Assuming spheres of uniform diameter D, we can write

$$S_V = 6V_V/D \text{ mm}^{-1}, \tag{6.53}$$

which, when substituted into Eq. (6.52), gives

$$V_V = A'_A\left(\frac{2D}{2D + 3t}\right). \tag{6.52a}$$

This result shows that the volume fraction of spherical particles in the absence of overlap, but with truncation at foil surfaces, is related to the observed projected area fraction by a simple geometrical correction term. When $t \longrightarrow 0$, $V_V \longrightarrow A'_A$; and when $t \gg D$, Eq. (6.52a) simplifies to Eq. (6.48a), where $\bar{H}' = D$ for spheres.

Overlap and truncation. The preceding equations embody the proviso that image overlap is negligible. In more realistic cases in which overlap becomes appreciable, a correction [19] must be applied to the measured areas on the projection plane. What is needed, of course, is the total projection of surface areas A''_A, but what is measured is the observed (or apparent) projected area A'_A.

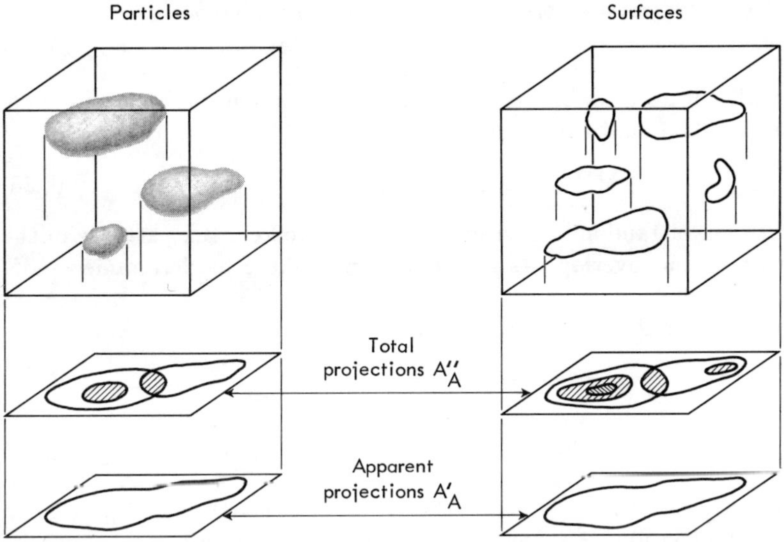

Fig. 6.11. Multiple overlaps on the projection plane from particles or surfaces.

Overlap of projected areas on the projection plane can be caused either by open surfaces or particles in the foil. The sketches in Fig. 6.11 illustrate the ambiguities inherent in the apparent projection: neither the amount of overlap nor the morphology of the structural features in the foil can be deduced readily.

On the other hand, the total projection of images on the projection plane is related unambiguously to the surface area of the features in the foil. All elements of the surface contribute to the total projection, regardless of whether they originate from planar, curved, or reentrant surfaces; convex or concave particles; or single or multiple overlaps. Thus the exact expression for surface area due to any source per unit volume in the foil is

$$\boxed{S_V = 2A''_A/t} \text{ mm}^{-1}, \tag{6.54}$$

where the factor 2 always applies so long as the surfaces are randomly oriented in the foil, or randomly projected to the projection plane. (Compare with Eqs. 6.13a and 6.47.)

We now derive an expression for the volume fraction of particles from their projected images that includes both truncation and overlap corrections. In terms of the total projection, we write the exact expression

$$A''_A = (A''_A)_{\text{out}} + (A''_A)_{\text{in}}, \tag{6.55}$$

where the subscripts "out" and "in" refer to the location of the particle centers with respect to the foil. It is apparent from Fig. 6.10 that

$$(A''_A)_{\text{out}} = A_A = V_V \tag{6.56}$$

for convex particles regardless of overlap. From Eq. (6.54) we have

$$(A''_A)_{\text{in}} = S_V t/2, \tag{6.57}$$

which applies to all surfaces within the foil, regardless of truncation or overlap. Substituting into Eq. (6.55) and rearranging gives

$$\boxed{V_V = A''_A - S_V t/2} \tag{6.58}$$

for the volume fraction of randomly positioned and oriented convex particles in the presence of truncation and overlap. As before, S_V and t must be determined independently.

For spheres of diameter D, Eq. (6.58) becomes

$$V_V = A''_A - \tfrac{1}{2}6V_V t/D \tag{6.59}$$

or

$$V_V = A''_A\left(\frac{D}{D + 3t}\right) \tag{6.58a}$$

which allows for both the effects. Note that both Eq. (6.58a) and Eq. (6.52a) reduce to Eq. (6.48a) when $t \gg D$, since $A''_A = 2A'_A$ for convex particles.

In order to use Eq. (6.58), a more accessible quantity, such as the observed projection A'_A, is desired instead of the total projection. However, if A'_A is used, an estimation of the overlap correction is required. In the case of convex particles, three separate types of overlap corrections are obtained, *viz.*, (1) when the particles lie completely within the foil, (2) when the centers of truncated particles lie outside the foil, and (3) when the centers of truncated particles lie inside the foil. In all these cases, for randomly oriented surface elements, $A''_A = kA'_A$, where $k = 2$, $k = 1$, and $1 < k < 2$, respectively. Thus the calculation of overlap involves a knowledge of type and amount of truncation.

There is another basic difficulty in devising an overlap correction for particles. This arises because of the physical impossibility of distributing discrete volume elements of any sort in space with true statistical randomness. If particles are added to a foil one at a time, the part of the foil occupied by previously added particles is unavailable for newly added particles. For this reason, the particles cannot be added randomly (in the usual sense), although they can be added *uniformly*. The mathematics of three-dimensional overlap have been examined in connection with the random nucleation of particles in an initially homogeneous matrix [58].

The problem of prior occupancy is circumvented in Mack's [27, 28] analysis, which considers the overlapping of laminae in a plane. The projected areas of all laminae are weighted equally, (that is, $k = 1$), and, by implication, when the results are extended to particles in a foil [4], the particles are also weighted equally. Thus truncation effects are not included, nor is anything done about the three-dimensional-overlap problem.

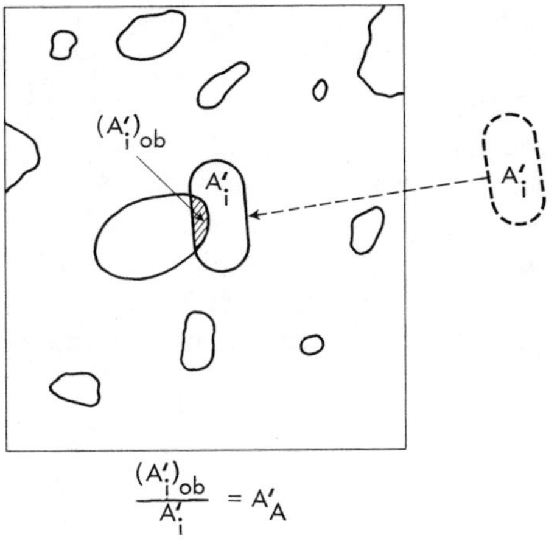

Fig. 6.12. Relationship between area of new particle A_i', area obscured $(A_i')_{\text{ob}}$, and area fraction of projected areas A_A'.

The model we develop here for particle overlap correction is based on the total projection of surfaces, and consequently is valid, in principle, for particles of any shape. The procedure is similar to that used by Cahn and Nutting [3] and Hilliard [4]. We start with a foil containing a small volume fraction of convex particles, and for which $(t/\overline{H}') \gg 1$, so that truncation is relatively unimportant. A new particle is added "randomly" to the foil. The fraction of the new particle image obscured by the original particle images is expected to equal their projected area fraction, A_A'. That is, if A_i' represents the projected area of the new (the ith) particle, and $(A_i')_{\text{ob}}$ the obscured part of the same particle, then the fraction obscured is given by

$$\frac{(A_i')_{\text{ob}}}{A_i'} = A_A'. \tag{6.60}$$

This situation is depicted in Fig. 6.12.

The increase in apparent projected area of all particles is

$$\Delta A' = A_i' - (A_i')_{\text{ob}}, \tag{6.61}$$

and the increase in the total projection of the particles is

$$\Delta A'' = 2A_i', \tag{6.62}$$

since both the upper and lower surfaces of the particle contribute to the total projection. We may then write

$$\frac{\Delta A'_A}{\Delta A''_A} = \frac{A'_i - (A'_i)_{\text{ob}}}{2A'_i} = (1 - A'_A)/2, \tag{6.63}$$

or, if there are many particles,

$$\frac{d(A'_A)}{d(A''_A)} = (1 - A'_A)/2. \tag{6.64}$$

Integration of Eq. (6.64) to obtain the total projection in the presence of particle overlap gives

$$A''_A = -2 \ln (1 - A'_A), \tag{6.65}$$

which is valid for small volume fractions of convex particles with negligible truncation. Hilliard [4] obtains a similar expression without the factor 2 because he corrects only for the overlap of *apparent* particle images and not their total projections. The interpretation given Eq. (6.65) is that for A'_A small with respect to unity, $A''_A \approx 2A'_A$; that is, overlap is negligible. As the volume fraction increases from zero, overlap increases and $A''_A > 2A'_A$. For example, when $A'_A = 0.1$, A''_A is 0.21072, and the excess over 0.2 is attributed to overlap.

Substitution of Eq. (6.65) into Eq. (6.58) yields

$$\boxed{V_V = -2 \ln(1 - A'_A) - S_V\, t/2} \tag{6.66}$$

for the volume fraction of convex particles from their projected images with a correction for overlap. For spherical particles of diameter D, Eq. (6.66) becomes

$$V_V = [-2 \ln(1 - A'_A)] \left(\frac{D}{D + 3t}\right), \tag{6.66a}$$

where t must be determined independently. Estimates can be made of a truncation correction that can be incorporated in the overlap correction of Eq. (6.65), using the three categories of particles discussed in the paragraph following Eq. (6.58a). Based on the relative number of particles in each category, we determine a weighted value of the constant k to replace the factor 2 in Eq. (6.62). For $t/D = 10$, the bracketed term in Eq. (6.66a) should be reduced by about 7.5% (see Problem 6.5).

6.7.2 Area of Surfaces in Space

In some cases all internal surfaces in the foil are visible on the projection plane regardless of overlap. Subgrain boundaries in a metal offer an example of surfaces that may be transparent to one another when viewed by electron transmission. Analysis of such structures should cause relatively little trouble.

In other cases, surfaces in the foil may obscure each other on the projection plane, and special techniques must be adapted to the special circumstances. For example, the mean projected area in the absence of overlap is related to spatial surfaces without reentrant folds by a factor of $\frac{1}{2}$ (Eq. 6.13), and to the surface area of convex particles by a factor of $\frac{1}{4}$ (Eq. 6.12a). An inequality results when there is an unspecified number of folds or overlapping surfaces or particles. Therefore corrections derived previously for overlap of areas on the projection plane are largely developed only for convex particles.

We wish to estimate here the overlap correction for surfaces which have reentrant curvatures or which obscure one another in the foil. As before, the required quantity is the total projection, and exact equations can be written in such terms. Unfortunately the total projection cannot always be obtained from measurements on the projection plane alone.

The general equation pertaining to surface area per unit volume, for surfaces of any degree of complexity, is

$$S_V = 2A_A''/t \text{ mm}^{-1} \tag{6.54}$$

where A_A'' is the mean value of the total projection averaged over all orientations. Allowances for both the overlap due to surfaces shadowing one another and to self-overlap due to reentrant curvatures are included by virtue of the total projection term. In order to express the exact term A_A'' by the observed term A_A', we must account for both kinds of overlap. A correction for surface-to-surface overlap can be derived under simplified conditions, using a procedure similar to that followed in deriving Eq. (6.65), only instead of adding convex particles to a foil, we insert simple surfaces of finite area without reentrant curvature.

When there is only overlap of open surfaces (but not self-overlap), Eq. (6.62) becomes

$$\Delta A'' = A_i'. \tag{6.67}$$

This leads to

$$A_A'' = -\ln(1 - A_A'), \tag{6.68}$$

which corrects for double overlap of surfaces without reentrant curvature. The same equation is derived by Mack [28] for flat, convex laminae randomly dispersed in a plane. It correctly allows for all double overlaps and some, but not all, cases of multiple overlap [27].

If Eq. (6.68) is substituted in the general expression Eq. (6.54), there results

$$\boxed{S_V = -2\ln(1 - A_A')/t} \text{ mm}^{-1} \tag{6.69}$$

for overlap of a system of simple open surfaces under the conditions specified above.

The dilemma raised by multiple overlaps in the interpretation of areas on the projection plane is illustrated in Fig. 6.11. Obviously, measurements on the projection plane alone are not sufficient. If the images are obtained by electron transmission microscopy, for example, then tilting the foil should give additional insight into the structure. In some cases the images of superimposed surfaces can be detected on the projection plane. In other cases, the relative density of the image may be used as a guide to the number of overlaps contributing to the image. As a final resource, however, one may have to prepare test sections and employ the equations given previously for opaque materials. Of course, this alternative is of doubtful value if the requisite magnifications cannot be achieved.

Bearing in mind the reservations enumerated above, one may use the overlap correction of Eq. (6.65) or (6.68) in simple cases. Reference to Mack's [27, 28] and Hilliard's [4] papers will yield additional insight into the problems associated with multiple overlaps greater than 2 or 3 in number. The analysis of this subject is far from complete at the present time [68], so improvisations and approximations are still required in order to obtain the true surface area from projected images.

6.7.3 Length of Lines in Space

Overlap on the projection plane of true linear features in space is a highly improbable event. The relationships for lines given previously in Eqs. (6.5) through (6.10) may be used without correction in such a case. Overlap of thick linear elements is possible, however, so each structure must be examined on its own merits [40] when one is deciding on the appropriate equations to use.

Another type of linear system in space consists of the perimeters of discrete surface elements in a thin foil. Overlap is commonly encountered with such a system and part of the perimeters may be obscured [4]. As a generalization of the results derived in Section 2.3.2, we know that the expected fractional length of a one-dimensional feature obscured by a planar figure is equal to the areal fraction of the figure [21].

We assume that discrete surface elements (plane figures) are randomly distributed throughout a thin foil, giving a projected area fraction of A'_A. An additional plane figure is added randomly to the foil. The expected fraction of the perimeter length that is obscured by an existing plane figure will equal A'_A. That is,

$$\frac{(L'_p)_{\text{ob}}}{(L'_p)_i} = A'_A \tag{6.70}$$

where $(L'_p)_i$ represents the projected perimeter length of the new plane figure and $(L'_p)_{\text{ob}}$ is the obscured part of its perimeter. The apparent increase of perimeter length in the projection plane is

$$\Delta L'_p = (L'_p)_i - (L'_p)_{\text{ob}} \tag{6.71}$$

and the increase in the total projected perimeter length of the system is

$$\Delta L''_p = (L'_p)_i, \tag{6.72}$$

since we assume that there is no self-overlap of the plane figures. We may now write

$$\frac{\Delta L'_p}{\Delta L''_p} = \frac{(L'_p)_i - (L'_p)_{\text{ob}}}{(L'_p)_i} = 1 - A'_A \tag{6.73}$$

and, for many plane figures

$$\frac{d(L'_p)}{d(L''_p)} = 1 - A'_A. \tag{6.74}$$

Integration of Eq. (6.74) and rearrangement gives the total projection of perimeter lengths for the case of not more than double overlaps

$$L''_p = L'_p/(1 - A'_A), \tag{6.75}$$

in terms of the measured projected perimeter lengths and area fraction. This result is similar to the equation derived by Hilliard [4] for planar laminae in a plane.

Equation (6.75) can be written more usefully in another form. Dividing both sides of the equation by the test area on the projection plane, we obtain

$$L''_A = L'_A/(1 - A'_A) \text{ mm/mm}^2, \tag{6.75a}$$

which corrects the measured projected perimeter lengths per unit area for random areal overlaps.

The mean perimeter length per plane figure is given exactly by

$$\overline{L''_p} = L''_A/N'_A \text{ mm}, \tag{6.76}$$

provided that N'_A can be counted in spite of overlap. Substituting Eq. (6.75a) gives

$$\boxed{\overline{L''_p} = L'_A/N'_A(1 - A'_A)} \text{ mm}, \tag{6.77}$$

which contains a perimeter overlap correction valid for small projected area fractions.

6.7.4 Points in Space

In this section, let us consider three types of point features in a foil and their relationships in the projection plane. These three systems involve (1) points, some of which are obscured on the projection plane by projected surfaces; (2) points that are generated by the intersection of a system of lines and a system of surfaces in a foil; and (3) points that arise from the intersection of test lines with a system of surfaces in the foil.

First case. We have a simple system of random point features in a foil in which open surfaces or particles are also present. The problem is to allow for the number of points obscured by projected areas on the projection plane. The derivation of a relationship that will correct for obscured points is straightforward and follows the same scheme as employed for Eq. (6.65), (6.68), and (6.75a).

The total projection of points per unit area is expressed, accordingly, as

$$P''_A = P'_A/(1 - A'_A) \text{ mm}^{-2}, \tag{6.78}$$

where by the total projection of points we mean both the visible points on the projection plane and the points obscured by surfaces.

Rewriting Eq. (6.4) for the case of overlap by projected surfaces, we have the exact expression

$$P_V = P''_A/t \text{ mm}^{-3}. \tag{6.79}$$

Inserting Eq. (6.78), there follows, for random points in space,

$$\boxed{P_V = P'_A/(1 - A'_A)t} \text{ mm}^{-3}, \tag{6.80}$$

which contains an allowance for those points that are obscured by surfaces on the projection plane, provided that the projected area fraction is not large.

Second case. We deal here with a system of points that are generated by the intersection of lines and surfaces. Since part of the projected lines will be obscured by projected areas, in general it will not be possible to obtain P_V by simple counting. We see from Eq. (2.63), that is, $P_V = \frac{1}{2}L_V S_V$, that the problem consists of determining S_V and L_V from measurements on the projection plane.

From Section 6.7.2, we see that the surface area per unit volume can be obtained from the projection plane by means of Eq. (6.69), which includes a correction for overlap.

To calculate L_V from the projection plane, we must make a correction for the line length obscured by projected surfaces. Rewriting Eq. (6.5a) in terms of total projection yields

$$L_V = \left(\frac{4}{\pi}\right) L''_A/t \text{ mm/mm}^3, \tag{6.81}$$

and substituting Eq. (6.75a) to correct for areal overlap gives

$$L_V = \frac{4L'_A}{\pi(1 - A'_A)t} \text{ mm/mm}^3. \tag{6.82}$$

Note that in the overlap of lines by projected surfaces, the length of line obscured is independent of whether the line is obscured by one surface or by several overlapping surfaces.

We can now proceed to express P_V in terms of projected quantities. Starting with

$$P_V = \tfrac{1}{2} L_V S_V, \tag{2.63}$$

and substituting Eqs. (6.69) and (6.82), we obtain

$$P_V = \frac{4L'_A[-\ln(1 - A'_A)]}{\pi(1 - A'_A)t^2} \text{ mm}^{-3}. \tag{6.83}$$

When the area fraction is small, Eq. (6.83) can be simplified by means of the approximation $-\ln(1 - X) = X$ to

$$P_V = \frac{4L'_A A'_A}{\pi(1 - A'_A)t^2} \text{ mm}^{-3}, \tag{6.84}$$

and further simplified with $L'_A = (\pi/2)P'_L$ to

$$\boxed{P_V = \frac{2P'_L A'_A}{(1 - A'_A)t^2}} \text{ mm}^{-3}. \tag{6.85}$$

This equation estimates the number of points per unit volume generated by random lines intersecting random surfaces. In this case, P'_L refers to the intersections of test lines with the visible projected lines.

Third case. In this section we deal with points that are generated by test lines intersecting a system of surfaces in a foil. This is not the same as the preceding case, since there the projected images consisted of both lines and areas. Here only projected areas appear on the projection plane, and the test lines (or grid points on the projection plane) are introduced independently. The situation is depicted in Fig. 6.3a, except that here we are treating the additional complication of overlapping surfaces.

We wish to solve for P_L, so the quantity P'_P is measured on the projection plane. As before, we require a correction for the loss of projected area due to overlapping surfaces. This correction can be deduced by a procedure similar to that employed previously. Consequently, it is found that

$$P''_P = P'_P/(1 - P'_P), \tag{6.86}$$

where P''_P represents the total projection of points of intersection in the foil, per grid points on the projection plane.

This equation is inserted into the exact expression for intersections of test lines with any type of surfaces in a foil

$$P_L = P''_P/t \text{ mm}^{-1}, \tag{6.87}$$

giving

$$\boxed{P_L = \frac{P'_P}{(1 - P'_P)t}} \text{ mm}^{-1}, \tag{6.88}$$

which provides the value of P_L for a system of surfaces in a foil in the presence of overlap. Note the close relationship of Eq. (6.42) to Eq. (6.87), where, in effect, P_L is measured perpendicular to the projection plane.

6.7.5. Mean Free Distance

This important parameter is expressed accurately and with complete generality by $\lambda = (1 - V_V)/N_L$ (Eq. 4.7), which is based on quantities measured on opaque test sections. For a system of separated, convex particles in a thin foil, we can obtain the mean free distance on the projection plane by converting to projected quantities. We use

$$N'_L = N_A t \qquad \text{from Fig. 6.3(h) and Eq. (6.15),}$$
$$N_L = N_A \bar{d} \qquad \text{from Fig. 6.6 and Eq. (6.21a),}$$
$$\overline{L'_2} = \bar{d} \qquad \text{Eq. (6.29),}$$

and

$$V_V = A'_A \bar{L}_3/t, \qquad \text{Eq. (6.48),}$$

which, upon substitution into Eq. (4.7), yield

$$\boxed{\lambda = \frac{t - A'_A \bar{L}_3}{N'_L \bar{L}'_2}} \text{ mm.} \tag{6.89}$$

This expression is valid for small volume fraction, i.e., negligible overlap. Terms that cannot be measured directly from the projection plane are t and $\bar{L}_3$, the mean particle intercept length. If we assume a particle shape, $\bar{L}_3$ can be converted to $\bar{H}'$, the mean tangent height on the projection plane.

Assuming that the convex particles closely approximate spheres, then $\bar{L}_3 = \frac{2}{3}\bar{D} = \frac{2}{3}\bar{H}'$, and Eq. (6.89) becomes

$$\lambda = \frac{t - \frac{2}{3}A'_A \bar{H}'}{N'_L \bar{L}'_2} \text{ mm} \tag{6.89a}$$

for the mean free distance of spheroidal particles in terms of quantities measurable on the projection plane. We may incorporate overlap and truncation corrections by substituting Eq. (6.66a) (for spheres) instead of Eq. (6.48).

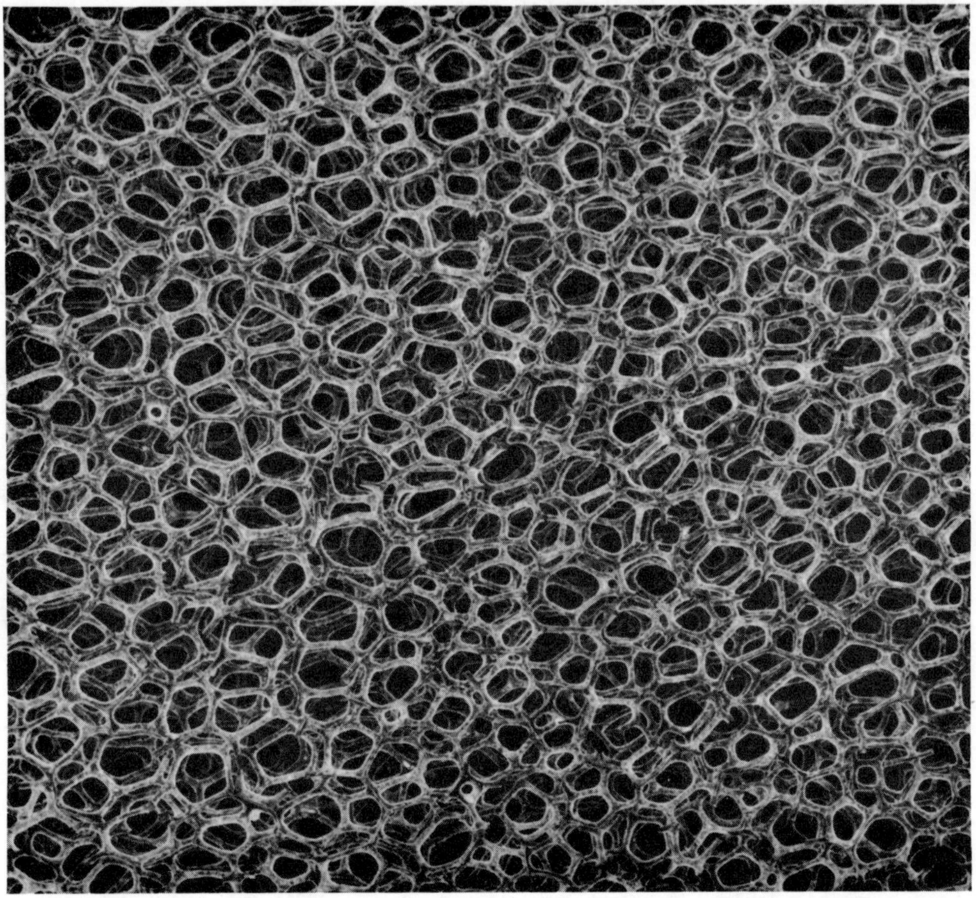

Fig. 6.13. Projected image of a polyurethane foam (courtesy of the Scott Paper Company).

The mean free distance between random lines in space is not calculated as readily as it is for particles. Since Eq. (4.7) involves a volume fraction term, either some assumptions about shape must be made (such as thin rods), or else another spacing parameter must be employed. The former assumption leads to the equations

$$\lambda = 1/\pi r N_A = t/\pi r N'_L \text{ mm} \tag{6.90}$$

for linear features approximated by thin straight rods with a constant radius r (see Table 4.1) and negligible volume fraction. We can see an example of a linear system in space in Fig. 6.13, which represents the projection of a polyurethane foam. This structure consists essentially of a topological network of edges and nodes (or corners) which outlines three-dimensional cellular spaces. If we consider the edges to have negligible cross section, Eq. (6.90) will give the mean free distance between edges.

On the other hand, if the linear features are truly one-dimensional, we can use the Δ_2 spacing parameter of Eq. (4.18a), which is based on the nearest-neighbor distance of random points in a plane. A vertical slice through the foil cuts the lines, giving the P_A count. On the projection plane, of course, we measure P'_L across the line images as depicted in Fig. 6.3(b). In order to convert to projected quantities, $P_A = P'_L/t$ from Fig. 6.3(b) is introduced into $\Delta_2 = \frac{1}{2}P_A^{-1/2}$, giving

$$\boxed{\Delta_2 = (t/4P'_L)^{1/2}} \text{ mm} \tag{6.91}$$

for the mean distance of separation of nearest-neighbor pairs of planar points originating from the intersection of a random system of lines in space by test planes. This spacing parameter is not exactly the same as λ, but serves much better for the purpose than the approximate spacing $\Lambda = P_A^{-1/2}$, which has been used for dislocation lines or particles intersected by a plane. The latter expression implicitly assumes a square array of points or particles in a plane, and can be seriously in error [51].

6.8 SIZE DISTRIBUTIONS OF CONVEX PARTICLES

The calculation of particle size distributions from the projection plane is an extremely difficult subject, which has been treated by only a few authors. Outstanding for its scope and depth of analysis is the paper by Hilliard [4], after which appeared papers by Ashby and Ebeling [52], Crompton, *et al.* [53], and Goldsmith [54]. The latter authors deal with relatively specialized aspects of particles and their projections.

Hilliard treats the problem of relating a projected image to the three-dimensional properties of the structure. He first gives a general analysis leading to N_V, the number of particles per unit volume, and D, the average tangent diameter, for convex particles in a thin foil. The determination of the size distribution is considered specifically for spherical particles.

As in Chapter 5, here we can consider only the simplest of particle shapes. Not only do particle truncation and overlap cause great complexities, but also the relation of particle diameter to foil thickness plays a critical role. When the number of overlaps extends beyond two, the mathematics become intractable. Another basic problem is that particles cannot be added randomly to a foil because of those regions already preempted by prior particles. In spite of these difficulties, appreciable progress has been made in defining the problems, in outlining their solutions, and in specifying the limits of applicability.

6.8.1 Direct Measurement of Size Distributions

Under favorable conditions, size distributions of convex particles may be measured directly from the projection plane. Provided that there are sufficient particles in each class interval, that orientations and locations are random, and that overlap and truncation at foil surfaces are negligible, then the problem resolves itself into one of simple measurements.

The tangent diameter (or height), H', of the projected image is the parameter of choice because of its relationships to other geometrical quantities and because of the ease with which it can be measured. Also, for convex particles in general, H' equals

D, the tangent diameter. For example, in determining the diameters of a system of disks [59] or circular loops, H' is the maximum dimension of the image on the projection plane, regardless of the eccentricity of the image.

It should be noted that there is a slight difference between $\bar{D}$ and $\bar{H}'$ because of truncation, with $\overline{H'} \lesssim \bar{D}$. However, this inequality becomes less important with increasing foil thickness, because the ratio of truncated particles to the total number decreases. Other difficulties arise when particle shapes become complex and when overlap becomes significant. Hilliard [4] has shown that the relative error in the number of particles counted if overlap is neglected is $3\ V_V t/2\bar{D}$. Thus, if $t/\bar{D}$ is 2 and V_V is 0.02, 94% accuracy or better is achieved.

6.8.2 Three-Parameter Representation of Size Distributions

For many purposes, the particle size distribution can be represented satisfactorily by three parameters instead of the complete distribution curve. The mean particle diameter $\bar{D}$, the standard deviation $\sigma(D)$, and the number of particles per unit volume N_V give all the information usually required. Hilliard has discussed this point [55, 56] in detail. DeHoff [29] has also shown how simplification can be achieved if the particle sizes adhere to the usual log-normal distribution. Let us describe these three quantities and their relationship to the projection plane.

Mean diameter. The arithmetic mean diameter has been discussed previously, and is defined in Eqs. (4.61) or (5.1). The "diameter" to be measured on the projection plane depends on the particles under study, but generally H', the tangent diameter, or L_2', the intercept length, are satisfactory. The measurements involved are illustrated in Fig. 6.5. An important relation given previously is

$$\overline{H'} = N_L'/N_A' \text{ mm}, \tag{6.21}$$

which is useful in simple cases.

Number of particles per unit volume. As indicated in Fig. 6.3(i), N_V and N_A' are simply related by $N_V = N_A'/t$ in the absence of overlap and truncation. When truncation effects are accounted for, we obtain the expression

$$\boxed{N_V = N_A'/(t + \overline{H'})} \text{ mm}^{-3} \tag{6.92}$$

by a procedure [3] analogous to that employed in deriving Eq. (6.52). Alternatively, consider a foil volume defined by an area A and thickness $t + \overline{H'}$. The number of particles having centers within this volume, $N_V(t + \overline{H'})A$, is equal to the number of images N' observed on the projection plane (within the same area). Equating the two numbers leads to $N_V = (N'/A)/(t + \overline{H'})$, or Eq. (6.92).

For systems of convex particles with larger volume fractions, or with $(t/\overline{H'}) \gg 1$, overlap becomes appreciable and a correction is required. Under these conditions, Eq. (6.92) is modified [4] to read

$$\boxed{N_V = (N_A' + M_A')/(t + \overline{H'})} \text{ mm}^{-3}, \tag{6.93}$$

where M_A' is the number of particle images per unit area lost by overlap. M_A' is a function of the degree of particle overlap, and the ability to resolve overlaps into the constituent particles. For spherical particles of diameter D and $(t/D) \gg 1$,

$$M'_A = \tfrac{3}{2} V_V \left(\frac{t}{D}\right) N'_A \text{ mm}^{-2}. \tag{6.94}$$

Assuming that $t/D = 10$ and $V_V = 0.02$, we see that $M'_A = 0.3\, N'_A$. However, a correction for overlap should be unnecessary if $V_V(t/D) < 0.04$.

The difficult problem of assigning a value to M'_A can be avoided if N'_A is known as a function of foil thickness t. In such a case, a graphical solution for N_V is available [4] through the equation

$$\ln [N'_A/(t + \overline{H'})] = \ln N_V - mt\, N_V, \tag{6,95}$$

where m is a constant dependent on overlap. A plot of $\ln [N'_A/(t + \overline{H'})]$ versus t gives a straight line intersecting the $t = 0$ axis at $\ln N_V$. The value of $\overline{H'}$ is established independently, using Eq. (6.21). In order to obtain N'_A measurements corresponding to various values of t, it is possible to use one wedge-shaped foil at different locations where the thicknesses are different.

Of course, if overlap is negligible, we may use Eq. (6.92) instead of Eq. (6.95). N'_A is plotted versus t, and N_V is calculated from the slope of the resulting straight line. Since the thickness is relatively difficult to measure, Eq. (6.92) is solved preferably with only one value of t, using direct substitution of N'_A and $\overline{H'}$.

Standard deviation. In order to determine the standard deviation $\sigma(D)$, we start with an expression for the total surface area S_V and number N_V per unit volume,

$$S_V = \pi N_V \overline{D^2} \text{ mm}^2/\text{mm}^3, \tag{6.96}$$

for a polydispersed system of spherical particles. Substitution of $\overline{D^2}$ from Eq. (5.3) gives

$$[\sigma(D)]^2 = \frac{S_V}{\pi N_V} - \bar{D}^2 \text{ mm}^2. \tag{6.97}$$

Introducing Eqs. (6.54) and (6.65) for S_V and A''_A, respectively, leads to

$$[\sigma(D)]^2 = \frac{-4 \ln (1 - A'_A)}{\pi N_V t} - \bar{D}^2 \text{ mm}^2. \tag{6.98}$$

This expression is valid for small volume fractions of spherical particles in the absence of truncation. Allowance for the latter can be made through the use of Eq. (6.58) instead of Eq. (6.65), if required.

Bach [30, 57] has analyzed the problems of obtaining the various parameters of a polydispersed system of spheres in a thin foil. Procedures are given for the determination of the mean radius, the variance, the mean sphere volume, and the number of spheres per unit volume. Similar parameters are discussed in the paper by Ashby and Ebeling [52] for polydispersed spherical particles *removed* from the sample surface by an extraction replica.

6.8.3 Overlap Correction for Size Distributions

In general, size distributions calculated from projected images overestimate the number of large particles because of unresolved overlap. The error may fall either way for small particles: truncation at the foil surfaces increases the number of small images, but, on the other hand, the loss of small particles by overlap is greater. These structural ambiguities may be resolved if one views the particles at more than one angle to the foil normal [8, 37].

Hilliard [4] gives an analysis by which the size distribution of spheres $N(D)$ may be obtained from their projected images. For the case of $(t/\bar{D}) \lesssim 2$, overlap is neglected and the true size distribution is calculated by Scheil's method (see Chapter 5). Modifications of Scheil's analysis are given by Crompton, *et al.* [53], who take into account explicitly (1) the contributions to the projection plane of truncated spherical particles with centers lying outside the foil, and (2) the effects of the relative foil-to-particle thickness.

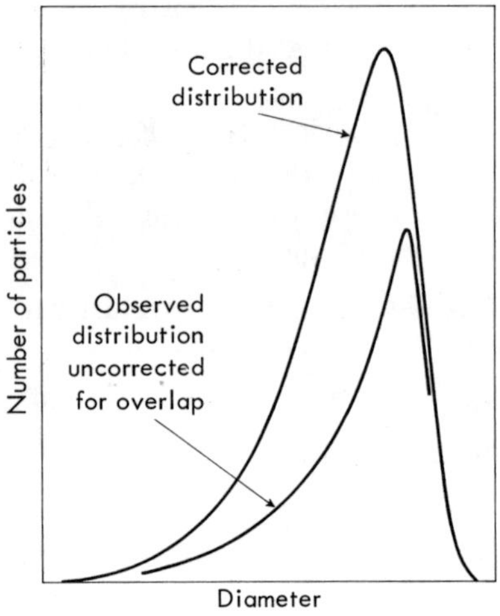

Fig. 6.14. Hypothetical distribution of spherical particles corrected for overlap on projection plane [4].

When $(t/\bar{D}) \gg 1$, an approximate numerical integration is available [4] which takes both truncation and overlap into account. Two separate size distributions are measured on the projection plane; that is, $N(H')$ and $N(L_2')$, using experimental procedures indicated in Fig. 6.5. An idea of the error introduced by neglecting the corrections is given in Fig. 6.14 for a hypothetical distribution of spherical particles.

These methods and equations are highly specialized and complicated, and for further information the reader is referred to the original publications.

6.9 DETERMINATION OF FOIL THICKNESS

One of the more important experimental quantities appearing in this chapter is the foil thickness. Because of its importance, we shall briefly review methods for obtaining foil thickness.

The measurement of thickness of foils cut by a microtome [31, 32, 33] does not present too great a problem, except that slight fluctuations in thickness are noted [15]. Metal foils produced by electrochemical methods are frequently tapered, or wedge-shaped [14, 34], and the thickness must be measured at every site of interest. Cleaved specimens are more uniform in thickness, and an average value can be determined that applies over a large area.

Standard methods for measuring the thickness of thin foils include optical interferometry [62], intensity reduction of transmitted beam [10, 63, 65, 66], chemical analysis [59], and direct weighing [59]. Since these methods are well documented elsewhere, they will not be discussed in detail here. Special techniques may be required for particular materials, however.

In petrography [13], the thickness of rock or mineral "sections" is determined by means of interference colors of a known mineral, such as quartz. Three methods for estimating the thickness of graphite foils are given by Thrower and Reynolds [9] and involve (a) the use of a Sigma Comparator, (b) folds in the graphite, and (c) measurements of optical density. Delavignette and coworkers developed analyses for determining the thickness of thin monocrystalline foils when flat [35] and when bent [36], using measurements of Kossel-Möllenstedt fringes. A method based on theoretical predictions of the number of periodic fringes from the edge of a tapered foil or inclined boundary, known as extinction contours, is described by Howie [5] and Hirsch, *et al.* [59]. Bachmann and Siegel [10] compare the absolute scattering distribution of electrons against samples of carbon or copper of known thickness.

Special methods that take advantage of structural features within a metal foil are based on (a) occurrence of slip traces [14, 69] or twin traces through the foil, (b) precipitate platelets or needles lying in a known crystallographic plane extending from top to bottom of the foil [7], and helical dislocations truncated by the foil surfaces [6]. Ishida [67] points out how foil thickness can be determined if random dislocation lines pierce both top and bottom surfaces of the foil. One uses the same micrographs to make both P'_L and P'_A counts. From the first count, $P'_L = P_A t$, and from the second, $P'_A = 2P_A$. Combining the two equations gives

$$t = 2P'_L/P'_A \text{ mm}, \tag{6.99}$$

which yields the foil thickness unambiguously from measurements on the projection plane. This is an average value, however, over the selected area.

Foil thickness can also be deduced from stereoscopic pairs of micrographs [64]. Sakaki and Eguchi [8] place two kinds of crystalline particles on opposite sides of the foil, then take photographs at two angles. Nankivell [60] gives an equation for foil thickness,

$$t = \frac{\Delta L}{2M \sin \theta} \text{ mm}, \tag{6.100}$$

where ΔL is the parallax, or difference in the separation of two objects on the two micrographs taken at a tilt angle (2θ), and M is the magnification of the micrographs.

The objects from which the parallax is measured must show sharp contrast and be on opposite surfaces of the foil.

It should be emphasized that in this method the inclination of an internal plane or line need not be known, and only the parallax between stereo pairs is measured. Hence objects such as grain boundaries, curved dislocations, and precipitates of unknown habit planes can be used [59]. The geometry of stereoscopic pairs is discussed elsewhere for the case of other applications [61] and other experimental techniques [37].

This listing is not complete, but should serve as an indication of the methods that have been used. Further work on this problem is required in order to produce a reliable, quick, and convenient method for determining foil thickness at any desired location.

REFERENCES

1. R. K. Ham and N. G. Sharpe, "A Systematic Error in the Determination of Dislocation Densities in Thin Films," *Phil. Mag.* **6** (Sept. 1961) 1193
2. U. Valdrè and P. B. Hirsch, "Rearrangements of Dislocations in Stainless Steel During Electropolishing," *Phil. Mag.* **8** (1963) 237
3. J. W. Cahn and J. Nutting, "Transmission Quantitative Metallography," *Trans. AIME* **215** (1959) 526
4. J. E. Hilliard, "The Counting and Sizing of Particles in Transmission Microscopy," *Trans. AIME* **224** (1962) 906
5. A. Howie, "Quantitative Experimental Study of Dislocation and Stacking Faults by Transmission Electron Microscopy," *Met. Rev.* **6**, 24 (1961) 467
6. E. J. Freise, A. Kelly, and R. B. Nicholson, "Guinier-Preston Zones in an Aluminum-Silver Alloy," *Acta Met.* **9** (1961) 250
7. M. von Heimendahl, "Determination of Metal Foil Thickness and Orientation in Electron Microscopy," *J. Appl. Phys.* **35** (1964) 457
8. Y. Sakaki and I. Eguchi, "A Method for Immediate Determination of Specimen Thickness," *Fifth Int. Cong. Electron Micros.* **1**, edited by S. S. Breese, Jr., New York: Academic Press (1962) EE-9
9. P. A. Thrower and W. N. Reynolds, "Problems in the Electron Microscopy of Irradiated Graphite," *Fifth Int. Cong. Electron Micros.* **1**, edited by S. S. Breese, Jr., New York: Academic Press (1962) F-9
10. L. Bachmann and B. M. Siegel, "Determination of Thickness of Thin Film by Inelastic Scattering of Electrons," *Proc. Europ. Reg. Conf. Electron Micros.* **1**, edited by A. L. Houwink and B. J. Spit, published by De Nederl. Veren. v. Electron., Delft (1960) 157
11. R. K. Ham, "On the Loss of Dislocations During the Preparation of a Thin Film," *Phil. Mag.* **7** (1962) 1177
12. P. B. Hirsch, "Dislocation Distributions and Hardening Mechanisms in Metals. 1. Techniques for Determining Dislocation Distributions in Metals. 2. Dislocations in Metals" (with J. W. Steeds), *The Relation Between the Structure and Mechanical*

Properties of Metals, Nat. Phys. Lab. Symp. No. 15, London: Her Majesty's Stationery Office **1** (1963) 40

13. W. W. Moorhouse, *The Study of Rocks in Thin Section*, New York: Harper, (1959)
14. G. Thomas, *Transmission Electron Microscopy of Metals*, New York: John Wiley (1962)
15. H. Haug, "Bedeutung und Grenzen der quantitativen Messmethoden in der Histologie," in *Medizinische Grundlagenforschung*, **IV**, edited by K. Fr. Bauer, Stuttgart, Germany: G. Thieme Verlag (1962) 299
16. A. Cauchy, *Oeuvres complètes d'Augustin Cauchy*, 1er Sér., Vol. II, Paris: Gauthier-Villars (1908) 167
17. P. S. Rudman, personal communication
18. V. Vouk, "Projected Area of Convex Bodies," *Nature* **162** (Aug. 28, 1948) 330
19. M. G. Kendall and P.A.P. Moran, *Geometrical Probability*, London: Charles Griffin, (1963) 74; 115
20. C. S. Smith, discussion on paper by P. Jordan, *Proc. Third Int. Conf. on Reactivity of Solids*, **1** Madrid: (1956) 442
21. C. S. Smith and L. Guttman, "Measurement of Internal Boundaries in Three-Dimentional Structures by Random Sectioning," *Trans. AIME* **197** (1953) 81
22. R. T. DeHoff, "The Determination of the Geometric Properties of Aggregates of Constant-Size Particles from Counting Measurements Made on Random Plane Sections," *Trans. AIME* **230** (1964) 764
23. E. J. Myers, "Sectioning of Polyhedrons," *Proc. First Int. Cong. for Stereology*, paper 15, Vienna, publ. by Vienna Medical Academy (1963)
24. H. Steinhaus, "Zur Praxis der Rektifikation und zum Längenbegriff," *Berichte Math. Phys. Klasse*, Akad. Wiss. Leipzig, 82 (1930) 120
25. S. A. Saltykov, *Stereometric Metallography*, second edition, Moscow: *Metallurgizdat* (1958)
26. J. E. Hilliard, "Specification and Measurement of Microstructural Anisotropy," *Trans. AIME* **224** (1962) 1201
27. C. Mack, "On Clumps Formed When Convex Laminae or Bodies Are Placed at Random in Two or Three Dimensions," *Proc. Cambridge Phil. Soc.* **52** (1956) 246
28. C. Mack, "The Expected Number of Clumps When Convex Laminae are Placed at Random and with Random Orientation on a Plane Area," *Proc. Cambridge Phil. Soc.* **50** (1954) 581
29. R. T. DeHoff, "The Estimation of Particle Size Distributions from Simple Counting Measurements Made on Random Plane Sections," *Trans. AIME* **233** (1965) 25
30. G. Bach, "Uber die Bestimmung von charakteristischen Grössen einer Kugelverteilung aus der Verteilung der Schnittkreise," *Proc. First Int. Cong. for Stereology*, paper 12, Vienna, publ. by Vienna Medical Academy (1963)
31. H. Fernandez-Moran, "A Diamond Knife for Ultrathin Sectioning," *Exptl. Cell Res.* **5** (1953) 255
32. V. A. Phillips, "Direct Observations on Structure of Ultramicrotomed Metals and Alloys," Symp. on *Advances in Electron Metallography and Electron Probe Microanalysis*, STP 317, ASTM, Philadelphia (1962) 34
33. W. C. Bigelow, "The Development of Electron Microscopic Methods for the Study of Metals," Symp. on *Advances in Electron Metallography and Electron Probe Microanalysis*, STP 317, ASTM, Philadelphia (1962) 58

34. P. M. Kelly and J. Nutting, "Techniques for the Direct Examination of Metals by Transmission in the Electron Microscope," *J. Inst. Met.* **87** (1958–59) 385
35. R. Siems, P. Delavignette, and S. Amelinckx, "The Buckling of a Thin Plate Due to the Presence of an Edge Dislocation," *Phys. Stat. Sol.* **2** (1962) 421
36. P. Delavignette and R. W. Vook, "A Method for Measuring the Thickness of Thin Bent Foils in Transmission Electron Microscopy," *Phys. Stat. Sol.* **3,** 4 (1963) 648
37. W. M. Williams, "Stereoscopic Microradiography," *Metallurgia* **63,** 376 (1961) 95
38. M. R. Spiegel, *Theory and Problems of Vector Analysis,* New York: Schaum (1959) 82
39. K. F. Hale and D. McLean, "Structure of Quench Aged Iron-Carbon and Iron-Nitrogen Alloys," *J. Iron Steel Inst.* **201** (1963) 337
40. J. E. Bailey and P. B. Hirsch, "The Dislocation Distribution, Flow Stress, and Stored Energy in Cold-Worked Polycrystalline Silver," *Phil. Mag.* **5** (1960) 485
41. *Techniques for Electron Microscopy,* edited by D. Kay, Oxford: Blackwell Scientific Publications (1961)
42. H. S. M. Coxeter, *The Real Projective Plane,* second edition, Cambridge: Cambridge University Press (1961) Students' Edition
43. D. Greenhood, *Mapping,* Chicago: University of Chicago Press (1964), Phoenix Science Series
44. W. H. Walton, "Feret's Statistical Diameter as a Measure of Particle Size," *Nature* **162** (Aug. 28, 1948) 329
45. W. Kaplan, *Advanced Calculus,* Reading, Mass.: Addison-Wesley (1953) 206
46. L. A. Lyusternik, *Convex Figures and Polyhedra,* New York: Dover Publications, (1963) 22, 140
47. *Quantitative Electron Microscopy,* edited by G. F. Bahr and E. H. Zeitler, Baltimore, Md.: Williams and Wilkins (1965)
48. O. Kallmes and H. Corte, "The Structure of Paper," *Tappi* **43,** 9 (1960) 737
49. H. Corte, "Paper and Board," in *Composite Materials,* edited by L. Holliday, New York: Elsevier (1966) 475
50. A. G. Ogston, "The Spaces in a Uniform Random Suspension of Fibers," *Trans. Faraday Soc.* **54,** 431 (1958) 1754
51. G. S. Ansell and J. S. Hirschhorn, "Reply to Comments on Dispersion Strengthening Models," *Acta Met.* **13** (1965) 574
52. M. F. Ashby and R. Ebeling, "On the Determination of the Number, Size, Spacing, and Volume Fraction of Spherical Second-Phase Particles from Extraction Replicas," *Trans. AIME* **236** (1966) 1396
53. J. M. G. Crompton, R. M. Waghorne, and G. B. Brook, "The Estimation of Size Distribution and Density of Precipitates from Electron Micrographs of Thin Foils," *Brit. J. Appl. Phys.* **17** (1966) 1301
54. P. L. Goldsmith, "The Calculation of True Particle Size Distributions from the Sizes Observed in a Thin Slice," *Brit. J. Appl. Phys.* **18** (1967) 813
55. J. E. Hilliard, "The Direct Determination of the Number of Convex Particles per Unit Volume and the Moments of their Size Distribution by an Intercept Analysis on a Section," *Stereology,* edited by H. Elias, *Proc. Second Int. Cong. for Stereology,* New York: Springer-Verlag (1967) 195
56. J. E. Hilliard, "Direct Determination of the Moments of the Size Distribution of Particles in an Opaque Sample," *Trans. AIME* **242** (1968) 1373

57. G. Bach, "Kugelgrössenverteilung und Verteilung der Schnittkreise: . . . ," *Quantitative Methods in Morphology*, edited by E. R. Weibel and H. Elias, New York: Springer-Verlag (1967) 23
58. J. E. Burke and D. Turnbull, "Recrystallization and Grain Growth," *Progress in Metal Physics*, **3**, edited by B. Chalmers, London: Pergamon Press (1952) 224–225
59. P. B. Hirsch, A. Howie, R. B. Nicholson, D. W. Pashley, and M. J. Whelan, *Electron Microscopy of Thin Crystals*, London: Butterworth (1965) 426
60. J. F. Nankivell, "Minimum Differences in Height Detectable in Electron Microscopy," *Brit. J. Appl. Phys.* **13** (1962) 126
61. J. F. Nankivell, "Stereoscopic Viewing of Metal Foils in the Electron Microscope," *Brit. J. Appl. Phys.* **13** (1962) 129
62. S. Tolansky, *Surface Microtopography*, London: Longmans Green (1960) 62
63. L. Reimer, "Contrast in Amorphous and Crystalline Objects," *Quantitative Electron Microscopy*, edited by G. F. Bahr and E. H. Zeitler, Baltimore, Md.: Williams and Wilkins (1965) 201
64. J. G. Helmcke, "Determination of the Third Dimension of Objects by Stereoscopy," *ibid.*, page 195
65. E. Zeitler and G. F. Bahr, "Contrast and Mass Thickness," *ibid.*, page 208
66. G. F. Bahr and E. Zeitler, "The Determination of the Dry Mass in Populations of Isolated Particles," *ibid.*, page 217
67. Y. Ishida, "Determination of Foil Thickness," *Phil. Mag.* **14,** 128 (August 1966) 411
68. S. A. Roach, *The Theory of Random Clumping*, London: Methuen (1968) 26
69. B. Vigeholm and J. Lindbo, "Thickness of Transmission Electron Microscope Specimens," *Trans. AIME* **245** (1969) 883

PROBLEMS

6.1. Derive Eq. (6.6): $\bar{L}' = (2/\pi)L$.

6.2. Express the following quantities in terms of D, the diameter of a monodispersed system of spheres, using the equations from Fig. 6.6.
Spatial: $\bar{V}, \bar{S}, \bar{L}_3, \bar{D}$
Section: $\bar{A}, \bar{L}_2, \bar{d}, \bar{L}_p$
Projection: $\bar{A}', \bar{L}_2', \bar{H}', \bar{L}_p'$

6.3. The surface area per unit volume of untruncated spheres of diameter D is correctly described by the equation $S_V = 6V_V/D$. Is the expression still valid when it is applied to randomly distributed spheres within a thin foil, as in Fig. 6.10? Why?

6.4. Consider a thin foil of thickness t containing N_V randomly located spherical particles of diameter D per unit volume. Assuming that $t/D = 10$ and that

$$(N_A')_{\text{total}} = (N_A')_{\text{untrunc}} + (N_A')_{\text{trunc-in}} + (N_A')_{\text{trunc-out}}$$

(according to the three categories discussed after Eq. 6.58a), what is the relative number of particles with centers lying in each of the above three categories?

6.5. Determine the weighted value of the constant k in $A''_A = kA'_A$ (see also Eq. 6.62), based on the results of Problem 6.4. (Assume that $k = \frac{2}{3} \times 2$ for truncated particles with centers lying inside the foil.) What is the modified version of Eq. (6.65)? Of Eq. (6.66a)?

6.6. Derive the following equations:
a) Eq. (6.68), $A''_A = -\ln(1 - A'_A)$,
b) Eq. (6.78), $P''_A = P'_A/(1 - A'_A)$,
c) Eq. (6.86), $P''_P = P'_P/(1 - P'_P)$,
d) Eq. (6.92), $N_V = N'_A/(t + \overline{H}')$.
Use a procedure analogous to that employed in deriving Eq. (6.52).

6.7. Show that

$$V_V = A'_A\left(\frac{\bar{L}_3}{\bar{L}_3 + t}\right)$$

can be obtained either for convex particles from Eq. (6.52) or for spherical particles from Eq. (6.52a).

6.8. Starting with Eq. (6.52), express N_V in terms of $\bar{L}_3$ instead of $\overline{H}'$ in a form comparable to Eq. (6.92). Is the result compatible with Eq. (6.92)?

6.9. What is the value of $(S_V)_{TP}$ for a sphere of unit radius embedded in a cube tangent to the sphere at all six faces? Use Eq. (6.44).

6.10. Calculate the mean edge length $\bar{L}_e$ of the network shown in Fig. 6.13, assuming the edges to be true lines. Use the definition $\bar{L}_e = L_V/E_V$, where $E_V = 2C_V$ (see Table 7.3). Note that C_V is the same as P_V.

6.11. Compare values of λ and Δ_2 from Eqs. (6.90) and (6.91) for the structure reproduced in Fig. 6.13. Do the two distance parameters have the correct magnitudes with respect to each other?

6.12. How would you estimate the volume fraction of the network structure in Fig. 6.13?

7

SPECIFICATION OF SHAPE

"To terms of magnitude, and of direction, must we refer all our conceptions of form. For the form of an object is defined when we know its magnitude, actual or relative, in various directions."
Sir D'Arcy W. Thompson

7.1 INTRODUCTION

We turn now to the most indeterminate subject of all—that of shape, or form—and its quantitative description. The word "form" is found extensively in the literature, especially in connection with the growth of an object. Here we are more interested in the geometrical aspects of the problem, so the synonym "shape" is used to emphasize this difference.

The classical book on the subject is D'Arcy W. Thompson's *On Growth and Form* [1]. His approach was unique in that he attempted to treat shape mathematically, to the fullest possible extent. Other books that deal with various aspects of shape are those by Hertel [13] on structure, form, and movement, by Weyl [2] on symmetry, by Wells [3] on three-dimensional chemistry, by Sinnott [4] on organic form, and by Whyte [5] on the esthetic to the mathematical. When one examines the tremendously variegated objects of nature, it is not surprising that most discussions of shape wax philosophical or artistic, rather than quantitative.

From the microstructural point of view, the publications of Cyril S. Smith (6a–d) are outstanding. They provide an excellent summary, both of previous work and of the latest developments in the field.

Shape (or form) has been defined as ". . . any definable combination of geometrical elements, whether real or imaginary." Only for the most regular objects can the shape be expressed in this quantitative sense. For the complex and irregularly shaped objects encountered in nature, one must be content with descriptions that are largely indirect and approximate.

The concept of shape also arises with geometrical elements other than solids. An assemblage of points presents collectively an impression of shape, as does a system of lines. For this reason, we include systems of points and lines, along with areal and volume elements, in a study of shape.

Definitions of "shape" in the four elemental cases are offered below. Here we place emphasis on the metric characteristics.

- The shape of a system of points in space is defined when the relative positions of the points are known in terms of distance and direction.
- For a system of lines, we must know not only the distances and directions between lines, but also the geometrical attributes of the lines themselves.
- The essential geometrical elements of a planar figure are the perimeter and the enclosed surface. These must be described quantitatively in order to fix the shape.
- For a three-dimensional object the bounding surface is the pertinent geometrical element. If we know the bounding surface, we can define the shape of the object.

Without doubt, studies of shape will involve various combinations of the above simple cases. In an assemblage of grains, for example, we find points, lines, surfaces, and volume elements.

Ideally, we would like to specify the shape of objects of any degree of complexity in exact mathematical terms. To handle shape quantitatively requires analytical expressions for lines (either planar or skewed) and for surfaces (either planar or curved). The vector-based definition by D'Arcy Thompson, quoted at the beginning of this chapter, is useful in describing the positions of points and lines.

For real objects, however, indirect or approximate methods are almost obligatory at the present time. A great deal of information is obtained from planar sections or projections instead of the three-dimensional object itself. Tedious experimental techniques, such as serial sectioning, are employed to build up the third dimension. Frequently, average shape parameters are obtained for an aggregate of objects rather than for the objects individually.

It is readily apparent that the subject of shape must approached on a broad front. Here we proceed from the exact to the approximate. We shall treat, in turn, the pertinent basic mathematical concepts, parametric representations, approximation methods, and visual devices. First, however, to place the subject in better perspective, let us survey the better-known methods for specifying shape from a historical point of view.

7.2 METHODS OF SPECIFYING SHAPE

Numerous proposals have been made to express the shape of particles or grains according to some arbitrary standard, such as a graded series of rectangles or ellipses [7, 8, 9, 12]. Calculations of the dimensions of imaginary grains that are uniform spheres or polyhedra [109] have some practical value, but our present concern is with the mathematical expression of shape of irregular bodies. A few examples of efforts in this direction are as follows: Scheil and Wurst [10], using laborious serial sectioning, derived a statistical grain shape from planar measurements on successive grain sections in ingot iron. Later, Hull and Houk [11] measured the areas and number of edges of the polygons obtained by systematic sectioning of simple polyhedra, and calculated distribution curves of areas or diameters for the various polyhedra. C. S. Smith [6b] studied the frequency of edges, sides, and corners in separated metal grains and in two-dimensional soap bubbles. These results compare well with those from human fat cells [21]. The topological properties of systems of cellular or interpenetrating structures are also being studied on a broad front [69, 85, 86].

Other mathematical approaches can be used if simplifying assumptions are permissible. For simple particles of the same size and shape, statistical shape factors can be defined which lead to a quantitative estimate of the shape [37]. Green [14] has discussed the problems inherent in this type of analysis when shape and size vary. Statistical shape factors are of doubtful value when applied to a real aggregate, since the averaging of surfaces and volumes is scale dependent and practically unanalyzable for nonuniform particles.

Of necessity, much work has been performed on two-dimensional sections through aggregates of cells or grains or particles. C. S. Smith [15] suggested a simple ratio for characterizing the two-dimensional grain shape. His dimensionless "grain shape index" is the ratio of the average perimeter length to the average intercept length. In his book [7], Saltykov discusses the problems of characterizing shapes, either by two- or three-dimensional form factors. He defines the two-dimensional factor in terms of two lengths; i.e., as the ratio of the length of the perimeter of a circle with area equal to the measured area, to the length of the actual perimeter. The three-dimensional factor is proportional to the ratio of the cube root of the volume to the square root of the surface area. These two form factors are claimed to be completely equivalent numerically.

We should also mention a very promising shape-sensing parameter based on the "tangent count" of Rhines [16]. The number of tangents produced by a test line sweeping across the outlines of irregular cells in a two-dimensional section depends on the complexity of the outlines and is related quantitatively to the average curvature. Independently, DeHoff [36] and Cahn [35] have been able to show that the average of the mean curvatures of closed surfaces in space is proportional to a simple ratio of N_A/N_L measured on the plane of polish.

The preceding discussion does not attempt to be complete. It is representative, however, of the more mathematically oriented methods of expressing shape quantitatively. We proceed now to the basic mathematical concepts needed for a quantitative description of the shape of lines and surfaces.

7.3 BASIC CONCEPTS IN THE THEORY OF CURVES

From elementary analytic geometry we learn that *every curve can be represented by an equation*, or, conversely, that *every equation in the two variables x and y represents a curve in the coordinate plane* [17]. Similarly, the coordinates of surfaces are given by the equations $z = f(x, y)$ or $F(x, y, z) = 0$. In general, the coordinate method, by establishing a close connection between elementary geometry and analysis, enables us to define many different curves and surfaces.

But analytic geometry, being restricted to the methods of algebra and elementary geometry, does not go further than the investigation of certain specific types of figures. The study of arbitrary curves and surfaces represents another branch of mathematics, known as metric *differential geometry* [18]. The basic tool is the differential calculus, whereby the differential properties of the curves and surfaces are investigated primarily "at a point." Thus the direction of a planar curve at a point is determined by its tangent at that point and the amount by which it curves is described by its curvature. Differential geometry investigates the properties of small segments of curves and surfaces, and is now being developed to include the properties of curves and surfaces "in the large," i.e., over their entire extent. However, only when curves have been defined by equations can their properties be investigated by mathematical analysis.

In the differential geometry of planar curves (linear elements in a flat surface) there are three basic concepts: length, tangency, and curvature. For space curves (linear elements in space) there are in addition the osculating plane and the torsion.

7.3.1 Length

Consider the planar curve $y = f(x)$ shown in Fig. 7.1. Its length between points $A(x_1, y_1)$ and $B(x_n, y_n)$ is obtained by the following procedure [17].

Let A and B be points on the curve and choose points $A = p_1, p_2, p_3, \ldots, p_i \ldots, p_n = B$, starting at A and moving along the curve to B. If the sum of the lengths of the chords joining successive points,

$$\overline{p_1p_2} + \overline{p_2p_3} + \overline{p_3p_4} + \cdots + \overline{p_ip_{i+1}} + \cdots + \overline{p_{n-1}p_n},$$

approaches a limit as the number of chords increases, in such a way that the length of the longest chord approaches zero, this is said to be the length of the curve between A and B.

Let p_i, p_{i+1} be any two adjacent points of intersection having coordinates (x_i, y_i), (x_{i+1}, y_{i+1}). The chord $\overline{p_ip_{i+1}}$ is the hypotenuse of a right triangle, the sides of which are equal to

$$\Delta x_i = |x_{i+1} - x_i|, \qquad \Delta y_i = |y_{i+1} - y_i|.$$

The length of the hypotenuse

$$\overline{p_ip_{i+1}} = [(\Delta x_i)^2 + (\Delta y_i)^2]^{1/2} = [1 + (\Delta y_i/\Delta x_i)^2]^{1/2}\Delta x_i. \tag{7.1}$$

It is apparent that there is a tangent parallel to $\overline{p_ip_{i+1}}$ at some point b on the curve between p_i and p_{i+1}, and the ratio $(\Delta y_i/\Delta x_i)$ may be replaced by the slope of the tangent at b; i.e., by the derivative $y'(x_b)$, where x_b is the abscissa of point b.

Equation (7.1) may now be expressed as

$$\overline{p_ip_{i+1}} = [1 + y'(x_b)^2]^{1/2}\,\Delta x_i. \tag{7.2}$$

The sum of all lengths of the chords between points A and B is

$$L_n = \sum_{i=1}^{n} [1 + y'(x_i)^2]^{1/2}\Delta x_i. \tag{7.3}$$

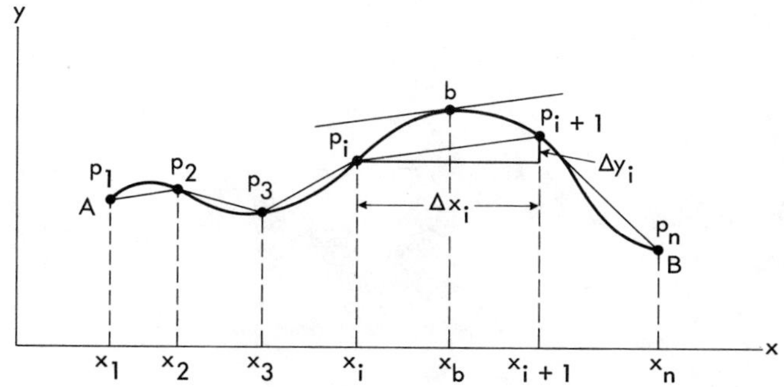

Fig. 7.1. Length of a planar curve.

Passing to the limit, where the greatest of the values Δx_i tends to zero, we obtain

$$L = \int_{x_1}^{x_n} [1 + (dy/dx)^2]^{1/2} dx. \tag{7.4}$$

One can derive the corresponding formula for the length of a space curve in basically the same way.

Equation (7.4) is important not only for computing the length of a curve, but also for investigating the general properties of length, its relations with other concepts, etc. The value of the length depends on the function $f(x)$, and its computation may not always be simple.

7.3.2 Tangency

In order to define the tangent at a point b on a curve, we choose a point x distinct from b, and consider the secant $\overline{bx}$. Then we allow x to approach b along the curve. If the secant $\overline{bx}$ converges to some limiting position, then the straight line in this limiting position is called the tangent at the point b. The meaning of a tangent for a space curve is completely analogous.

The tangent has the important geometric property that near the point of tangency the curve departs less, in a well-defined sense, from this straight line than from any other. Therefore a small segment of the curve may be replaced by a corresponding segment of the tangent with an error which is small in comparison with the length of the segment. This procedure often permits certain simplifications, since in the limit it gives completely exact results.

7.3.3 Curvature

In order to determine whether a line is more or less curved, we need an exact quantitative description of its curvature. Intuitively, of course, our impression of the curvature of a line is based on the rapidity of change of direction of the curve.

We can obtain a more precise definition of curvature by referring to Fig. 7.2. Let a and b be nearby points on the curve. The angle $\Delta\phi$ between the tangents at these points expresses how much the curve has changed direction in the segment Δl between a and b. The average rate of change of direction is given by $\Delta\phi/\Delta l$.

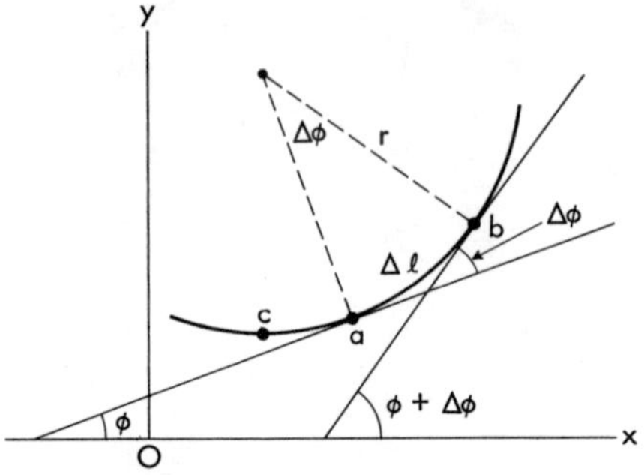

Fig. 7.2. Curvature of a planar curve.

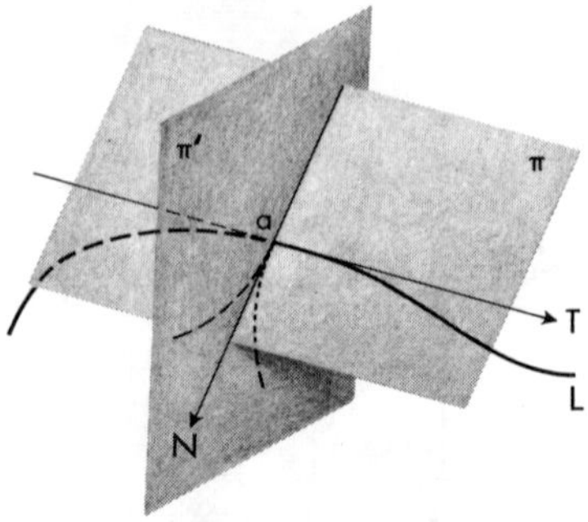

Fig. 7.3. The relationship of the osculating plane π to a space curve L.

Then the curvature at a is naturally defined as the limit of the ratio $\Delta\phi/\Delta l$ as $b \longrightarrow a$; in other words, as $\Delta l \longrightarrow 0$. Thus the curvature is defined by the formula

$$\kappa = \lim_{\Delta l \to 0} \left(\frac{\Delta\phi}{\Delta l}\right) = \frac{d\phi}{dl}\,\text{mm}^{-1}. \tag{7.5}$$

Note that in general we may define the curvature of a curve at a point by comparing the curve with the circumference of a certain circle, the circle of curvature, which plays the role of a model or standard for the curvature. In fact, the curvature of the given curve proves to be equal to the reciprocal of the radius of the (unique) circle which fits the curve most closely in the neighborhood of the point.

Since, for a circle of radius r, the arc subtending the angle $\Delta\phi$ has length $\Delta l = \Delta\phi r$, then

$$\Delta\phi/\Delta l = 1/r,$$

and the ratio $\Delta\phi/\Delta l$ is constant. The curvature at a point can then be written as

$$\kappa = 1/r\,\text{mm}^{-1}, \tag{7.6}$$

where r is the radius of the curve at that point. In more general terms, the curvature of a planar curve at a point may be expressed as

$$\kappa = y''/(1 + y'^2)^{3/2}, \tag{7.7}$$

where y' and y'' are the first and second derivatives of y with respect to x at the chosen point. The derivation of this equation may be found in standard texts [19]. Note that when $y' = 0$, $\kappa = y''$ (see, for example, point c in Fig. 7.2).

7.3.4 Osculating Plane

Although a space curve does not lie completely within one plane, it is generally possible to associate a certain plane with each point in the curve such that this plane lies closer to the curve than any other plane. This closest plane is called the *osculating plane* of the curve at that point. The relationship of the curve L to the osculating plane π is depicted in Fig. 7.3.

The osculating plane passes through the chosen point a and contains the line T tangent to the curve. However, there are many planes that contain both the point and the tangent line. In order to choose the one plane that deviates least from the curve, we investigate the deviation of the curve from the tangent [17]. To do this, we project the curve near a onto the normal plane π', a plane which is perpendicular to T at the point a. The projection on π' of a segment of the curve containing a forms a new curve, usually having a cusp at point a. If the projected curve has a tangent N at the point a, then the plane π determined by T and N will be closest to the original curve in the neighborhood of the point a, that is, it will be the osculating plane at the point a. It can be shown that when the functions defining the original curve have second derivatives (and the curvature of the curve at point a is not zero), then the osculating plane necessarily exists and its equation may be expressed simply in terms of the first and second derivatives of the functions defining the curve.

Of the many possible straight lines in space that are perpendicular to the tangent, only one, the line N, lies in the osculating plane. This line is called the *principal normal* to the curve, with its direction toward the concave side of the curve (as determined by the projection of the curve on the osculating plane). Thus the principal normal plays the same role for a space curve as the ordinary (unique) normal does for a plane curve.

7.3.5 Torsion

From point to point along a space curve, the position and orientation of the osculating plane will probably change. Just as the rate of change of direction of the tangent characterizes the curvature, so the rate of change of direction of the osculating plane characterizes another quantity: the *torsion* τ of the curve. Here, as in the case of curvature, the rate is taken with respect to arc length. If $\Delta\psi$ is the angle between the osculating planes at a fixed point a and at a nearby point b, and if Δl is the length of the arc $\overline{ab}$, then the torsion at point a is defined as the limit

$$\tau = \lim_{\Delta l \to 0} \left(\frac{\Delta\psi}{\Delta l} \right) = \frac{d\psi}{dl} \ \text{mm}^{-1}. \tag{7.8}$$

The sign of τ depends on the side of the curve toward which the osculating plane turns as it moves along the curve. Just as the circle of curvature is related to the curvature of a planar curve, we may define the torsion of a curve in terms of the (unique) helix* that best approximates the curve in the neighborhood of the given point. With a certain analogy to curvature, the torsion characterizes the rate at which the curve leaves its osculating plane.

Fundamental results from the theory of differential equations help define curves in terms of the above concepts. The curvature and torsion as functions of arc length define a curve completely except for its position in space; they describe all the properties of a curve by stating the relationship between its length, its curvature, and its torsion [20]. With the help of the latter three parameters, we can also express the simplest concepts in the theory of surfaces, to which we turn next.

---

*The helix has the same torsion at all its points.

Of course, the theory of curves has not been exhausted by the above remarks. There are many other concepts relating to a curve, and their utilization requires almost every branch of mathematics. Progress in describing shape quantitatively depends on advances in the appropriate mathematical fields.

7.4 BASIC CONCEPTS IN THE THEORY OF SURFACES

In order to study surfaces by means of mathematical analysis, we must be able to express them analytically. The simplest way to represent a surface is by an equation of the form

$$z = f(x, y), \tag{7.9}$$

in which x, y, and z are Cartesian coordinates of a point lying in the surface. Surfaces given by equations in this form are familiar from analytic geometry; for example, $z = Ax + By + C$ represents a plane and $z = x^2 + y^2$ a paraboloid of revolution.

For the application of differential calculus, it is necessary that the function $f(x, y)$ have first, second, and sometimes even higher derivatives. A surface given by such an equation is called *regular*, and conforms geometrically to a surface that curves continuously (without breaks or other singularities). If we are concerned only with small segments of the surface, then the definition of a surface by Eq. (7.9) is perfectly general, and the methods of classical differential geometry may be applied.

7.4.1 Tangent Plane

At each point of a smooth curve there is a tangent *line* which is closest to the curve in the neighborhood of the point; surfaces may also have at each of their points a tangent *plane* (see Fig. 7.4). A plane π passing through the point a on a surface S is said to be tangent to the surface at a if the angle α between the plane and the secant $\overline{ab}$, drawn from a to another point b on the surface, converges to zero as the point b approaches the point a. Moreover, the tangents to all curves that lie in the surface S and pass through the point a lie in the tangent plane π.

A surface is called *smooth* if it has a tangent plane at each point and if, as we pass from point to point, the position of this plane varies continuously. As a first approximation, a small segment (or element) of the surface may be replaced by a segment of the tangent plane near the point of tangency. The perpendicular to the tangent plane which passes through the point of tangency acts as a perpendicular to the surface at this point and is called a *normal* (see N in Fig. 7.4).

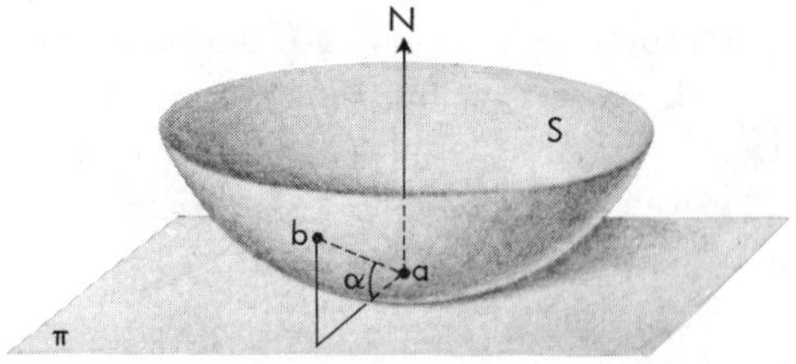

Fig. 7.4. Tangent plane to a surface S at the point a.

7.4.2 Area of a Surface

The possibility of replacing an element of the surface by a segment of the tangent plane is useful in many situations. This procedure may also be used as a basis for defining the area of various surfaces. The curved surface is divided into elementary areas, each of which is projected onto a plane tangent to the surface at some point within the elementary area. We thus obtain a number of plane regions or segments, the sum of whose areas gives an approximation to the area of the surface. The area of the curved surface is defined exactly as the limit of the sum of the areas of the segments under the condition that the partitions of the surface become extremely small. In this way, we can derive an exact expression for the total area S in the form of a double integral,

$$S = \iint_B \left[1 + \left(\frac{\partial z}{\partial x}\right)^2 + \left(\frac{\partial z}{\partial y}\right)^2\right]^{1/2} dy\, dx, \tag{7.10}$$

where the limits of integration depend on the projection B on the xy-plane of the region whose area we wish to calculate.

7.4.3 Curvature of a Surface

The curvature of a surface is characterized by the rate at which the surface leaves its tangent plane. Depending on the direction, the surface may leave its tangent plane at different rates. Thus the curvature of a surface at a given point depends on the curvatures of the whole set of curves lying in the surface and passing through the given point in different directions.

In Fig. 7.5, a tangent plane π touches the surface S at the point a. The curves L', L'', ... formed by intersections of the surface with the planes π', π'', ... that pass through the normal N are called *normal sections*. The curvature of a normal section is considered positive if the curve is concave in the direction of the normal, and negative if it is concave in the opposite direction. Thus, for the surface shown in Fig. 7.5, with the normal pointing in the vertical direction, the curvatures of curves L' and L'' are negative.

A normal section is defined by the angle ϕ (Fig. 7.5) by which its plane is rotated from some initial direction in the tangent plane. Knowing the curvature of the normal

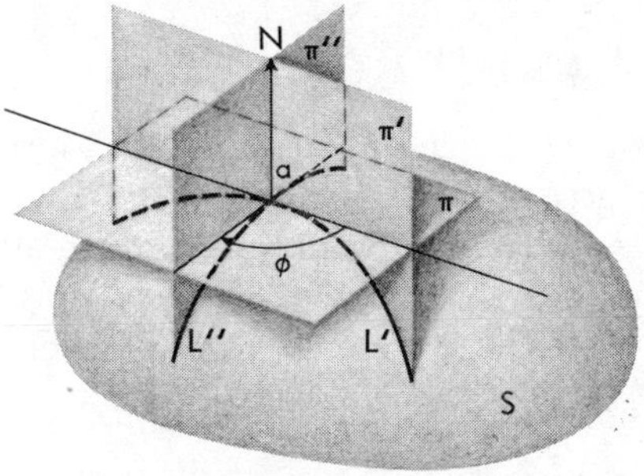

Fig. 7.5. Curvature of a surface S at the point a.

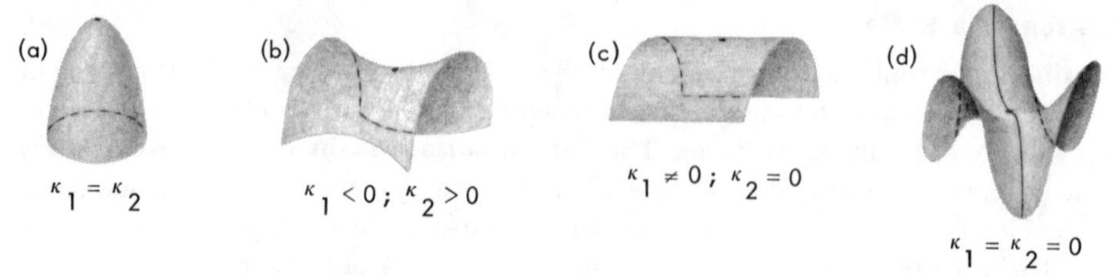

Fig. 7.6. Surfaces with different curvatures. (a) Elliptic paraboloid. (b) Hyperbolic paraboloid. (c) Parabolic cylinder. (d) Monkey saddle.

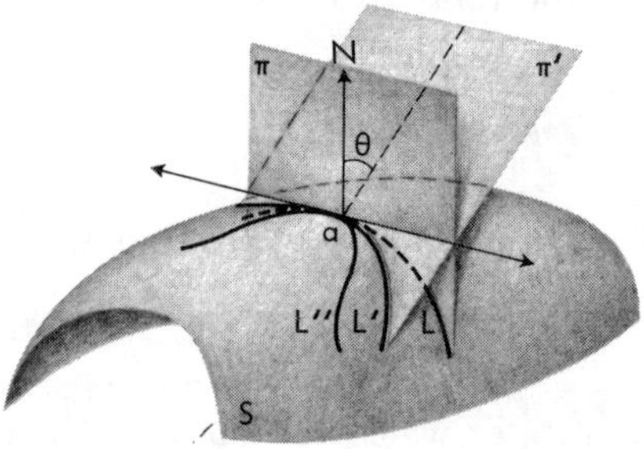

Fig. 7.7. Curvature of lines on a surface S at the point a.

section $\kappa(\phi)$ in terms of the angle ϕ gives a rather complete picture of the behavior of the surface in the vicinity of the point a.

Although a surface may be curved in many different ways, the dependence of the curvature κ on the angle ϕ is not arbitrary [17]. In fact, for the surfaces studied in differential geometry, a simple relationship establishes the connection between the curvatures of the normal sections passing through a given point in various directions.

At each point of a surface there exist two particular directions such that:

1) The two directions are mutually perpendicular.
2) The curvatures κ_1 and κ_2 of the normal sections in these directions are the smallest and largest values of the curvatures of all normal sections.*
3) The curvature $\kappa(\phi)$ of the normal section rotated from the section with curvature κ_1 by the angle ϕ is expressed by the formula

$$\kappa(\phi) = \kappa_1 \cos^2 \phi + \kappa_2 \sin^2 \phi. \tag{7.11}$$

Such directions are called the *principal directions* and the curvatures κ_1 and κ_2 are called the *principal curvatures* of the surface at the given point.

This theorem of Euler shows that, in spite of the diversity of surfaces, the form of the surfaces in the neighborhood of each point must be one of a very few completely defined types. Figure 7.6 portrays various types of surfaces and the signs of their principal curvatures at a point.

---

*For the case in which $\kappa_1 = \kappa_2$, the curvature of all normal sections is the same; as, for example, on a sphere.

If one of the section planes is inclined arbitrarily to the normal, we have the situation depicted in Fig. 7.7. The curvature of the normal section L is related simply to the curvature of L', provided that the normal plane π and the arbitrary plane π' have the same direction, i.e., that they intersect the tangent plane in the same straight line. According to a formula due to Meusnier [17],

$$\kappa_{L'} = |\kappa_L|/\cos\theta, \tag{7.12}$$

where θ is the angle between the normal N and the plane π'.

Moreover, the curvature of *any* curve L'' lying in the surface and having the plane π' as its osculating plane may be shown to be identical with the curvature of the curve L' formed by the intersection of π' with the surface. Thus the curvature of any curve in the surface is defined if we know the principal curvatures κ_1 and κ_2, the direction of its tangent, and the angle between its osculating plane and the normal to the surface.

As a consequence of these results, the curvature of a surface at a point is characterized by the two numbers κ_1 and κ_2. Their absolute values are equal to the curvatures of two mutually perpendicular normal sections, and their signs show the direction of the concavity of the respective normal sections with respect to a chosen direction on the normal.

7.4.4 Mean Curvature

Frequently, in studies of surfaces, one does not require the principal curvatures themselves, but certain quantities dependent on them; for example, the mean curvature or the total (Gaussian) curvature.

The mean curvature of a surface at a given point may be defined as the average of the principal curvatures

$$\mathscr{K}_m = \tfrac{1}{2}(\kappa_1 + \kappa_2)\ \text{mm}^{-1}. \tag{7.13}$$

It can be either negative, positive, or zero, depending on the relative magnitudes and signs of κ_1 and κ_2.

The mean curvature is useful in physical problems involving the pressure exerted by a thin film, its surface tension, and the surface of least area bounded by a given contour. The latter problem has been considered extensively in mathematical studies of minimal surfaces.

7.4.5 Total Curvature

The total (or Gaussian) curvature of a surface at a point is the product of the principal curvatures

$$\mathscr{K}_T = \kappa_1\kappa_2\ \text{mm}^{-2}. \tag{7.14}$$

The sign of the total curvature defines the character of the surface near the point under consideration. For $\mathscr{K}_T$ negative, the surface is like a saddle (Fig. 7.6b), that is, the surface lies partly on one side and partly on the other side of the tangent plane in the neighborhood of a point. Other examples of $\mathscr{K}_T$ negative are the inner surface of a torus, or the hyperboloid of revolution of one sheet.

For $\mathscr{K}_T$ positive, the surface at a point has the form of a bowl (Fig. 7.6a). A surface with positive total curvature at every point is the sphere or the ellipsoid. Surfaces of zero total curvature are cylinders, as are, in fact, all developable surfaces.

The absolute value of the total curvature, although more of an abstraction, gives the degree of curvature of the surface in general.

The most remarkable property of the total curvature pertains to a flexible but inextensible surface whose shape can be changed by bending or twisting, but not by stretching or tearing. Although the principal curvatures are changed, the value of $\mathscr{K}_T$ remains unchanged at every point. This fundamental result shows that two surfaces with different total curvatures are inherently distinct from each other. No matter how they are deformed (without stretching or tearing), they can never be superposed on each other. A well-known example of this is the impossibility of bending or twisting the surface of a sphere so as to make it lie on a plane.

The above discussion has dealt with certain basic concepts in the theory of surfaces. We have not gone into the details of this theory because our intent has been only to indicate the mathematical basis for a quantitative study of the shapes of lines and surfaces. Other subjects, such as intrinsic geometry, deformation of surfaces, geodesic lines, curvilinear coordinates, and vector notation are not reviewed here, but may prove useful to the student who wishes to pursue the quantitative specification of surfaces further.

Particularly important are the methods applicable to geometry "in the large" (as opposed to those methods mentioned above for geometry "in the small"), which considers the properties of an entire closed surface. For such an analysis, new mathematical tools are required, and these include the theory of differential equations, studies of nonregular functions, the calculus of variations, and the theory of convex bodies. Although the subject of geometry is still being developed, it has not yet reached the stage of development at which useful applications can be made readily.

7.5 MATHEMATICAL METHODS

In this section we shall summarize various approaches that have been, or can be, adapted to the mathematical expression of shape. Of necessity, the more precise methods frequently apply only to the simplest or most regular of figures or objects. Other procedures, such as the topological analyses, apply to structures of any degree of complexity. The difficulties in this case are largely experimental, however. The impact of rapidly developing computer and computer-graphic techniques promises to revolutionize the entire problem of shape description.

7.5.1 Simple Geometrical Figures

Exact representation of the shape of a planar figure frequently depends on an analytical expression for the outline of the figure. Equations for simple planar curves are given in Table 7.1. The general second-degree equation in two variables [22] from which these equations can be obtained is

$$Ax^2 + By^2 + 2Cxy + 2Dx + 2Ey + F = 0, \tag{7.15}$$

where the capital letters are constants, not all equal to zero.

TABLE 7.1 *Simple curves and their equations* [22]

$\frac{x^2}{a^2} + \frac{y^2}{a^2} - 1 = 0$		Circle
$\frac{x^2}{a^2} + \frac{y^2}{b^2} - 1 = 0$		Ellipse
$\frac{x^2}{a^2} + \frac{y^2}{b^2} + 1 = 0$		Imaginary ellipse
$\frac{x^2}{a^2} + \frac{y^2}{b^2} = 0$		Point (a pair of imaginary lines intersecting in a real point)
$\frac{x^2}{a^2} - \frac{y^2}{b^2} - 1 = 0$	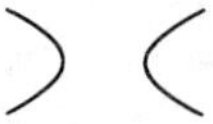	Hyperbola
$\frac{x^2}{a^2} - \frac{y^2}{b^2} = 0$		A pair of intersecting lines
$y^2 - 2px = 0$		Parabola
$x^2 - a^2 = 0$		A pair of parallel lines
$x^2 + a^2 = 0$		A pair of imaginary parallel lines
$x^2 = 0$		A pair of coincident straight lines

The constants a, b, and p are not all equal to zero.

Table 7.2 represents simple geometrical figures and their equations. The general second-degree equation in three variables that applies here is

$$A_1x^2 + A_2y^2 + A_3z^2 + 2B_1yz + 2B_2xz + 2B_3xy + 2C_1x + 2C_2y + 2C_3z + D = 0, \tag{7.16}$$

where the capital letters are constants, not all equal to zero. The specific equations (canonical forms) for the figures shown in Tables 7.1 and 7.2 can be derived from Eqs. (7.15) and (7.16) by appropriate rotations of the coordinate system.

Of course, real objects seldom have the simple shapes shown in the tables. The practical value of this compilation, and other collections of more complicated curves [23], is that shapes of real figures can often be approximated by these simpler curves.

7.5.2 Polar Plots

Figures with some degree of symmetry can be represented conveniently by polar equations using polar coordinates. Familiar curves are the cycloids, spirals, and roses based on logarithmic, trigonometric, and exponential functions. The first four curves of Fig. 7.8 are simple roses based on sine or cosine functions. The next three curves of Fig. 7.8 represent more complicated examples that closely approximate real objects [1].

Another technique for handling more complicated shapes is illustrated in the last curve of Fig. 7.8, in which one curve has been combined with another. Onto a

TABLE 7.2

Simple surfaces and their equations [22]

$\frac{x^2}{a^2}+\frac{y^2}{a^2}+\frac{z^2}{a^2}-1=0$		Sphere
$\frac{x^2}{a^2}+\frac{y^2}{b^2}+\frac{z^2}{c^2}-1=0$		Ellipsoid
$\frac{x^2}{a^2}+\frac{y^2}{b^2}+\frac{z^2}{c^2}+1=0$		Imaginary ellipsoid
$\frac{x^2}{a^2}+\frac{y^2}{b^2}-\frac{z^2}{c^2}-1=0$		Hyperboloid of one sheet
$\frac{x^2}{a^2}+\frac{y^2}{b^2}-\frac{z^2}{c^2}+1=0$		Hyperboloid of two sheets
$\frac{x^2}{a^2}+\frac{y^2}{b^2}-\frac{z^2}{c^2}=0$		Second-order cone
$\frac{x^2}{a^2}+\frac{y^2}{b^2}+\frac{z^2}{c^2}=0$		Imaginary second-order cone
$\frac{x^2}{a^2}+\frac{y^2}{b^2}-2cz=0$		Elliptic paraboloid
$\frac{x^2}{a^2}-\frac{y^2}{b^2}-2cz=0$		Hyperbolic paraboloid
$\frac{x^2}{a^2}+\frac{y^2}{b^2}-1=0$		Elliptic cylinder
$\frac{x^2}{a^2}+\frac{y^2}{b^2}+1=0$		Imaginary elliptic cylinder
$\frac{x^2}{a^2}+\frac{y^2}{b^2}=0$		A pair of intersecting imaginary planes
$\frac{x^2}{a^2}-\frac{y^2}{b^2}-1=0$		Hyperbolic cylinder
$\frac{x^2}{a^2}-\frac{y^2}{b^2}=0$		A pair of intersecting planes
$y^2-2px=0$		Parabolic cylinder
$x^2-a^2=0$	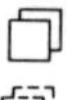	A pair of parallel planes
$x^2+a^2=0$		A pair of imaginary parallel planes
$x^2=0$		A pair of coincident planes

The constants a, b, c, and p are not all equal to zero.

simple outline curve we superpose secondary sine curves of lesser period and varying amplitude, after the fashion of a Fourier series.

7.5.3 Curve Fitting

In addition to the equations quoted above, which describe simple curves and surfaces exactly, there are approximation methods for describing less regular curves. Approximation functions may consist of algebraic polynomials or trigonometric polynomials, and may utilize interpolation formulas or the mean square method.

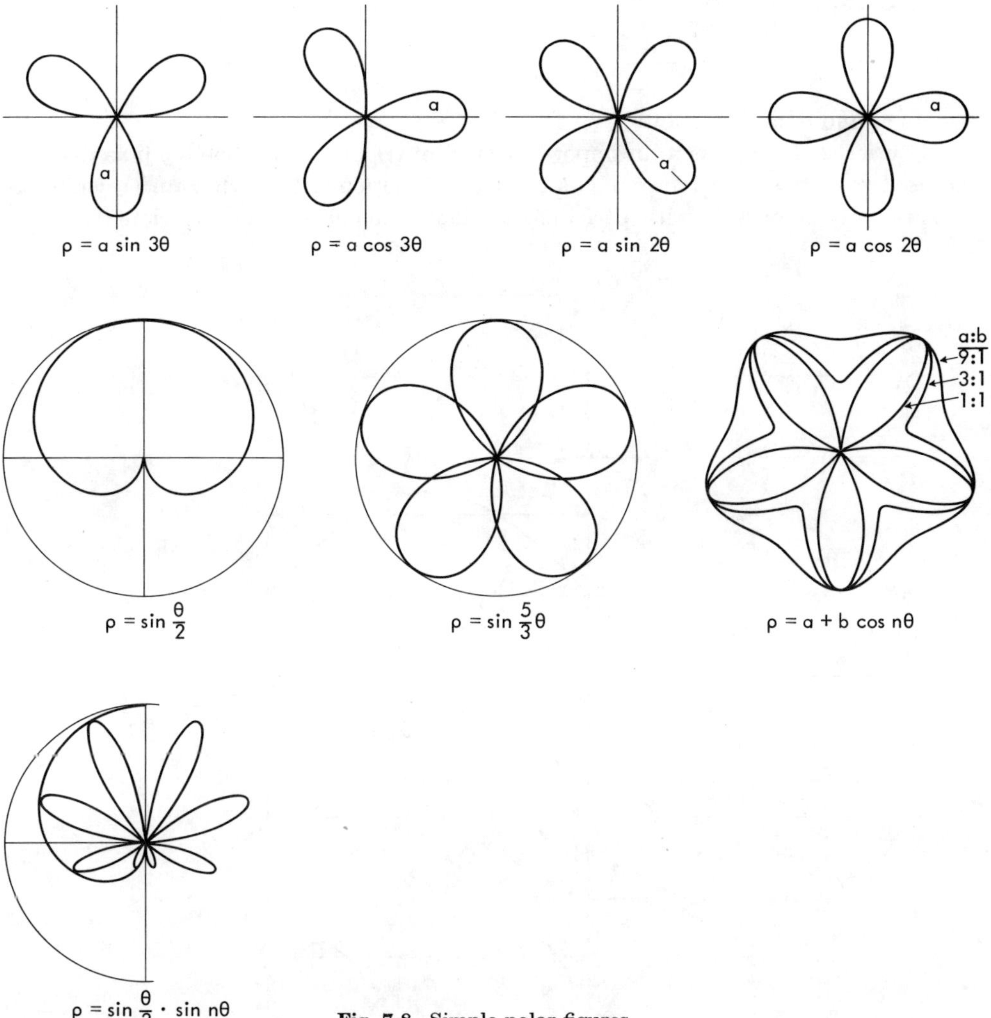

Fig. 7.8. Simple polar figures.

Taylor's formula [19] in one variable

$$f(x) \approx f(a) + f'(a)(x-a) + f''(a)(x-a)^2/2! + \cdots + f^{(n)}(a)(x-a)^n/n! \qquad (7.17)$$

is an example of an algebraic polynomial used to approximate a given function or graph. Also useful is Taylor's formula in two variables, $f(x, y)$, as an approximation function for surfaces.

Another important means of approximation consists of the trigonometric polynomials. An example of the latter is a function of the form

$$\begin{aligned} F_n(x) = a_0 + a_1 \cos x + b_1 \sin x + a_2 \cos 2x + b_2 \sin 2x + \cdots \\ + a_n \cos nx + b_n \sin nx, \end{aligned} \qquad (7.18)$$

or more concisely,

$$F_n(x) = a_0 + \sum_{k=1}^{n} (a_k \cos kx + b_k \sin kx), \tag{7.19}$$

where a_k and b_k are constants.

Fourier series represent an important class of trigonometric polynomials that can be used to represent a more or less arbitrary function. Although Fourier series are traditionally associated with periodic functions, their use is not restricted to peri-

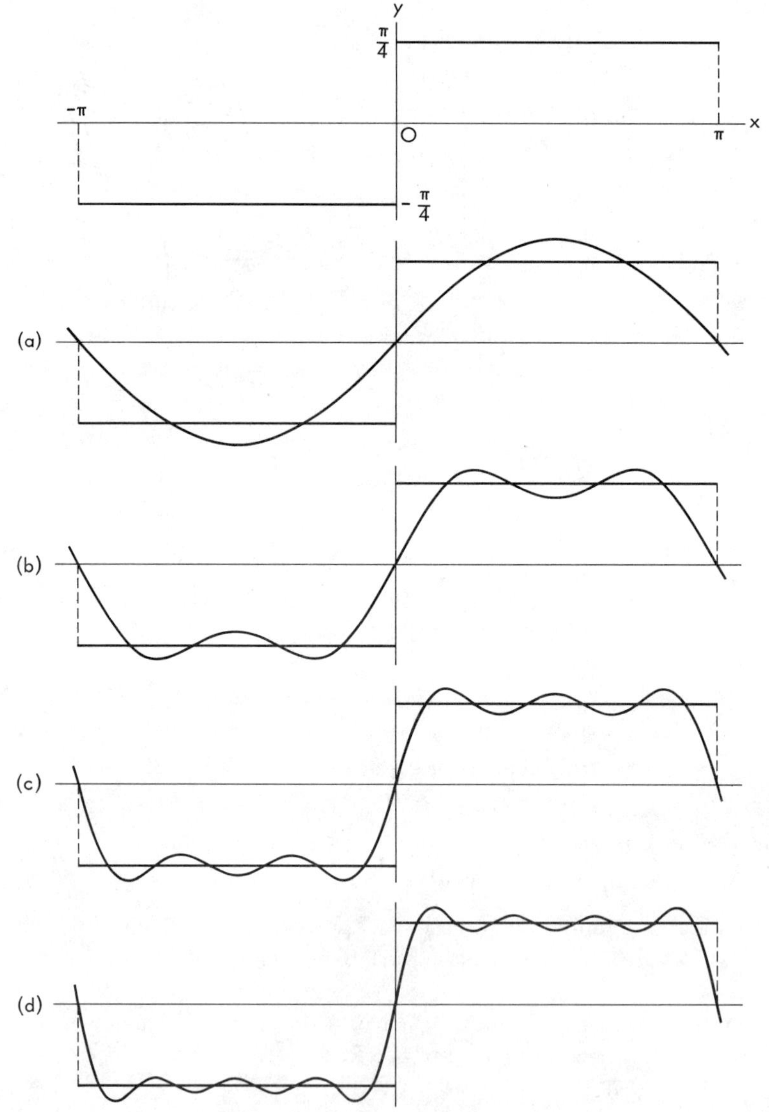

Fig. 7.9. Fourier series approximation to a square wave [24].

odic functions. If the function $f(t)$ is defined only over a finite range of t, for example, from $-\pi \leq t \leq +\pi$, then the Fourier series would represent the function within these limits.

Figure 7.9 illustrates how a trigonometric Fourier series approximates a square wave $f(x)$ with increasing number of terms [24]. The equations corresponding to the four plotted curves consist of partial sums of the Fourier series:

$$\left.\begin{aligned} &\text{a) } S_1(x) = \sin x \\ &\text{b) } S_3(x) = \sin x + \frac{\sin 3x}{3} \\ &\text{c) } S_5(x) = \sin x + \frac{\sin 3x}{3} + \frac{\sin 5x}{5} \\ &\text{d) } S_7(x) = \sin x + \frac{\sin 3x}{3} + \frac{\sin 5x}{5} + \frac{\sin 7x}{7} \end{aligned}\right\} \tag{7.20}$$

By a suitable choice of origin, one can expand the Fourier series either as a sine series or a cosine series. In general, of course, the series would contain both sine and cosine terms.

The subject of approximation functions is voluminous and can become very difficult. For further details, the reader is referred to standard mathematical texts [25]. These methods may become of increasing importance in the future, when they are coupled with automatic scanning devices, as well as computer curve-fitting techniques.

7.5.4 Transformations of Coordinates

In one of the most celebrated chapters in his book [1], D'Arcy Thompson makes startling comparisons of the shapes of living creatures by means of simple deformation of coordinates. Although his goal was more one of comparison than of precise definition, his method suggests a useful technique for expressing shape quantitatively.

As D'Arcy Thompson looked at it, the problem is closely akin to that of the cartographer who transfers identical data, in each small unit of area, to the corresponding area in some other projection [26, 27]. In our case, we might wish to transform a particular curve from rectilinear coordinates into curvilinear coordinates in order to attain the desired shape. For example, transformations into polar, elliptic, bipolar, or other curvilinear coordinates have been worked out for many practical applications [28, 29, 30]. The transformations can be performed analytically, provided one knows the equation of the curve, or graphically, if that is adequate for the requirement at hand.

A few simple examples [1] of transformations of planar curves will illustrate the idea involved here. Part (a) of Fig. 7.10 shows a circle inscribed in a square net in an orthogonal system of coordinates. When we alter or deform this network in various ways, the figure inscribed in the original net will be deformed in proportion to the deformation of the entire configuration. Part (b) shows how a linear contraction along the x-direction changes the circle into an ellipse. In mathematical language, for the

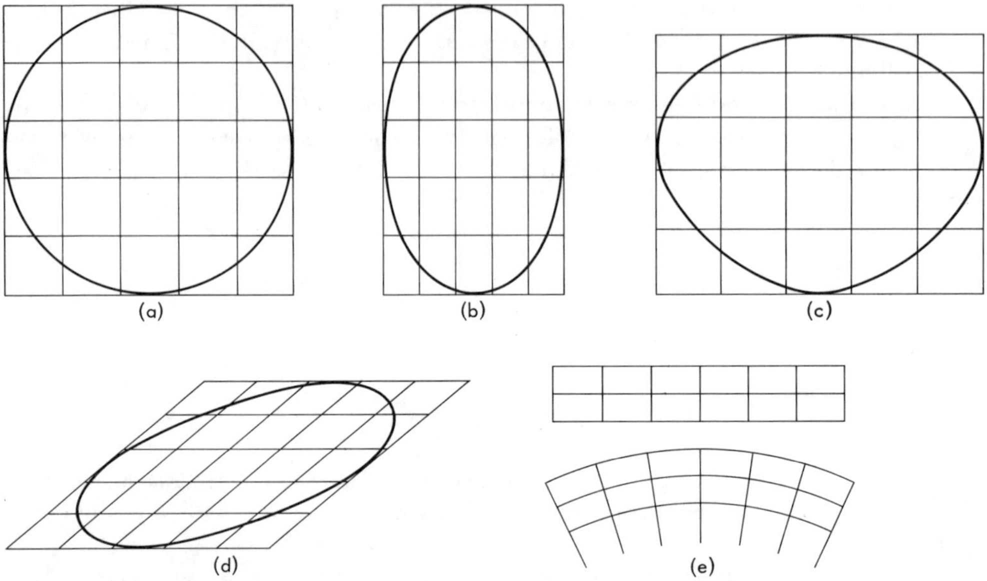

Fig. 7.10. Transformations of planar curves [1].

original x we have substituted Cx, and the equation of the original circle $x^2 + y^2 = a^2$ becomes that of the ellipse, $C^2x^2 + y^2 = a^2$.

If we deform the original net unequally, we obtain other shapes from the circle. For example, the ordinate can be spaced logarithmically, giving the shape shown in part (c) of Fig. 7.10. In this case, we have substituted e^y for y, giving $x^2 + e^{2y} = a^2$. It is obvious that both the abscissa and ordinate may be plotted logarithmically.

A third type of deformation is the simple shear, in which the rectangular coordinates become oblique, i.e., the axes are inclined to one another at a certain angle ω. The original net now appears as in part (d) of Fig. 7.10. The system may be described in terms of oblique axes, or may be referred to new rectangular coordinates α, β by the transpositions $x = \alpha - \beta \cot \omega$ and $y = \beta \operatorname{cosec} \omega$.

Another important class of deformation may be represented by the use of radial coordinates, in which one set of lines radiates outward from a point or "focus," while the other set is transformed into circular arcs cutting the radii orthogonally. Radial coordinates are especially applicable to cases in which no deformation occurs in a particular direction, such as the beam in simple flexure shown in part (e) of Fig. 7.10.

We may need spiral curves to define the shape of a figure. The example shown in part (a) of Fig. 7.11 represents a system of straight lines transformed into a conformal system of logarithmic spirals. The straight line $y - ax = b$ corresponds to the logarithmic spiral $\theta - c \log \rho = d$, where a, b, c, and d are constants.

Most of the transformations considered above (other than that of simple shear) are particular cases of a general transformation obtainable by the method of conjugate

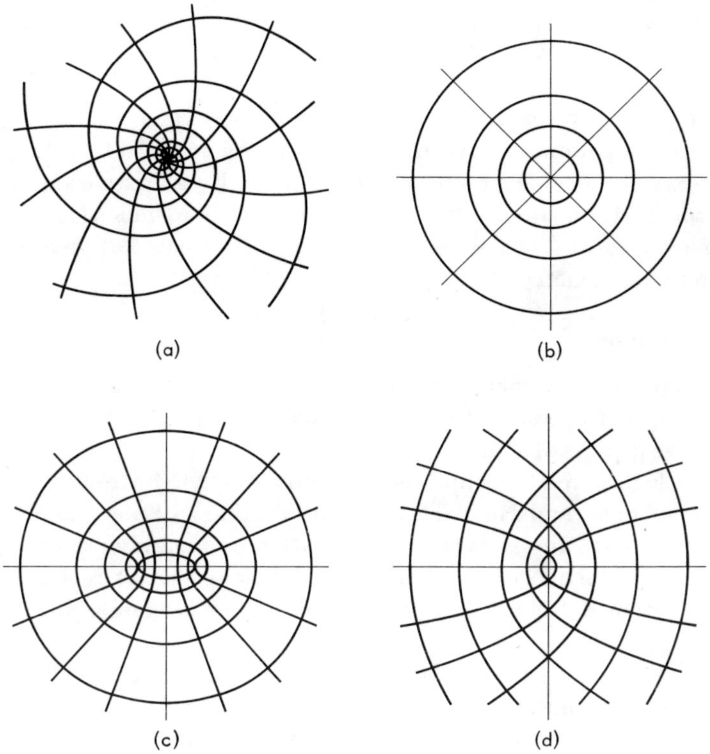

Fig. 7.11. Systems of coordinates: (a) Logarithmic spiral coordinates, (b) polar coordinates, (c) elliptic coordinates, (d) parabolic coordinates.

functions and equivalent to the projection of the original figure on a new plane. Examples of such transformations are a coaxial system in which the Cartesian coordinates are replaced by coaxial circles, or a confocal system in which they are replaced by confocal ellipses or parabolas. Some of these coordinate systems are illustrated in parts (b), (c), and (d) of Fig. 7.11.

Similar transformations may be found useful for characterizing surfaces. Morse and Feshbach [31] illustrate 12 types of three-dimensional coordinate systems in stereographic pairs of line drawings. These systems include elliptic and parabolic cylinder coordinates; conical, ellipsoidal and toroidal coordinates; and spherical, prolate, and oblate spheroidal coordinates. A full mathematical account of these separable coordinate systems is given in their book.

The method of curvilinear coordinates, introduced by Gauss, is a natural generalization of the idea of Cartesian coordinates in the plane [17]. Instead of expressing the surface by an equation of the form $z = f(x, y)$, we choose coordinates u and v on the surface itself. This device eliminates the inconvenience of having the coordinates of a point in the surface change when the surface is deformed. The space coordinates x, y, and z of a point are in every case functions of two numbers u and v, which are called its *curvilinear coordinates*. The functions can be expressed by the vector $\mathbf{r} = \mathbf{r}(u, v)$, issuing from some fixed origin to the points in the surface and called the *radius vector* of the surface.

To define a curve lying in a given surface, it is convenient to express the coordinates u, v as functions of one parameter t; then the radius vector to a point moving along the curve is expressed as a composite function $\mathbf{r}[u(t), v(t)]$. By an extension of this approach we can express, for example, the length of a curve on the surface or the formula for area in general form. The details of this subject are treated at greater length in books on differential geometry [18, 32].

7.5.5 Topological Relationships

There are some characteristics of a system of grains or cells that are not dependent on the size or shape of the cells. Nevertheless the information to be gained does shed some light on the qualitative aspects of shape.

We refer to those topological relationships for polygons and polyhedrons that link the *numbers* of corners, edges, and faces. These equations, due to Euler, do not depend on magnitudes (i.e., length, area, or volume), but rather on *numerical* quantities, and are independent of distortion of any kind, provided connectivity is maintained. Thus parameters obtained from topological considerations alone are shape insensitive.

This situation appears to create a paradox. How, on the one hand, can we invoke topology to characterize shape, when the equations are independent of shape?

The answer lies, to some extent, in the awareness that topology is not used alone. Merely stating that the average cell in an aggregate controlled by surface tension considerations has 5.1043 edges per face, 13.394 faces and 22.789 corners does not fix a shape. However, if we combine shape-sensitive quantities (such as mean diameter, surface area, volume-to-surface ratio, etc.) with topological results, then useful shape-sensitive parameters are possible.

Let us consider a space-filling aggregate of polyhedral cells or grains. Their surfaces consist of a number of polygonal faces bounded by edges where adjacent faces intersect. Edges terminate at corners, or points, where several edges intersect. For an aggregate of such polyhedrons, Euler's law [6a] may be stated in the simple but general form

$$\boxed{+C - E + F - B = 1}\,, \tag{7.21}$$

which relates the number of point corners C, lineal edges E, polygonal faces F, and polyhedral bodies B.* Note especially that these quantities represent the actual, observable features that could be counted in a transparent aggregate of grains.

*A useful device for remembering the various forms of Euler's law is the following. We start with $C - E + F - B = 1$, which applies to an aggregate of polyhedrons. For one polyhedron, $B = 1$, and the basic equation becomes $C - E + F = 2$. If we are considering a 2-dimensional array of polygons, $B = 0$, and we have $C - E + F = 1$. Finally, for isolated polygons, $F = 1$, and the equation reduces to $C - E = 0$.

TABLE 7.3

Topological relationships for polyhedrons

Feature	Separate polyhedrons*		Aggregate of polyhedrons†		Averaged aggregate values		
					Number per cell		Number per unit volume
	$C - E + F = 2$		$C - E + F - B = 1$		Exact	Approximate (large number)	
Number of corners, C	$C = \frac{2}{3}E$	$C = \frac{4\bar{n}}{6 - \bar{n}} = \frac{\sum nF_n}{3} = \frac{\bar{n}F}{3}$	$C = \frac{1}{2}E$	$C = \frac{\sum nF_n}{6} = \frac{\bar{n}F}{6}$	$\bar{C} = \frac{4C}{B} =$	$\frac{4\bar{n}}{(6 - \bar{n}) - 6/F} \approx \frac{4\bar{n}}{6 - \bar{n}}$	$C_V = \frac{1}{4}\bar{C}N_V \approx \frac{\bar{n}N_V}{6 - \bar{n}}$
Number of edges, E	$E = 3F - 6$	$E = \frac{6\bar{n}}{6 - \bar{n}} = \frac{\sum nF_n}{2} = \frac{\bar{n}F}{2}$	$E = 2(F - B - 1)$	$E = \frac{\sum nF_n}{3} = \frac{\bar{n}F}{3}$	$\bar{E} = \frac{3E}{B} =$	$\frac{6\bar{n}}{(6 - \bar{n}) - 6/F} \approx \frac{6\bar{n}}{6 - \bar{n}}$	$E_V = \frac{1}{3}\bar{E}N_V \approx \frac{2\bar{n}N_V}{6 - \bar{n}}$
Number of polygonal faces, F	$F = \frac{C + 4}{2}$	$F = \frac{12}{6 - \bar{n}} = \sum F_n$	$F = C + E + 1$	$F = \frac{\sum nF_n}{\bar{n}}$	$\bar{F} = \frac{2F}{B} =$	$\frac{12}{(6 - \bar{n}) - \bar{n}/C} \approx \frac{12}{6 - \bar{n}}$	$F_V = \frac{1}{2}\bar{F}N_V \approx \frac{6N_V}{6 - \bar{n}}$
Number of polyhedral bodies, B	$B = 1$	—	$B = F - C - 1$	$B = \frac{\sum (6 - n)F_n - 6}{6} = \frac{(6 - \bar{n})F - 6}{6}$		—	$B_V = \frac{B}{V} = N_V$
	$3C = 2E = \bar{n}F$		$2C = E = \bar{n}F/3$			$F - \bar{C} \approx 1$	$C_V - E_V + F_V \approx B_V$

*Three faces meet at a corner. Two faces meet at an edge.
†Four cells meet at a corner. Three cells meet at an edge. Two cells meet at a face.

(Later on we shall calculate fictitious numbers, based on the observed values that would apply if the aggregate were exploded into separate cells.)

The law expressed by Eq. (7.21) is independent of curvature of edges and faces, so we can apply it to situations controlled by surface tension requirements. If we inject the ordinary conditions that each corner is shared by 4 cells, each edge is shared by 3 cells, and each face is shared by 2 cells, we can obtain additional relationships.

Table 7.3 lists important relationships for separate polyhedrons and for aggregates of polyhedrons. The important difference between these two cases is the relationships for number of corners to edges, $C = \frac{2}{3}E$ and $C = \frac{1}{2}E$, respectively. The first equation (for $B = 1$) results from the conditions that 3 edges meet at every corner, and that each edge contains 2 corners. The second equation (for B large) stems from the requirement that 4 edges meet at every corner, and that each edge contains 2 corners.

Note that the equations in Table 7.3 are specialized according to the assumptions stated at the bottom of the table. The second and fourth columns of equations are useful for polyhedrons with mixed types of faces. The applicable equations contain an important parameter, n, which is the number of edges around each polygonal face, regardless of sharing; i.e., whether the edge was counted previously with an adjacent face. The average value, $\bar{n}$, is defined as

$$\bar{n} = (1/F) \sum^{n} nF_n, \tag{7.22}$$

where F_n is the number of faces with n edges. Thus the truncated octahedron, with six 4-sided and eight 6-sided faces, has a mean number of edges per face of

$$\bar{n} = \frac{4 \times 6 + 6 \times 8}{14} = \frac{36}{7} \approx 5.143.$$

The parameter $\bar{n}$ is also useful in defining the interrelationship of terms for separate polyhedrons

$$3C = 2E = \bar{n}F \tag{7.23}$$

and for an aggregate of polyhedrons

$$2C = E = \bar{n}F/3. \tag{7.24}$$

The last two columns in Table 7.3 list equations for the average number of corners, edges, or faces, per cell or per unit volume. As indicated for $\bar{n}$ above, the terms $\bar{C}$, $\bar{E}$, and $\bar{F}$ are defined without regard to duplication in counting corners, edges, or faces. For example, $\bar{C}$, $\bar{E}$, and $\bar{F}$ for an aggregate of truncated octahedrons equals 24, 36, and 14, respectively. It is interesting to note that the three approximate equations (for $\bar{C}$, $\bar{E}$, and $\bar{F}$) have the same forms as the exact equations (for C, E, and F) for separate polyhedrons.

When we investigate the shapes of three-dimensional grains or cells, we are primarily interested in shape parameters that refer to the three-dimensional properties. Except for the rare case in which the grains can be separated and examined individually, we are mostly restricted to observations made on the plane of polish through the grains. Unfortunately, the topological data obtained from random two-

TABLE 7.4

Topological relationships for polygons

Feature	Separate polygons		Array of polygons*		Finite array of polygons	
	$C - E = 0$		$C - E + F = 1$		Boundary edges with corners	Boundary edges rounded off
Number of corners, C	$C = E$	$C = \sum nF_n$	$C = \frac{2}{3}E$	$C = \frac{\sum nF_n}{3} = \frac{\bar{n}F}{3}$	$C = \frac{\sum nF_n}{3} + \frac{2C_1' + C_2'}{3}$	$C = \frac{\sum nF_n}{3} + \frac{E_b}{3}$
Number of edges, E	$E = n$ (one polygon)	$E = \sum nF_n$ (many separate polygons)	$E = 3F - 3$	$E = \frac{\sum nF_n}{2} = \frac{\bar{n}F}{2}$	$E = \frac{\sum nF_n}{2} + \frac{C_1' + C_2'}{2}$	$E = \frac{\sum nF_n}{2} + \frac{E_b}{2}$
Number of polygonal faces, F	$F = 1$	$F = \sum F_n$	$F = \frac{C + 2}{2}$	$F = \frac{\sum nF_n}{\bar{n}}$	$\sum (6 - n)F_n = E_b + 6 - 2C_2'$	$\sum (6 - n)F_n = E_b + 6$
	$C = E = n$		$3C = 2E = \bar{n}F$		$E_b = C_1' + C_2'$	$E_b = C_2'$

*Three polygons meet at a corner. Two polygons meet at an edge.

dimensional sections do not refer directly to the three-dimensional grains [110]. For example, when we see a triple point in the two-dimensional section, we know that this originates from an edge in three-dimensional space. Similarly, an edge lying in the section comes from the intersection of a grain face. This point should be kept clearly in mind in the ensuing discussion.

The two-dimensional form of Euler's law for an array of polygons is

$$\boxed{C - E + F = 1}, \tag{7.25}$$

where the terms are as before. As a consequence of the "rubber geometry" of topology, the polygonal network does not need to lie on a flat surface, but could be, for example, the map on a globe or on a polyhedron, or it could be the irregular topography of an intergranular fracture surface.

Table 7.4 summarizes the more important two-dimensional topological relations for isolated polygons and arrays of polygons. The equations incorporate the usual observations that three polygons meet at a corner and two polygons meet at an edge. The second and fourth columns of equations are useful when one is dealing with polygons of mixed type.

The two basic equations that relate the terms in Euler's law are

$$C = E = n, \tag{7.26}$$

for separate polygons, and

$$3C = 2E = \bar{n}F, \tag{7.27}$$

for a large array of polygons. Note that Eq. (7.23) is the same as Eq. (7.27).

Special problems of a more practical nature arise when finite numbers of grains in an aggregate are examined on the plane of polish. The boundary polygons do not contain their full complement of edges and corners, and so allowances must be made for these deficiencies. Two major cases exist: (1) When the boundary edges are rounded off, as they are in a group of bubbles, (2) when the boundary edges retain their corners (see Fig. 7.12). Since normal sharing does not exist at the boundary, correction terms are required, and are given in the equations in the last two columns of Table 7.4. Three new terms are introduced. C_1' and C_2' refer to the number of incomplete boundary corners abutting one and two polygons, respectively, and E_b is the number of unshared edges at the boundary. Thus, with C_1' and C_2' corners, $E_b = C_1' + C_2'$; while without C_1' corners (rounded-off edges), $E_b = C_2'$.

An example of the application of the equations to the case in which boundary edges have corners is given in Fig. 7.12. The unshared boundary edges are outlined; C_1' corners and C_2' corners are indicated by solid and open circles, respectively. Although the polygons within the area selected conform to the equation in the figure, the grain sections do not obey the criterion $C_1' - C_2' = 6$, mentioned by Smith [6b], which is based on an average value for interior polygons of $\bar{n} = 6$. The figure shows, however, that all grains have not completely attained their equilibrium shapes, and $\bar{n} = 5.7$.

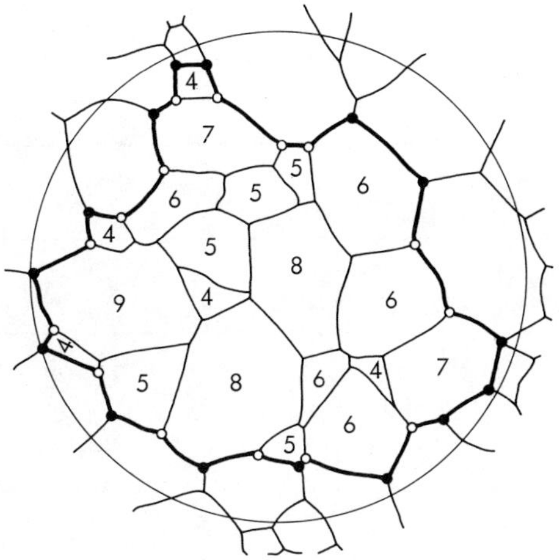

— $E_b = 30$

• $C_1' = 15$

○ $C_2' = 15$

n	4	5	6	7	8	9
F_n	5	5	5	2	2	1
6 − n	2	1	0	−1	−2	−3
$F_n(6-n)$	10	5	0	−2	−4	−3

$\sum F_n = 20$

$$\sum (6-n)F_n = E_b - 2C_1' + 6$$

$$6 = 30 - 30 + 6$$

Fig. 7.12. Topological analysis of typical section [8] through metal grains.

Further details on the various Euler equations for two- and three-dimensional structures may be obtained from the papers by Smith [6], Coxeter [33], and Meijering [34]. Experimental applications of these equations will be discussed in Section 7.6.

7.5.6 Curvature

In this section we shall develop in greater detail the subject of curvature, especially as it pertains to the geometrical relationships between a three-dimensional system of curved surfaces and the corresponding two-dimensional traces obtained by sectioning through the surfaces. Basic terms defined previously include the curvature κ, mean curvature $\mathscr{K}_m$, and total curvature $\mathscr{K}_T$, while important equations for the curvature of surfaces are given in Eqs. (7.11) and (7.12).

Four new terms [35, 36] that define quantities required in subsequent derivations are introduced in Table 7.5. They are denoted by k and $\bar{k}$, K_m and $\bar{K}_m$. Note that the lower-case k and κ apply to planar curves and the capital K and $\mathscr{K}$ refer to surfaces.

The general relationship of the new terms to the curvature or mean curvature is evident in Table 7.5. Their explicit analytical form can be worked out with the aid of the supplementary equations in the last column. For example, it is seen that

$$k = \int \kappa \, dl = \frac{y''}{1 + (y')^2} \, dx, \tag{7.28}$$

where y' and y'' are the first and second derivatives of y with respect to x. Unless the equation of the curve or surface is available, however, direct evaluation of the curvature parameters in Table 7.5 is not always possible.

There are certain specific cases in which numerical results can be obtained without specifying the equation of the curve. For planar curves, either individually or on the

TABLE 7.5

Interrelationships and definitions of curvature parameters

Planar Curves		
Curvature at a point: $\kappa = 1/r$	mm^{-1}	$\kappa = y''[1 + (y')^2]^{-3/2}$
Local curvature: $\kappa = d\theta/dl$	mm^{-1}	$dl = [1 + (y')^2]^{1/2}\,dx$
Integral curvature along a curve: $k = \int \kappa\,dl$	mm^0	
Average curvature along a curve: $\bar{k} = \int \kappa\,dl \Big/ \int dl$	mm^{-1}	$L = \int dl$
Curved Surfaces		
Mean curvature: $\mathscr{K}_m = \frac{1}{2}(\kappa_1 + \kappa_2)$	mm^{-1}	
Total curvature: $\mathscr{K}_T = \kappa_1\kappa_2$	mm^{-2}	
Integral mean curvature over a surface: $K_m = \iint \mathscr{K}_m\,ds$	mm	$ds = \left[1 + \left(\frac{\partial z}{\partial x}\right)^2 + \left(\frac{\partial z}{\partial y}\right)^2\right]^{1/2} dy\,dx$
Average mean curvature over a surface: $\bar{K}_m = \iint \mathscr{K}_m\,ds \Big/ \iint ds$	mm^{-1}	$S = \iint ds$

average, we can make a convenient classification into (1) open curves, and (2) closed curves. A few basic relationships will be developed for these two cases [35, 36, 111].

Open curves. For an open curve of length L, we substitute $\kappa = d\theta/dl$ in $\bar{k} = (1/L)\int \kappa\,dl$, giving the average curvature

$$\bar{k} = \Delta\theta/L\ \text{mm}^{-1}, \tag{7.29}$$

where $\Delta\theta$ is the net change in direction of tangents at the ends of the curve (see Fig. 7.13a). It should be noted that Eq. (7.29) is applicable in the limit of sharp corners, even though the curvature there is infinite, that is, $\int \kappa\,dl$ remains integrable and equal

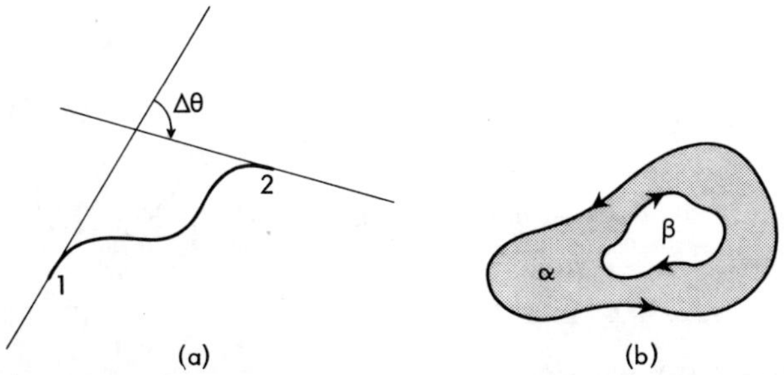

Fig. 7.13. Conventions adopted in measuring angles and signs of two-dimensional curves. (a) Net angle $\Delta\theta$ of an open curve. (b) Closed curves assigned positive and negative angles. α-phase: $\Delta\theta = +2\pi$; β-phase: $\Delta\theta = -2\pi$.

to the change in angle. For many open curves

$$\boxed{\bar{k} = \Delta\theta_A / L_A} \text{ mm}^{-1}, \tag{7.30}$$

where the ratio of $\Delta\theta_A$ to L_A represents the average curvature of all curves within a unit test area.

DeHoff [36, 111] suggests a way to evaluate $\Delta\theta_A$ by use of the tangent count, introduced by Rhines [16]. If a test line is swept across the planar system of open (or closed) curves, tangents will be formed with some curves. The total number of tangents, divided by the total area swept by test lines from all directions, is designated T_A. This number includes tangents with both convex and concave elements of arc.

Given that N is the number of test lines swept across an area $A = L^2$, the total area traversed by all test lines is NL^2. Moreover, the fraction of test lines that form a tangent with an element of arc dl is equal to $d\theta/\pi$. Thus the number of tangents formed with the arc dl per unit area of structure sampled is

$$dT_A = N\frac{d\theta}{\pi}(1/NL^2) = d\theta/\pi A.$$

For a curve of finite length, we find that the *total* number of tangents formed is

$$T_A = \int \frac{d\theta}{\pi A} = \frac{\theta}{\pi A} = \frac{\theta_A}{\pi} \text{ mm}^{-2}, \tag{7.31}$$

where θ_A is the *total* angle subtended by all arcs in a unit area, regardless of convexity or concavity.

In order to get the net change in angle, we need to distinguish between segments of convex arc (denoted by $+$) and concave arc (denoted by $-$). A separate tangent count must be made for each type. For all convex arcs in the sample

$$T_{A+} = \theta_{A+}/\pi, \tag{7.32}$$

while for all concave arcs,

$$T_{A-} = \theta_{A-}/\pi. \tag{7.33}$$

Combining Eqs. (7.32) and (7.33) gives the desired expression for the net change in angle, $(\theta_{A+}) - (\theta_{A-})$, or

$$\boxed{\Delta\theta_A = \pi \Delta T_A} \text{ mm}^{-2}, \tag{7.34}$$

where $\Delta T_A = T_{A+} - T_{A-}$.

The average curvature is now obtained by substituting Eq. (7.34) in Eq. (7.30), giving

$$\bar{k} = \pi \Delta T_A / L_A \text{ mm}^{-1}. \tag{7.35}$$

If we evaluate L_A by means of Eq. (2.49), we get

$$\boxed{\bar{k} = 2\Delta T_A / P_L} \text{ mm}^{-1}, \tag{7.36}$$

where ΔT_A is the net number of tangents (convex minus concave) per unit area, and P_L is the number of intersections with the system of open curves per unit length of a randomly placed test line. Note that the latter test line is not the same as the moving or "swept" test line required for the tangent count.

It is sometimes desirable to express the average curvatures of convex and concave arcs separately. As indicated above, we can write

$$\bar{k}_+ = 2T_{A+}/P_{L+} \tag{7.37}$$

for convex arcs and

$$\bar{k}_- = 2T_{A-}/P_{L-} \tag{7.38}$$

for concave arcs. These equations are important because they give the local curvature at specific locations as opposed to an average curvature for the entire system of curves.

It should be mentioned that the above equations require particularly laborious experimental measurements. Not only must each concave and convex segment be identified, but their lengths must also be marked off. Moreover, it is not particularly clear how to distinguish the concave and convex sides of a planar open curve. Struik [18] discusses means for assigning a sense to oriented open curves based on a counter-clockwise direction from the OX-axis to the OY-axis (see his Sections 1.2 and 1.4). Fortunately, these difficulties do not exist for closed curves, such as the traces found separating one phase from another in two-phase materials.

Closed curves. For a closed curve in a plane, $\Delta\theta = \pm 2\pi$, regardless of the extent of concavity or convexity of the loop. Adopting the convention that $\Delta\theta$ is positive when the loop encloses an area of interest (such as a particle of α-phase surrounded by the matrix, see Fig. 7.13b), the average positive curvature is given by

$$\bar{k}_+ = 2\pi / L \text{ mm}^{-1}. \tag{7.39}$$

For the closed loops around many particles,

$$\bar{k}_+ = 2\pi N_{A+} / L_A \text{ mm}^{-1} \tag{7.40}$$

or

$$\bar{k}_+ = 4 N_{A+} / P_L \text{ mm}^{-1}, \tag{7.41}$$

where P_L has been defined above and N_{A+} is the number of solid particles intercepted per unit area of the sectioning plane. Since $2N_L = P_L$, Eq. (7.41) can be written as

$$\boxed{\bar{k}_+ = 2N_{A+}/N_L} \text{ mm}^{-1} \tag{7.42}$$

where N_L is the number of intersections with particles per unit length of test lines.

In case of voids or other phases entrapped within the particles of α-phase, or when α-phase surrounds the matrix phase, the negative term is invoked in $\Delta\theta = \pm 2\pi$. Then the last four equations are written with a negative sign, and the average negative curvature is denoted by $\bar{k}_-$.

When both positive and negative loops occur in the same structure, the average curvature is expressed by

$$\bar{k} = 2\pi(N_{A+} - N_{A-})/L_A \text{ mm}^{-1} \tag{7.43}$$

where N_{A+} represents the number of loops per unit area with positive curvature (e.g., particles surrounded by matrix) and N_{A-} stands for the number of loops per unit area with negative curvature (e.g., holes within particles). Letting $\Delta N_A = (N_{A+}) - (N_{A-})$, we have

$$\bar{k} = 2\pi\Delta N_A/L_A \text{ mm}^{-1} \tag{7.44a}$$

$$\bar{k} = 4\Delta N_A/P_L \text{ mm}^{-1} \tag{7.44b}$$

$$\bar{k} = 2\Delta N_A/N_L \text{ mm}^{-1} \tag{7.44c}$$

for microstructures consisting of both convex and concave loops. An interesting consequence arises in the case of solid particles with irregular outlines. By comparing Eqs. (7.36) and (7.41), we obtain $\Delta T_A = 2N_{A+}$, which requires, on the average, that the number of convexities exceed the number of concavities by two, per particle. Frequently this condition is not found in individual particles, so care must be taken to ensure adequate sampling.

Cahn [35] also treats the case of clumped particles. The entire clump can be considered as one, if desired, or as separate particles, provided that their interfaces are resolvable under the microscope. The curvature is calculated much as discussed above, except that allowance must be made for sharing of interfaces if particles are considered separately. The principles involved have been set forth in Section 4.5.

Curved surfaces. When curved surfaces are cut by a sectioning plane, the traces form planar curves on the section. We consider three distinct cases of planar curves, based on the position of the sectioning plane:

1) A plane that passes through both the normal to the surface and a principal direction (see π', Fig. 7.5).
2) A plane that passes through the normal to the surface and is rotated by the angle ϕ from a principal direction (see π'', Fig. 7.5).
3) A plane that is rotated by an angle ϕ from a principal direction and tilted from the normal by the angle θ (see π', Fig. 7.7).

The curvature at a point on the planar curve (trace) may be expressed for these three cases as:

1) $\kappa = \kappa_1$,
2) $\kappa(\phi) = \kappa_1 \cos^2 \phi + \kappa_2 \sin^2 \phi$ (Eq. 7.11),
3) $\kappa(\phi, \theta) = (\kappa_1 \cos^2 \phi + \kappa_2 \sin^2 \phi)/\cos \theta$ (Eq. 7.12),

respectively.

We wish to relate the average mean curvature of a system of curved surfaces to the average curvature of their traces seen on random sectioning planes through the surfaces. To do this, we express the average curvature $\bar{k}$ in terms of the function $\kappa(\phi, \theta)$, averaged over the total surface area, and weighted according to the probabilities of the random sectioning plane making intersections at arbitrary angles of rotation and tilt. Thus [35, 111]

$$\bar{k} = \frac{1}{S} \iiiint \kappa(\phi, \theta) \mathscr{P}(\phi)\, d\phi \mathscr{P}(\theta)\, d\theta\, ds, \tag{7.45}$$

where $\mathscr{P}(\phi)\, d\phi$ and $\mathscr{P}(\theta)\, d\theta$ are the probabilities that the sectioning plane lies at angles between ϕ and $\phi + d\phi$, and θ and $\theta + d\theta$, respectively.

For a range of ϕ between $-\pi/2$ and $+\pi/2$, $\mathscr{P}(\phi) = 1/\pi$, a constant. For θ ranging from 0 to $\pi/2$, $\mathscr{P}(\theta) = (4/\pi) \cos^2 \theta$. One of these $\cos \theta$ terms originates in the probability that a surface element lies at an angle θ to the random sectioning plane, and the other $\cos \theta$ factor relates to the probability that the surface element is sectioned by the random plane.

Substituting into Eq. (7.45), from above, we obtain

$$\begin{aligned} \bar{k} &= \frac{1}{S} \iiiint \left(\frac{\kappa_1 \cos^2 \phi + \kappa_2 \sin^2 \phi}{\cos \theta}\right) \frac{d\phi}{\pi} \left(\frac{4}{\pi} \cos^2 \theta\, d\theta\right) ds \\ &= \frac{1}{S} \iiiint (\kappa_1 \cos^2 \phi + \kappa_2 \sin^2 \phi) \left(\frac{4 \cos \theta}{\pi^2}\right) d\phi\, d\theta\, ds \\ &= \frac{2}{\pi S} \iint (\kappa_1 + \kappa_2)\, ds = \frac{4}{\pi} \bar{K}_m. \end{aligned} \tag{7.46}$$

Thus the average mean curvature of the surfaces is given by

$$\boxed{\bar{K}_m = \frac{\pi}{4} \bar{k}} \quad \text{mm}^{-1}. \tag{7.47}$$

This important result was obtained independently by both Cahn [35] and DeHoff [36, 111], using somewhat different procedures. Note that, on the average, the curvature of the traces is about 27% greater than the average curvature of the surfaces.

By substituting the appropriate equation for $\bar{k}$ from Eqs. (7.29) through (7.44), we can evaluate $\bar{K}_m$. An important relation results when we insert Eq. (7.42) in Eq. (7.47), giving

$$\boxed{\bar{K}_m = \frac{\pi N_A}{2 N_L}} \quad \text{mm}^{-1} \tag{7.48}$$

for closed loops with positive curvature only. Thus $\bar{K}_m$ is obtained by simple counting measurements on the plane of polish (N_A and N_L) that are independent of particle concavity or convexity. It should be recalled, however, that N_L is the number of *interceptions* with particles per unit length, and not the number of *particles* intercepted per unit length. An interesting observation is that $\bar{K}_m$ in Eq. (7.48) is the inverse of r in the expression given for uniform spheres of radius r in Table 4.1: $r = 2N_L/\pi N_A$.

A useful relationship can be shown between the average curvature and the mean tangent diameter of convex particles [35, 36]. To Eq. (7.48), we bring $N_A = N_V\bar{D}$ (see Chapter 5) and $N_L = S_V/4$ (see Table 2.2), giving

$$\bar{K}_m = 2\pi\bar{D}/\bar{S}\ \text{mm}^{-1}, \tag{7.49}$$

where $\bar{S} = S_V/N_V$. This equation offers one way of obtaining $\bar{D}$, provided $\bar{S}$ can be determined. Further manipulation of Eq. (7.49) with the relationships tabulated in Fig. 6.6 leads to interesting results; for example,

$$\bar{K}_m = \pi/2\bar{d}$$

and

$$\bar{K}_m = \pi^2/2\bar{L}_p,$$

where $\bar{d}$ is the mean tangent diameter of the planar loops, and $\bar{L}_p$ is the mean perimeter length of the particle loops.

An elegant and general treatment of quantitative characterization of microstructure has been developed independently by Matheron [118, 119] and others [120, 121, 122] in the French school of stereologists. In addition to such basic quantities as volume fraction, specific surface area, spacings, and numbers per unit volume, they provide a basis for determining the curvature of interfaces from measurements on two-dimensional sections. Even though the constituents have complicated shapes, one can obtain the distribution function of the radii of curvature—positive and negative—the mean values, and the degree of angularity, roundness, and sphericity.

7.5.7 Numerical Specification of Form

In a particularly noteworthy paper, Meltzer, Searle, and Brown [117] outline a procedure for defining the shape of two-dimensional figures.

In their method, a two-dimensional figure is represented by a function taking the values 1 and 0. This function is analyzed into a sum of weighted basic pattern functions that are particular examples of the class of orthonormal functions. Generalization to three-dimensional objects is stated to be straightforward. Moreover, it is possible to devise numerical measures of qualitative features of the pattern such as degree of symmetry or degree of fine structure.

Figure 7.14 illustrates the essential steps of this method. Figure 7.14(a) depicts a typical two-dimensional figure; Fig. 7.14(b) shows examples of the basic pattern (Walsh) functions; and Fig. 7.14(c) gives the histograms representing the particular shape shown. It can be seen that the peaks in the histograms become more accentuated for the more palmate figures. The simplicity of the experimental method should permit widespread use, as well as adaptation to automated equipment.

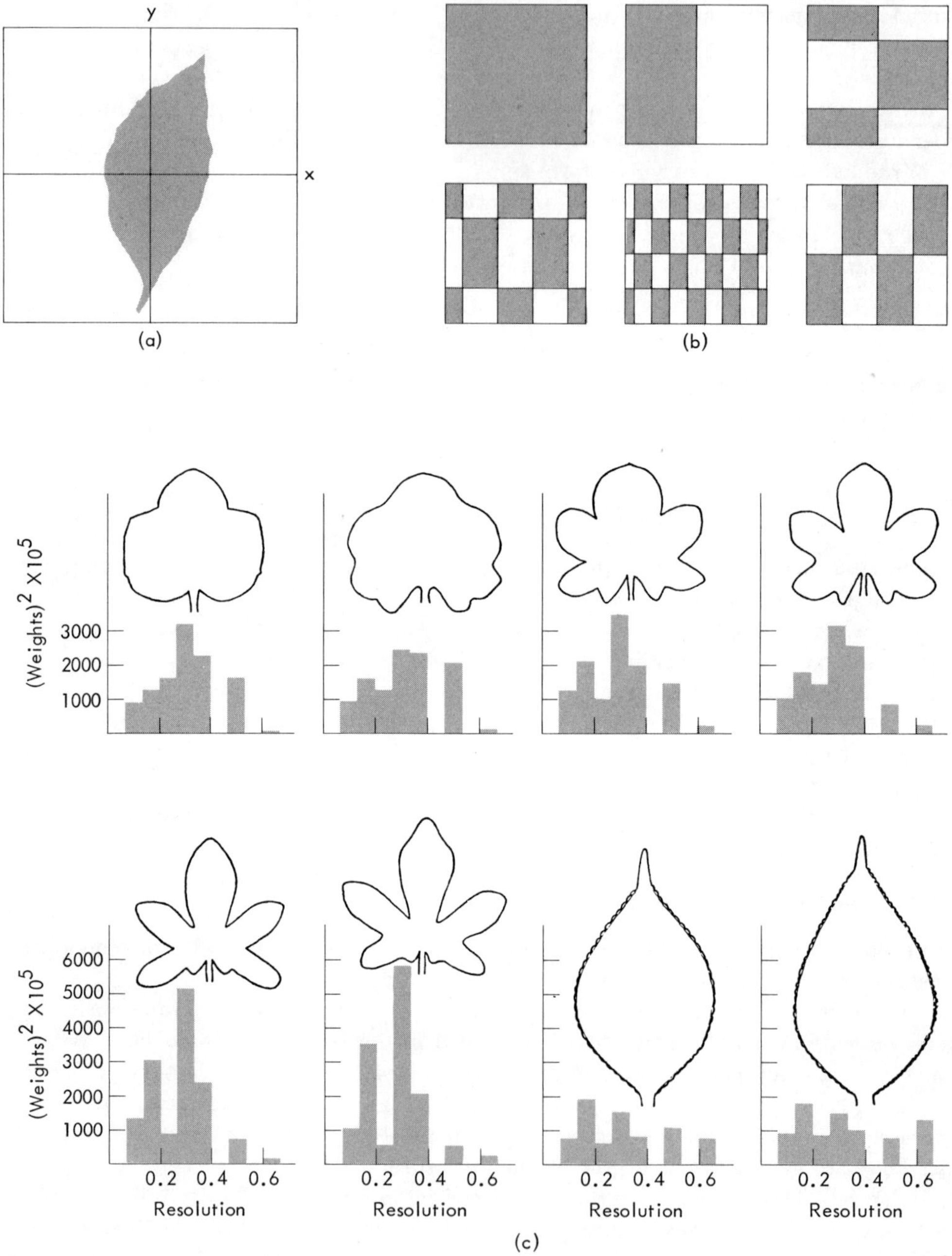

Fig. 7.14. Numerical description of two-dimensional figures [117]. (a) Location of figure in coordinate frame. (b) Selection of basic Walsh patterns. (c) Outlines and corresponding histograms of selected types of leaves.

7.6 EXPERIMENTAL METHODS

In this section we shall survey typical procedures used to characterize either the total shape or some particular aspects of it. Included here are methods based on random sections or projections, on serial sectioning, and on the three-dimensional cell, grain, or particle. Finally, we shall briefly review techniques for stereoscopic reconstruction.

7.6.1 Random Sections or Projections

Statistical methods are frequently employed to express the average particle or grain shape of an aggregate. Measurements and calculations based on random two-dimensional sections through the aggregate may then be extrapolated to encompass the three-dimensional quantities.

Form factors. One of the oldest ways to characterize shape numerically is through the statistical methods developed for small-particle technology. Dimensionless parameters, called *form (or shape) factors*, are related to the length, surface, or volume characteristics of the aggregate.

The basic measurement required for these form factors is some consistently applied linear measure of particle magnitude. The choice of "length" or "diameter" is, however, quite arbitrary. Various kinds of statistical diameters were defined in Section 4.6. In addition, there are the tangent diameter, the equivalent diameters, the projected diameters, the intercept diameter, etc. [114]. The outstanding virtues of the mean intercept length $\bar{L}_3(=\bar{L}_2)$ have been emphasized in previous chapters.

Having chosen a particular length measure, we can proceed to calculate the form factors. Let L represent this arbitrary length. Then, in an aggregate of cells or particles, we can express the mean tangent diameter $\bar{D}$, the mean surface area $\bar{S}$, and the mean volume $\bar{V}$, through the defining equations [37]

$$\bar{D} = \alpha\bar{L}, \tag{7.50}$$

$$\bar{S} = \beta\overline{L^2}, \tag{7.51}$$

$$\bar{V} = \gamma\overline{L^3}, \tag{7.52}$$

where α, β, and γ are the dimensionless, shape-sensitive statistical form factors. The latter also represent [38]:

a) proportionality factors, between particle diameters determined by different methods,
b) conversion factors, for expressing measurements in terms of equivalent figures, and
c) transformation coefficients, for converting diameters into surface areas or volumes.

For a system of uniform-sized spheres of radius r, we know that $D = 2r$, $S = 4\pi r^2$, and $V = 4\pi r^3/3$. If we let r equal the arbitrary length L, the shape factors have the values

$$\left.\begin{aligned} \alpha &= 2, \\ \beta &= 4\pi, \\ \gamma &= 4\pi/3. \end{aligned}\right\} \tag{7.53}$$

When the particle size is expressed as the *projected diameter*, d_p (the diameter of a circle having the same area as the projected image of the particle), the shape factors may be defined [38] according to

$$\bar{D} = \alpha' \bar{d}_p, \tag{7.54}$$

$$\bar{S} = \beta' \overline{d_p^2}, \tag{7.55}$$

$$\bar{V} = \gamma' \overline{d_p^3}, \tag{7.56}$$

where the primes signify shape factors based on projected diameters.

Another linear measure is the *equivalent diameter* d_e (the diameter of a sphere having the same volume as the particle). The average volume is then

$$\bar{V} = (\pi/6)\overline{d_e^3}. \tag{7.57}$$

Combining Eqs. (7.56) and (7.57), we can write

$$\gamma' \overline{d_p^3} = (\pi/6)\overline{d_e^3}, \tag{7.58}$$

which provides a basis for conversion between d_p and d_e. Similar relationships are required for the interconversion of other diameters and form factors.

Shape indices. Closely tied in with the subject of form factors are the so-called *shape indices*. Generally, these indices are shape-sensitive parameters independent of the number and size of particles. Shape indices may range from simple dimensionless ratios to complicated combinations of form factors.

In order to give an idea of the scope and diversity of shape indices that have been proposed, we list various representative types.

1) $Q = \dfrac{\text{length}}{\text{width}}$

2) $q = \dfrac{D_{\parallel}}{D_{\perp}} = \dfrac{(P_L)_{\perp}}{(P_L)_{\parallel}}$

3) $I_s = \dfrac{\bar{L}_p}{\bar{L}_2} = \dfrac{\pi P_L^2}{2N_A}$

4) $\text{C.I.} = \dfrac{\bar{L}_p}{\bar{A}} = \dfrac{\pi P_L}{2A_A} = \dfrac{\pi}{\bar{L}_2}$

5) $F_2 = \dfrac{(\bar{L}_p)_{\text{equivalent circle}}}{(\bar{L}_p)_{\text{actual}}}$

6) $\psi = \left[\dfrac{a_2^2 + b_2^2}{a_0^2}\right]^{1/2}$

7) $\Omega_{12} = \dfrac{(P_L)_{\perp} - (P_L)_{\parallel}}{(P_L)_{\perp} + 0.571(P_L)_{\parallel}}$

8) $K = \dfrac{N_L N_V}{N_A^2} = \dfrac{\bar{d}}{\bar{D}} = \dfrac{4\bar{L}_3}{\bar{S}}$

9) $\phi = \dfrac{\alpha\gamma}{\beta^2} = \dfrac{N_A P_P}{16N_L^2}$

10) $f_1 = \dfrac{\bar{V}^{1/3}}{\bar{S}^{1/2}} = \dfrac{N_A^{1/2}}{N_V^{1/3}}$

11) $f_2 = \dfrac{\bar{V}^{1/3}}{\bar{L}} = \dfrac{N_L}{N_V^{1/3}}$

12) $f_3 = \dfrac{\bar{S}^{1/2}}{\bar{L}} = \dfrac{N_L}{N_A^{1/2}}$

The above form factors require some explanation. Number (1) is very common, and is used whenever the grains or cells are appreciably elongated. Similarly, number (2) is also an index of nonequiaxiality [7]. The symbols $D_{\parallel}$ and $D_{\perp}$ refer to diameter or intercept lengths measured parallel ($\parallel$) and perpendicular ($\perp$), respectively, to an arbitrary orientation axis (see Chapter 3). Similarly, $(P_L)_{\perp}$ and $(P_L)_{\parallel}$ are the boundary intersections made per unit length in the perpendicular and parallel directions.

The I_s in number (3) is a *grain shape index* proposed by Smith [15, 39]. It is simply the mean perimeter length, $\bar{L}_p$, divided by the mean intercept length, $\bar{L}_2$. Both terms can, of course, be determined rigorously and without assumptions. The letters C.I. in number (4) refer to the *complexity index*, designed to reflect an increased jaggedness of the perimeter per area of the particle or grain [40]. It, too, can be determined without assumptions.

The shape factor F_2, in number (5), employs the common device of referring to an equivalent circle (see Eqs. 7.54, 7.55, 7.56). In this case, F_2 has a maximum value of unity [7]. Number (6) is an anisotropy parameter based on an analysis by Hilliard [41] of the orientation distribution of a system of lines in a plane. The constants a_0, a_2, and b_2 are Fourier coefficients that are quite sensitive to shape changes; ψ has a maximum value of unity.

The Ω_{12} in number (7) expresses the degree of orientation [7] of a planar system of lines, the subscript 12 referring to lines (1) in a plane (2). The denominator represents the total line length, and the numerator the oriented portion of the lines along an arbitrary orientation axis. For a completely random system of lines, $(P_L)_\perp = (P_L)_\parallel$ and $\Omega_{12} = 0$. The reader should consult Chapter 3 for further details.

The K in number (8) represents a dimensionless form factor that appears to have great generality in the equations for convex particles. Inspection of Tables 4.2 and 4.3 show that K, the constant relating N_V and N_A^2/N_L, has different values depending on the particle shape. Equivalent forms for K are given in terms of $\bar{d}$, the tangent diameter of particle sections and $\bar{D}$, the tangent diameter in space; or in $\bar{L}_3$, the mean intercept length and $\bar{S}$, the mean surface area per particle. A similar shape-sensitive constant, defined by $k = N_A/N_V$, has been documented [42] for simple polyhedrons.

In number (9), the ϕ is a dimensionless form factor, for particles of the same size and shape, which is related to simple counting measurements on the plane of polish [37]. The reciprocal of ϕ has been given by Braun [43], but was developed from his three form factors defined by numbers (10), (11), and (12).

The above compilation has not gone into three-dimensional form factors explicitly, but direct relationships between the two-dimensional and three-dimensional factors are possible. Their use is a matter of choice or modification, depending on the particular requirements at hand.

Applications. A typical development of shape indices for particles of the same shape and size is given by DeHoff [37]. Starting with the basic particle equation $\bar{D} = N_A/N_V$ and the definitions $\bar{S} = S_V/N_V$ and $\bar{V} = V_V/N_V$, from Eqs. (7.50), (7.51), and (7.52), we have

$$N_V = \frac{N_A}{\alpha \bar{L}} = \frac{S_V}{\beta \bar{L}^2} = \frac{V_V}{\gamma \bar{L}^3}. \tag{7.59}$$

We can eliminate the linear dimension L in several ways. For example, we can write

$$N_V = \frac{\beta N_A^2}{4\alpha^2 N_L} = \left[\frac{\gamma N_A^3}{\alpha^3 P_P}\right]^{1/2} = \frac{8\gamma^2 P_L^3}{\beta^3 P_P^2}, \tag{7.60}$$

where $S_V = 2P_L$ and $V_V = P_P$. The relationships given in Eq. (7.60) make it possible to obtain N_V, provided that we know the appropriate form factors (i.e., that we know the particle shape exactly). From Eq. (7.59), we see that L, the characteristic particle dimension, equals $4\alpha N_L/\beta N_A$. Thus we can readily determine both the particle density and "size."

If one knows the shape only qualitatively, then one has recourse to the shape index, ϕ, shown as number (9) in the above list of shape indices. Simple counting measurements—N_A, P_P, and N_L—give the value of ϕ. Master curves of ϕ versus particle axial ratio permit the appropriate axial ratio, as well as the values of α, β, and γ, to be read off directly. Then N_V and L are determined as outlined above.

Bach [44] has taken DeHoff's analysis a step further to include size distributions of equi-shaped particles. He derives the mean characteristic dimension, the variance, and the number of particles per unit volume for symmetrical and nonsymmetrical distribution curves in terms of form factors and the three counting measurements given above.

Fullman [45] derives expressions for the mean sizes of spherical or disk-shaped particles (see Section 4.3.3). For a nonuniform size distribution of spheres, he gives

$$\left.\begin{aligned} \bar{r} &= \pi/4\bar{m}, \\ \bar{S} &= 2\pi N_L/\bar{m}N_A, \\ \bar{V} &= \pi V_V/2\bar{m}N_A, \\ N_V &= 2\bar{m}N_A/\pi, \end{aligned}\right\} \tag{7.61}$$

where $\bar{r}$ is the mean particle radius, $\bar{m}$ is the mean value of the reciprocals of the diameters of the intersected circles, $\bar{S}$ is the average surface area of the spheres, and $\bar{V}$ their average volume. The symbols N_L, N_A, V_V $(=P_P)$ represent simple counting measurements.

An expression for N_V is also given by DeHoff [46] for a size distribution of particles, all of the same shape. We modify Eq. (7.59) to read

$$N_A = \alpha\bar{L}N_V, \tag{7.62}$$

$$S_V = \beta\overline{L^2}N_V, \tag{7.63}$$

$$V_V = \gamma\overline{L^3}N_V. \tag{7.64}$$

Dividing Eq. (7.63) by Eq. (7.62) squared gives

$$\frac{S_V}{N_A^2} = \frac{\beta\overline{L^2}}{\alpha^2 N_V\bar{L}^2}. \tag{7.65}$$

Substituting $2P_L$ for S_V and rearranging, we obtain

$$N_V = \frac{\beta N_A^2\overline{L^2}}{2\alpha^2 P_L\bar{L}^2}, \tag{7.66}$$

where $\overline{L^2}$ and $\bar{L}^2$ are related to first and second moments of the size distribution, respectively, and to the coefficient of variation.

Other treatments of the same type may be found in the literature [43, 47]. However, most of these methods require that the particle shape be simple, that it be known in advance (at least qualitatively), and that the particles have only one shape. In spite of these analytical limitations, it has been recorded experimentally that the ratios $\bar{V}/\bar{D}^3$ or $\bar{S}/\bar{D}^2$ remain practically unchanged for different size grades of irregular or nonuniform discrete particles [48].

Only a few papers deal with the statistical analysis of systems of particles that vary both in size and shape. Green [14], as well as Stoeber [49], discuss nonuniformity in size and irregularity of shape in terms of shape factors and average particle diameters. Braun [43] considers the two cases of (1) all grains geometrically similar but with different volumes, and (2) all grains with the same volume but different shapes. The mathematics are complicated and outside the scope of this book. Further details can be found in the original papers and in standard texts [38, 50].

Also appropriate to a discussion of shape measurements from sections are the two-dimensional computer techniques [51] and the planar distribution studies from large models [11]. Interest in automated electronic or optical scanning methods has accelerated during the past few years [9, 52], and an excellent up-to-date summary is available in the Proceedings of the Second International Congress for Stereology [53]. The status of computer applications to quantitative analysis of microstructures can be seen in the publications of Moore [51, 54]. Procedures that apply to the expression of shape have also been developed. Examples given by Moore [54] include a shape complexity factor, $\bar{L}_p^2/\bar{A}$; a serpentine length, defined as the length of an equivalent rectangle having the same area and perimeter length as the figure; a compactness factor, which is the ratio of the actual area to the area of the circumscribed rectangle; and others.

Another type of shape analysis is based on the plane distribution curves obtained by random sectioning of grains or particles. In an effort to simplify the experimental

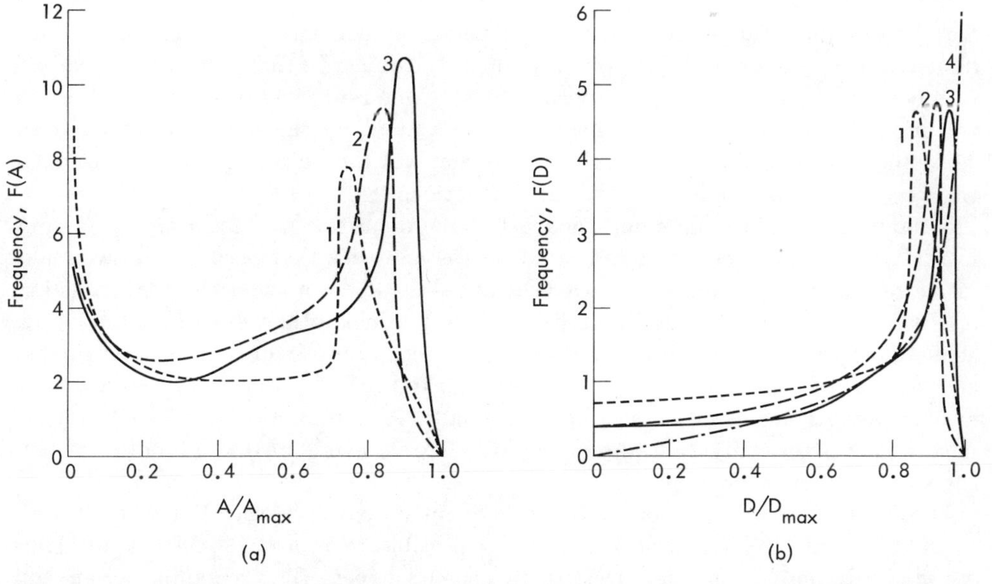

Fig. 7.15. Frequency curves of regular polyhedrons [11]. 1 = cube, 2 = truncated octahedron, 3 = pentagonal dodecahedron, 4 = sphere.

problems, Hull and Houk [11] elected to study four regular polyhedrons that approximate the shapes of metal grains. They measured the areas of random planar sections through a cube, a truncated octahedron, a pentagonal dodecahedron, and a special truncated cube. Frequency curves of planar areas and diameters are shown in Fig. 7.15 for some of these objects. The qualitative differences among curves of this type could be used as a basis for discrimination between particle shapes. Hull and Houk anticipate that the plane distribution curve for an average metal grain would exhibit the same general characteristics as the curves shown in Fig. 7.15. However, the maximum would be much broader because of the lack of symmetry in the grain shape, and would occur at smaller values of the ratio.

The tangent count [16, 36] is an important shape-sensitive parameter. We discussed an application of this measurement in Section 7.5.6. Consider, for example, a section through a two-phase material. A test line is swept through the test area, either by translation parallel to itself, or by rotation through the test area. The number of tangents produced by the sweeping test line per unit area is denoted by T_A. Obviously, if all second-phase particles were convex, $T_A = 2N_A$. The coefficient 2 can increase without limit as the complexity of the perimeter outline increases. Thus we have here a sensitive and powerful tool for characterizing shape. The complete ramifications of this type of measurement have not yet been fully explored. However, in the important equation for the average mean curvature of a system of surfaces (Eq. 7.47), the tangent count is used to evaluate $\bar{k}$, the average curvature of traces in the section. Further significant developments are to be anticipated using the tangent count.

7.6.2 Serial Sectioning

Serial sectioning usually consists of successive parallel, flat cuts through a three-dimensional object or space. Reconstruction of the object so that it may be visualized in three dimensions is then accomplished by stacking reproductions of the sections in the original order, either as transparencies, as a movie of the successive sections, or as profiles that collectively form the actual shape of the object. Sketches or calculations are frequently made from the reconstructions.

Undoubtedly anatomists and biologists have done the greatest amount of work in this area. The extremely complicated organs and cells that occur in nature make visualization from isolated sections a difficult task, and account for the heavy reliance on built-up models, movies, and drawings. Recently other experimenters, such as metallurgists, have adopted serial sectioning techniques in order to obtain further clarification of their microstructures [56, 68, 70, 75]. As more use is made of reconstructions from serial sections, especially in conjunction with quantitative measurements on individual planes, we can expect even greater advances in this shape-visualization technique.

Because the reconstruction methods frequently involve physical models, movies, or even optical imaging in space, it is not possible to reproduce them here. Thus we shall rely mainly on references to the original papers. The remaining discussion of this topic will consist of brief mention of typical reconstruction devices.

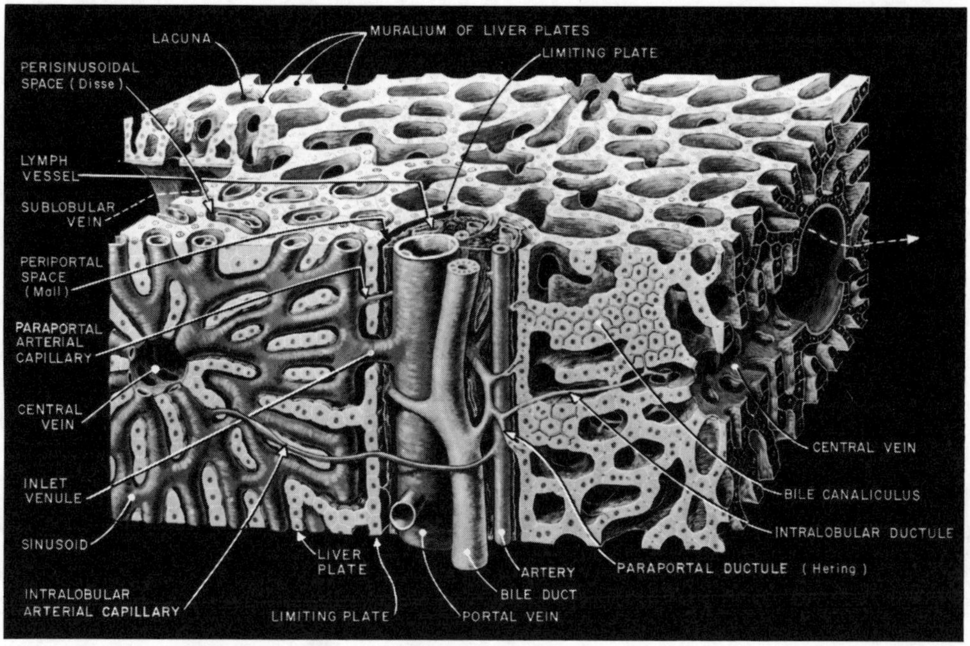

Fig. 7.16. Structure of the human liver [58].

Drawings. Included under this heading are sectional or cut-away drawings of selected portions of the subject being studied. Figure 7.16 shows a reconstruction by Elias [58], originally in color, of the internal structure of the normal human liver. The complexity of this organ, as well as the high degree of artistic skill required to make drawings of it, can be appreciated. Other three-dimensional drawings of great complexity appear in the books and articles by Krieg [59] and Elias [60, 63].

Klemm [61] presents sketches of successive polished surfaces in a steel, and compares them with the corresponding photomicrographs. The final picture consists of 12 superimposed contours in one drawing. Hines and Hugill [62], in a study of crack formation in a steel, sketch the three-dimensional appearance of the crack system from two series of sections taken at right angles to each other.

Models. Another common method of presentation consists of a model or drawing of the entire object. Examples are the drawings of voids in sintered compacts of copper particles [115], the different structures of graphite in cast irons [64] and the plastic model of the human bronchial tree [65] shown in Fig. 7.17.

A familiar type of three-dimensional model is the phase diagram, or "ternary diagram" resulting from admixtures of three metals. The influence of temperature on the various alloys creates solid-phase regions and a melting surface. Isothermal sections through the ternary diagram represent the "serial sections." Books by Masing [66] and Rhines [67] illustrate these models beautifully.

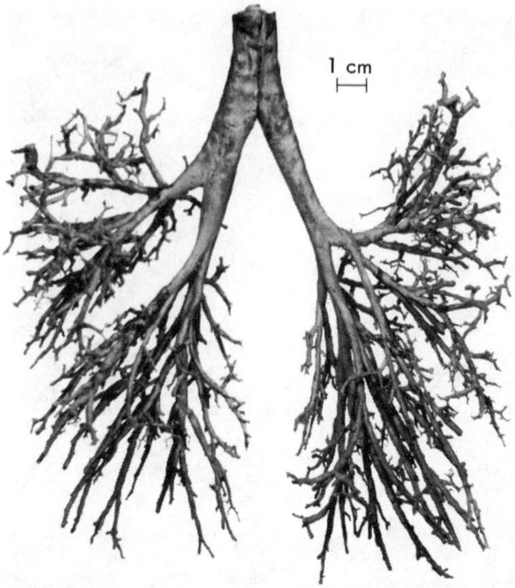

Fig. 7.17. Model of the human bronchial tree [65].

An unusually effective model of an embryo has been devised by Hegre [72]. A series of cardboard pages, each representing a sectional view, are mounted on vertical posts. The outer edges of the pages are trimmed to the sectional contour so that when the pages are stacked together, the embryo appears as a three-dimensional model when viewed from the front.

Figure 7.18 represents an oblique view of a three-dimensional map. In this case, the surface portrays the population density of central Kansas, but the principle is applicable to a broad range of subjects [116].

Sequential pictures. A wide variety of structures have been represented by means of a series of pictures or photographs taken from serial sections. No attempt is made necessarily to build them up into a three-dimensional representation. An example is given in Fig. 7.19(a), which is a sketch of a nerve cell in an animal cortex [57], showing by dashed line the location of the first cut. Part (b) presents the sequence of 8 sections observed after making cuts 3μ apart.

Figure 7.20 by Hahn [68] reconstructs the plastically deformed region ahead of a crack in steel, as revealed by serial sectioning and etching. A considerable amount of work has been done at the University of Florida on the serial sectioning of metallic materials, an account of which appears in a paper by Kronsbein, *et al.* [69].

Dyer [70] presents interesting pictures taken in series from horizontal and vertical sections under a hardness impression in copper. A selective etch reveals the dislocation density in each section, giving the appearance of an inverted, spongy tree. Underwood and Coons [71] employ serial sectioning techniques to follow the traces of twin lamellae in a single crystal of Mo-35 a/o Re.

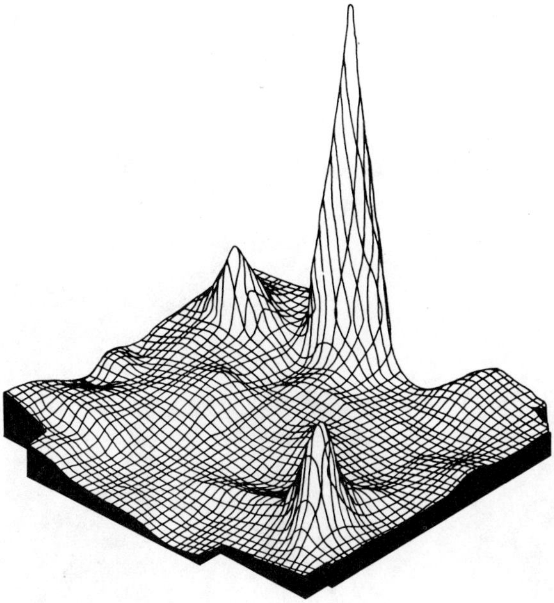

Fig. 7.18. Oblique view of a three-dimensional map [116].

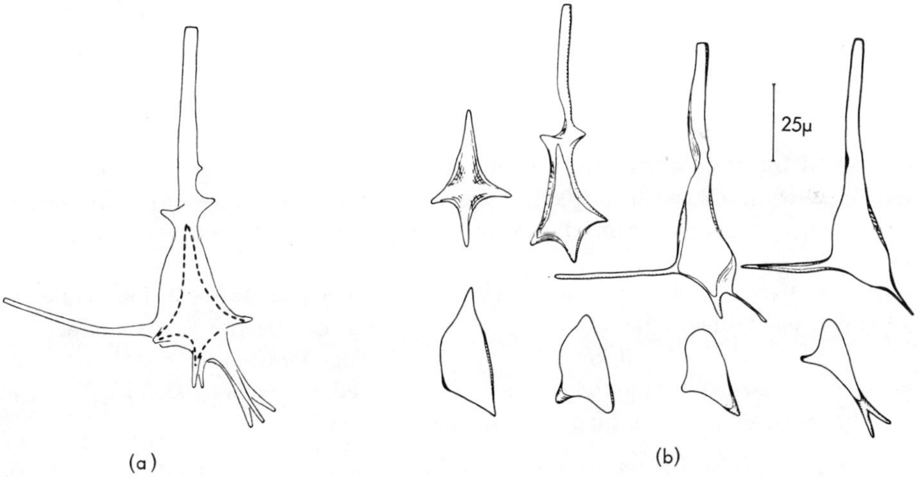

Fig. 7.19. Nerve cell in an animal cortex [57]. (a) Before sectioning. (Dashed line shows location of first cut.) (b) Sequence of eight slices, from left to right.

Transparent layers. This technique employs stacked transparencies made from the serial sections. The layers may be touching, or more commonly, may be separated vertically by appropriate spacers. Kendrew [73], in an x-ray analysis of a protein molecule, builds up the three-dimensional electron-density map with tracings on lucite sheets. From the superimposed images he is able to reconstruct pictorially the extremely complicated three-dimensional model of the molecular structure.

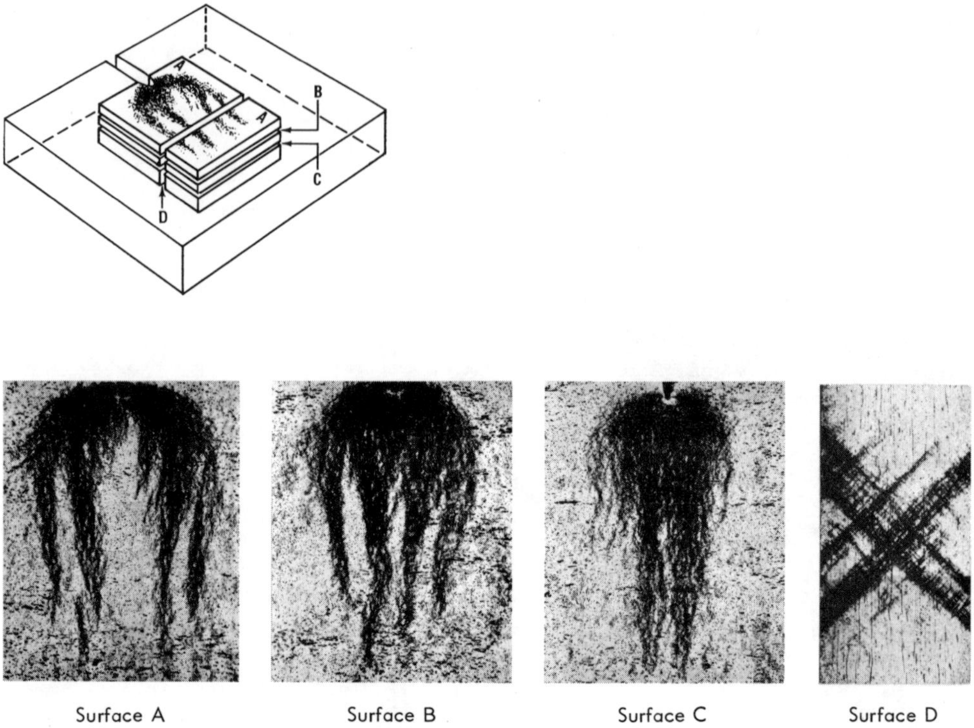

Fig. 7.20. Plastic zone near the tip of a notch in silicon iron [68].

Of especial interest is the study by Cahn and Charles [74] of the connectivity of a minor phase in a two-phase glass. Their space model is constructed from transparent sections, computed from theoretical considerations of precipitation in such systems.

Kraft [75] has consistently utilized three-dimensional concepts in his studies of unidirectionally solidified eutectic alloys. The locations of selected lineal microstructural features are determined by photography during continuous metal removal. These sites are marked on plexiglass sheets, connected by colored strings, and the sheets are stacked in sequence for three-dimensional viewing.

Movie techniques. Now let us speak briefly of some of the methods employed to reproduce serial sections by movie techniques (also called cinephotography, cinematography, cinephotomicrography, etc.). Surface preparation is an important factor. In general, microtoming usually provides an adequate surface for the soft biological specimens. Metals and ceramics, however, must be laboriously polished and etched after griding in order to remove scratches and any deformation introduced on the surface.

Automated devices have been constructed by Hegre [72] and Postlethwait [76] for photographing serial microscope sections of biological specimens. The specimens are automatically cut, stained, and photographed, giving a complete film record of sections through the specimen. The resulting photographic record can be used for projection or as the basis for making an accurate three-dimensional reconstruction.

Applications of movie techniques to metals are typified in studies of dislocations in zinc [77, 78], spheroidized structures in steel [79], and growth faults in an Al-Cu eutectic alloy [75]. In the latter investigation, the alloy surface was removed at a controlled rate in an electrolytic bath, and was continuously observed through a microscope while a movie camera recorded the changing microstructure. The magnification used was 650 times.

Optical visualization. We shall discuss here three methods for reconstructing a visual image of a three-dimensional object. The equipment required is rather elaborate and expensive, but wider use of these advanced methods is definitely indicated.

An instrument developed by Montebello [80] uses serial sections that have been recorded on a film strip. The images are projected rapidly in sequence into the empty viewing space, where the viewer's eye achieves a three-dimensional synthesis. This virtual solid is a duplicate of the original specimen, except that it has a transparent character akin to a radiograph. The image can be inspected from all angles, rotated, and dissected, making it possible to analyze the inner structural relationship in greater detail.

Another remarkable system for synthesizing objects in three dimensions is known as man-computer graphics [81]. A computer-controlled cathode ray tube allows an operator to "draw" pictures on an oscilloscope screen with a light-sensitive "pen." Although there are a myriad of potential applications of this system, the applications of interest here are the surface definition and molding capabilities described in review articles by Flanagan and Hefner [82] and by Chasen and Seitz [112].

Surface molding is a rapid pictorial method for creating free-formed surfaces. The designer generates the desired figure, line by line, by pointing the light pen to desired locations on the oscilloscope, then giving the appropriate command. As the surface is modified, an equation that defines the surface is automatically built, term by term. The first term of the equation defines the shape of the basic surface. Each additional step in shaping the surface adds one term to the equation. Much as a sculptor puts the finishing touches on a statue by adding or removing clay here and there, the additional terms build up or shave off portions of the surface until the final form is obtained.

Geometrical information about the surface, such as cross-sectional or longitudinal traces, surface area, and volume of the body, can also be obtained. If desired, the screen displays the three orthogonal views of the object and a perspective view. These can all be rotated simultaneously, expanded, or altered at the direction of the operator. A new mathematical method, described by Coons [113], makes possible the description of free-form doubly curved surfaces of a very general kind. All in all, the potentialities of the man-computer graphics system appear virtually unlimited.

Visual images are also formed by a technique called holography [83], in which a low-intensity laser beam reconstructs an object with its full three-dimensional appearance. The solid image is formed from a two-dimensional photographic transparency, and no stereo-optical viewing equipment is required. By moving the head from side to side while observing the image, the observer sees the complete three-dimensional phenomenon of shifting perspective, or parallax.

Mathematical analyses. Although considerable experimental work has been done on the qualitative aspects of shape as revealed by serial sectioning, relatively little is available on the quantitative aspects. In general, two types of experiments may be undertaken. One approach is typified by studies of large aggregates of grains. The other approach investigates the sectional properties of a single representative or prototype model. Mathematical analyses of the results have been essentially statistical in nature, giving the average properties or the distributive (frequency curve) properties. More recent analyses are concerned with the topological relationships discernible from serial sections.

Scheil and Wurst [10], continuing earlier research reported by Scheil [84], investigated the spatial shapes of 68 grains from a sample of ingot iron. Serial sections about 8μ apart were prepared and photographed for subsequent analysis of the intercepted planar grains. The successive intercept areas of each of the grains were measured by a planimeter. Since the authors knew the sectional areas and the section spacings, they could calculate the volume of each grain. As the measure of grain size, they selected the equivalent radius of a sphere of equal volume, and established an experimental frequency curve (1), based on grain radii.

This experimental curve (1) was checked by other means. A frequency curve (2), based on equivalent circular areas of the grain sections, was derived but found unsatisfactory. Therefore a statistical grain shape was calculated from further detailed measurements on 10 grains. First, the equivalent circle radius was determined for each of 20 sections equally spaced through each grain. Next the 20 radii were ordered according to magnitude, giving a cumulative distribution curve, from which a frequency distribution was obtained for each grain. The distribution curves for all 10 grains were then averaged, giving the curve representative of the statistical grain shape. Comparison of the frequency distribution curve (3) derived in this manner with the frequency curve (1) showed much better agreement than did curve (2).

Further analysis of the grain section shapes was made. Instead of assuming circular grain intercept areas, the authors determined an average chord intercept. Twenty parallel chords were drawn across each grain section and the average half-length ("radius") calculated. This procedure was repeated on 20 sections through the grain, giving twenty average half-chord lengths for each grain. As before, these average "radii" were ordered according to size, yielding another frequency distribution curve (4). The agreement with frequency curve (1) was not significantly improved, however.

We have described Scheil and Wurst's experimental procedure in some detail, primarily because of the uniqueness and thoroughness of their investigation. So far as grain shape is concerned, however, the results based on equivalent circles or spheres are rather unsatisfactory for a really quantitative description of shape. The use of chord intercepts (either random or parallel) looks interesting and should be investigated further, preferably on material more representative of equilibrium conditions.

Other mathematical analyses based on serial sectioning have been reported. Especially noteworthy are papers by Kronsbein, *et al.* [69], on the topological relationships of sinter bodies, and by Steele [85], on the serial section analysis of grain structures. In the latter paper, Steele describes the way to obtain the volume densities of grains, faces, and edges from serial sections. He calculates the average grain volume, tangent diameter, facet area, and triple line length by combining data obtained from ordinary two-dimensional measurements with those from serial sections. In another paper, Kronsbein and Steele [86] describe the topological properties of structural components embedded in ordinary three-space, based on measurements from serial sections.

7.6.3 Three-Dimensional Cell Studies

There has been considerable work on the geometric and topological properties of space-filling cells or grains, but the results have been mostly descriptive. Efforts in this area may be classified according to whether the cells or grains are:

1) Separate or extracted from an aggregate,
2) Observed *in situ* by optical or x-ray means,
3) Made artificially by compression of initially spherical pellets.

Let us take up each of these three categories in turn.

Separated cells. In a celebrated study, Desch [87] examined the exposed faces (or facets) of metal grains from a disintegrated brass and an embrittled chromium steel. It was found that the frequency of occurrence of polygon faces is almost identical with the frequency curve of foam cells. Five-sided faces are by far the most frequent. In fact, the individual grains tend to approximate pentagonal dodecahedrons with curved faces. Space cannot be filled completely by this shape alone, so the gaps are occupied by small tetrahedrons and hexahedrons. Later, Desch [88] noted that in β-brass grains the number of faces per grain ranged from 11 to 20 and that the number of edges per face varied from 3 to 8. Again the maximum occurred at 5 edges per face. Desch also tried to determine the relation between the form of a polyhedron and those of its plane sections. He compared distribution curves of the number of sides per grain section for the rhombic and pentagonal dodecahedrons, for the truncated octahedron, and for actual metal grains. However, he found that frequency curves based on the number of sides per grain are not sufficiently characteristic to indicate the polyhedron closest to metal grains.

Figure 7.21 shows extracted grains and cells that are remarkably similar in appearance. Part (a) portrays β-brass grains separated by mercury [6a]; part (b), after Lewis [89], shows plant cells; and part (c), also after Lewis [21], shows models of cells in human fat tissue. C. S. Smith [6b] has given an excellent summary of the literature dealing with extracted grains and cells. Despite the disparity in origins of these cells, it appears that the final structures must all conform, more or less closely, to the requirements of minimum energy configurations. Thus the equilibrium angles

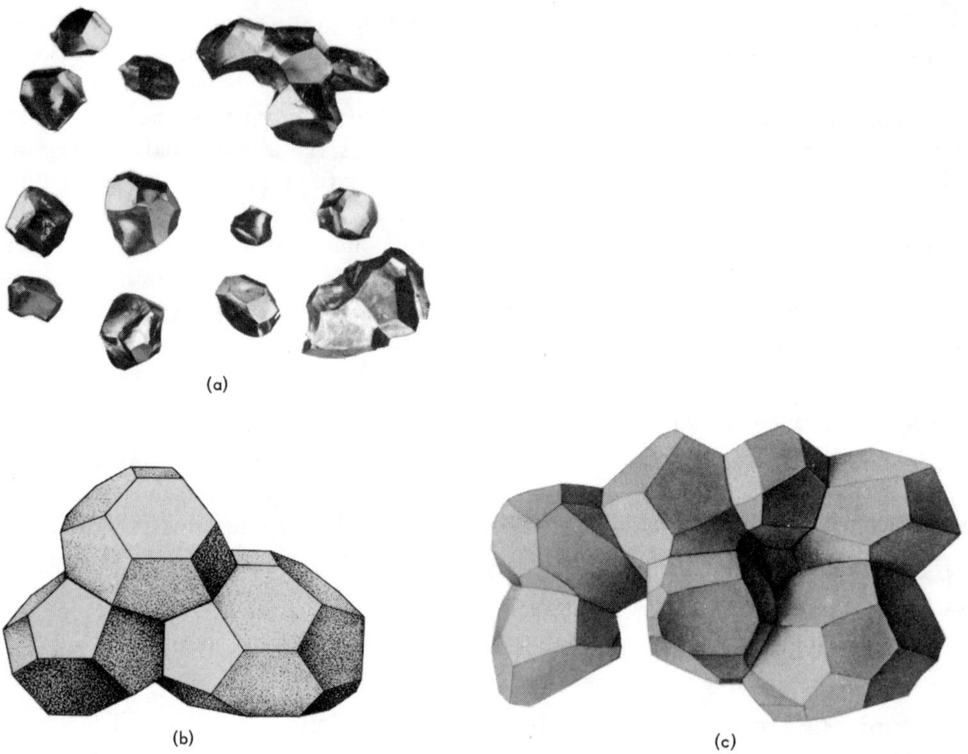

Fig. 7.21. Extracted grains and cells. (a) β-brass grains [6a]. (b) Plant cells [89]. (c) Cells in human fat tissue [21].

of junction, the relative surface energies, and the tendency to form minimum boundary area are factors common to all cases.

Other studies utilizing extraction methods have been reported. See, for example, Rhines and Timpe [96], who employed selective etching to separate components from binary eutectic alloys. These particles are not space-filling and do not readily lend themselves to the topological analyses described above under the heading of Mathematical Methods. Rhines and Timpe gained considerable insight into their true shape, however, merely by being able to observe them as a whole.

An analysis of the rather unusual cell shapes in liver walls was made by Elias [58]. These cells perform a function more like bricks in a tubular wallwork; consequently, only four general types of cells were observed, i.e., 5–, 8–, 10–, and 12-sided polyhedrons. Although Elias did not make a topological analysis, he determined average cell volumes and contact areas.

Grains and cells in situ. The appearance of grains and cells is also frequently depicted by sketches of the outlines, or edges, of the grain. Figure 7.22 reveals the striking similarities in the cell shapes of diverse materials. Part (a) represents bubbles in

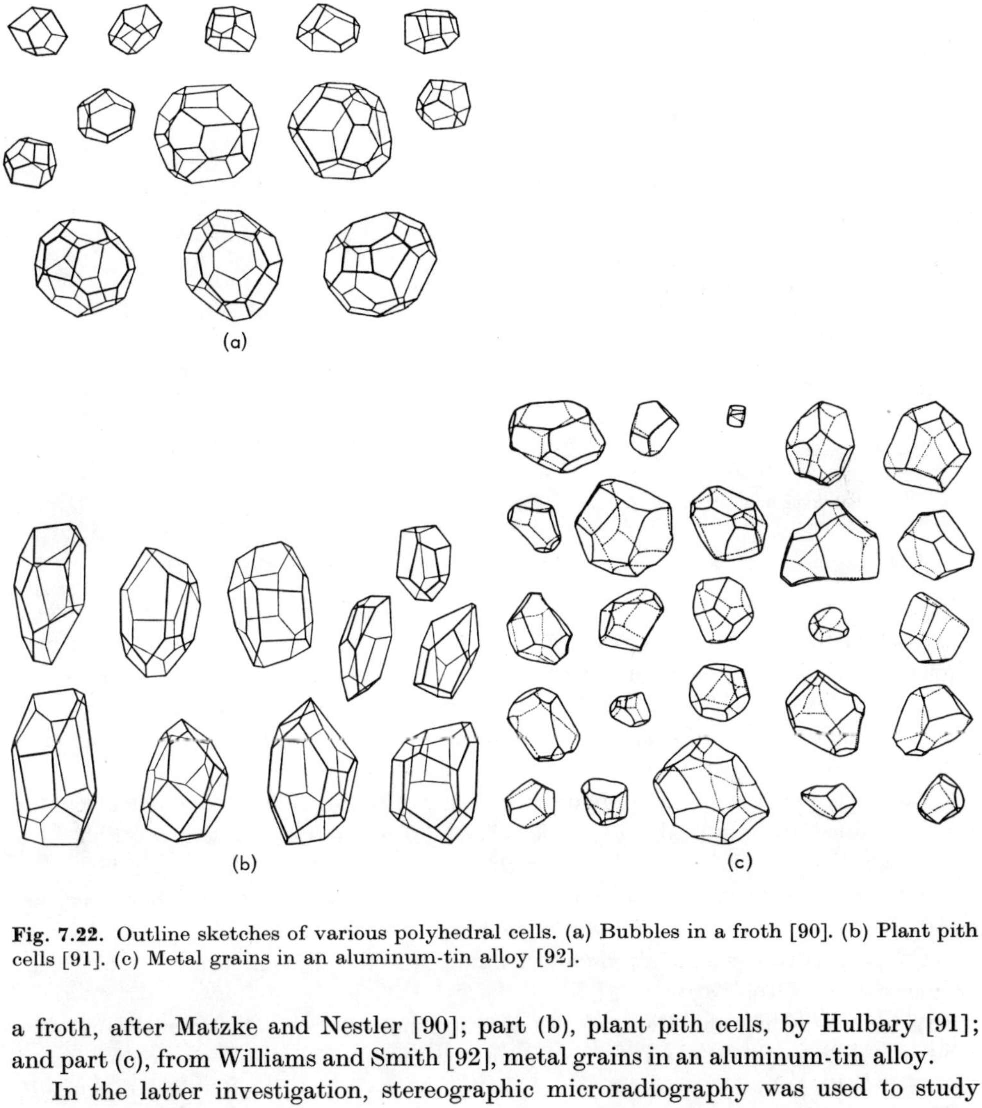

Fig. 7.22. Outline sketches of various polyhedral cells. (a) Bubbles in a froth [90]. (b) Plant pith cells [91]. (c) Metal grains in an aluminum-tin alloy [92].

a froth, after Matzke and Nestler [90]; part (b), plant pith cells, by Hulbary [91]; and part (c), from Williams and Smith [92], metal grains in an aluminum-tin alloy.

In the latter investigation, stereographic microradiography was used to study the shapes and topological properties of almost 100 contiguous grains, *in situ*. The number of faces per polyhedron and the number of edges per face were determined. Examination of the cells in part (c) of Fig. 7.22 reveals grain corners where two or four faces appear to meet. The existence of these anomalous configurations was later questioned because of poor viewing angles [6a].

Matzke [93] conducted a detailed examination of vegetable cells and of uniform and nonuniform soap bubbles. The latter system was composed of 50 large and 100 small bubbles [90]. Results from these three systems, along with data for β-brass [88], an aluminum-tin alloy [92] and the β-tetrakaidecahedron [124], are plotted in

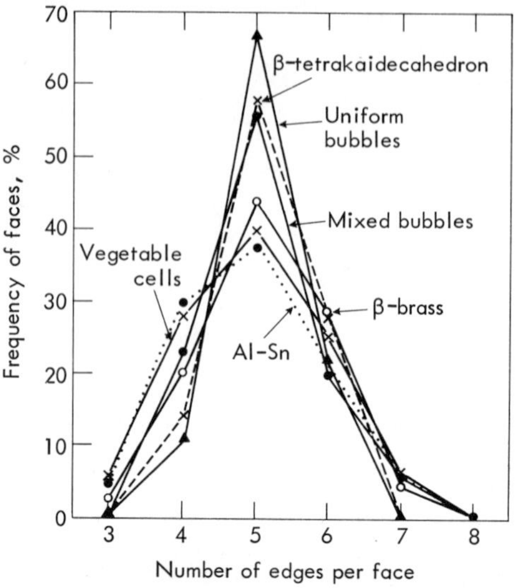

Fig. 7.23. Frequency of various polygonal faces in grains, cells, and bubbles [6a].

Fig 7.23 as the frequency of polygonal faces versus the number of edges per face. The similarity among frequency curves for the solid vegetable cells and the metal grains is noteworthy. The narrower and higher peaks for the two bubble systems attest to the greater uniformity of the bubbles, compared with the solid grains and cells. The most frequent number of edges per face is 5 in every case, however.

Table 7.6 makes a simple comparison of some of the significant topological parameters obtained from the above studies. Recall from our discussion of topological relationships in Section 7.5.5 that $\bar{n}$ is the average number of edges per face, $\bar{F}$ is the average number of faces per polyhedron, and $\bar{C}$ is the average number of corners per polyhedron, all for separated polyhedrons. Table 7.6 also gives values for an "ideal" polyhedron, which represents a minimum energy configuration based on a corner angle of 109°28′16″ [6a].

We see from this table that the values fall rather uniformly between those for the pentagonal dodecahedron and those for the truncated octahedron, except for irregularities in the β-brass grains. Any anomalies in the brass results are undoubtedly to be attributed to the small number of grains in the sample, as well as to the fact that all grains in the original volume were not examined. It does appear significant that the pentagonal dodecahedron and the truncated octahedron act as limits to the naturally occurring cells and grains, and that the ideal polyhedron properties lie close to the middle of the range.

The absolute predominance of pentagonal faces, in the above studies and others [95], suggests that the pentagonal dodecahedron approaches the average grain closer than any other simple polyhedron. The truncated octahedron, on the other hand, has never been observed in real systems [6c].

TABLE 7.6

Basic topological parameters for isolated bubbles, cells, and grains

	$\bar{n}$	$\bar{F}$	$\bar{C}$	References
Pentagonal dodecahedron	5	12	20	3, 6a
100 Al-Sn alloy grains	5.06	12.48	21.04	92
150 mixed bubbles	5.095	13.260	22.520	90
Ideal polyhedron	5.104	13.394	22.789	6a
600 uniform bubbles	5.111	13.702	23.404	93
450 vegetable cells	5.123	13.802	23.572	93
30 β-brass grains	5.142	14.500	24.852	88
β-tetrakaidecahedron	5.143	14	24	124
Truncated octahedron	5.143	14	24	3, 6a

The edges of the pentagonal dodecahedron meet at angles of 108°, not far from the characteristic angle of 109°28′, and the faces meet at an angle not far removed from 120°. A slight curvature of the faces is enough to turn the pentagonal dodecahedron into a possible figure of equilibrium for grains or cells. It cannot be close-packed so as to fill space, whether equal or unequal in size, but occurs commonly in real space-filling systems in the presence of other polyhedrons.

Indeed, as C. S. Smith [6a] remarked, apropos of a random froth of bubbles and an ideal array of truncated octahedrons, "... the introduction of faces with five sides, nearer to the average than the fours and sixes in the Kelvin array, results in there being fewer higher curved surfaces, and the cell edges are more clearly straight. ..." Further, Smith says [6b], "It is obvious that as far as local conditions at a single corner are concerned, the junction of six pentagons at a point represents a better conformity with surface-tension conditions than does the... tetrakaidecahedron."

Unfortunately, however, neither the pentagonal dodecahedron nor the truncated octahedron comply with all requirements of a space-filling, minimum-energy configuration. Even the Kelvin [123] "minimal tetrakaidecahedron" (with curved surfaces) is not satisfactory because of the absence of pentagonal faces. R. E. Williams [124] remedied this deficiency with his β-tetrakaidecahedron, so-called because of its close relationship to the minimal (or α-)tetrakaidecahedron. It can be mechanically derived from the α-tetrakaidecahedron by taking any edge common to two hexagons plus the edges that meet at each end of this edge (see Fig. 7.24a), rotating

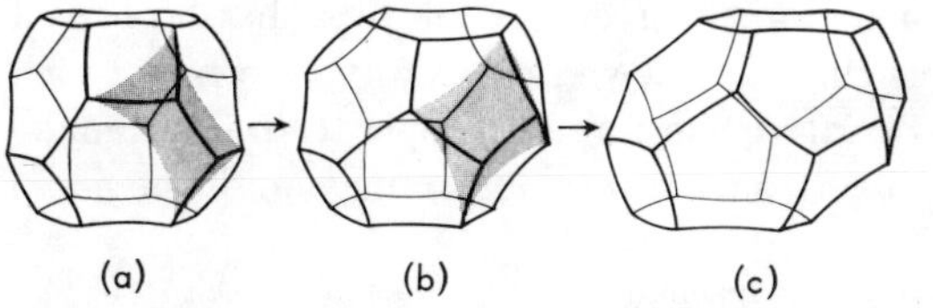

Fig. 7.24. The transformation from Kelvin's α-tetrakaidecahedron, part (a), to Williams' β-tetrakaidecahedron, part (c) [124].

them 90°, and reconnecting them (Fig. 7.24b). This operation transforms two hexagons and two quadrilaterals into four pentagons.* The same operation is then performed with the corresponding group of edges on the opposite side of the polyhedron, giving the β-tetrakaidecahedron (Fig. 7.24c).

This new polyhedron retains the same number of faces (14), corners (24), and edges (36) as the α-tetrakaidecahedron, and the corner juncture angles remain at 109°28′. Moreover, the β-tetrakaidecahedron has two quadrilateral, eight pentagonal, and four hexagonal faces, which give an average of 5.143 sides per face. The frequency distributions of faces for this new polyhedron and for the cells obtained from natural packings compare extremely well (see Fig. 7.23). Williams' new space-filling β-tetrakaidecahedron should serve both as the ideal polyhedron for theoretical studies and as the practical prototype for naturally occurring cells and grains.

Polyhedrons obtained by compression. Another approach to the study of grain shapes has been used to gain deeper insight into the laws governing space-filling structures. Spherical particles, stacked within a closed container, are compressed until the interstices are filled. A remarkable variety of shapes can be obtained in this way.

In 1939, Marvin [97, 33] repeated an experiment performed in 1727 with peas. He placed uniform lead shot in a steel cylinder, then compressed the shot to eliminate all interstices. When the shot was stacked in a square array, nearly perfect rhombic dodecahedrons were obtained. But if the shot were just poured into the cylinder, irregular-faced bodies were formed.

Thompson [1] did much the same experiment with clay pellets. When the pellets were dry, he obtained rhombic dodecahedrons. If the pellets were wetted thoroughly, so that they could slide easily on one another, they came out with square and hexagonal facets.

Bernal [95] compressed irregularly packed balls of Plasticine rolled in chalk. He determined the frequency of polyhedron and polygon types, which yielded an average of 13.3 faces per polyhedron. His central geometrical thesis postulated that irregular dense packing and pentagonal arrangements are necessarily related.

W. C. Graustein [21] calculated the relative surface area and volume of polyhedrons obtained by the imaginary compression of spheres arranged in various configurations. The simple cubic, hexagonal, and body-centered cubic packings were considered, and yielded, upon compression, cubes, rhombic dodecahedrons, and tetrakaidecahedrons, respectively.

7.6.4 Stereoscopic Reconstructions

Almost without exception, our previous quantitative relationships have derived from planar sections or projected images that are viewed uniaxially. We now touch briefly on the advantages afforded by biaxial viewing. Stereoscopic techniques make available new approaches to difficult problems, such as the determination of the mean

*Lewis [21] obtains the same result, but by a different mechanism. Observations of soap bubbles and segmenting ova reveal that the two quadrilaterals (connected by the shaded area in Fig. 7.24a) enlarge until they touch at a point, producing a tetrahedral angle. A further enlargement of the same facets makes four pentagons (as in Fig. 7.24b).

tangent diameter of cells or particles, and the characterization of unflat surfaces. The increased use of stereoscopy should accelerate the development of more quantitative methods of analysis.

Aside from the additional information offered by three-dimensional reconstructions, a major benefit from the use of stereoscopic pairs is the increase in resolution they afford, compared with the resolution afforded by individual pictures. In addition, imperfections in the photographic material itself tend to cancel out. There is also a twofold gain in the number of photographic grains that contribute to the net image.

A few examples have been selected from the literature to illustrate applications of stereoscopic pairs. Although the various methods may employ different radiation (such as light, x-ray, electron, or radar), the basic principles are much the same. Only the details may vary somewhat.

One of the most comprehensive and up-to-date treatises on stereoscopy appears in a book by Valyus [98]. In it, he surveys the principal factors contributing to binocular vision and describes the chief ways of forming and viewing spatial images. A wide range of applications are covered, including astronomy, microscopy, x-ray, aerial photography, and photogrammetry. Numerous examples of the use of optical stereo-pairs may be found in this book, as well as in references cited previously in other contexts [3, 26, 29].

References to x-ray stereographs and their applications appear frequently in texts [99, 114] and the literature [53, 55, 94]. Dislocation line arrays have been studied extensively [100, 101]. Williams and Smith [92] used stereoscopic microradiography successfully to reveal the shapes of metal grains and to determine the size distribution. Some of their results have been reported previously in this book. Williams [102] gives equations for determining from stereo-pairs the depth of a point within the sample body and the internal angle of intersecting lines. To illustrate the method, he analyzes the shape and dimensions of a large grain of aluminum-tin.

Rather more stereoscopic work has been done by electron microscopy than by x-rays. The symposia and proceedings of conferences on electron microscopy that refer to stereoscopy are numerous, and include both biological investigations and materials studies [53, 103, 104]. Helmcke [105] discusses the unique advantages of the Elmigraph I for quantitative stereoplotting and three-dimensional coordinate measurements. Simpler and cheaper apparatuses, the Elmitop [105] and the Stereopret [106], are also described.

A problem arises with linear dislocation arrays that is peculiar to stereo-electron microscopy. The usual method for forming stereo-pairs of dislocations in metal crystals fails because the image of the dislocation does not coincide with its position. By operating in different reflections, the displacement of the image relative to the actual dislocation position might be larger than the shift of the dislocation lines produced by tilting. Basinski [107] avoids this difficulty by choosing the axis of tilt that coincides with the normal to the reflecting plane, thus ensuring that dislocations in each photograph of the stereo-pair are imaged by the same reflection.

Stereoscopic devices based on aerial or radar photography may prove useful in structural studies of shape. The mathematics of photogrammetry and the instrumentation for obtaining true elevations from overlapping radar photographs are covered extensively [108]. Typical projection systems for aerial photography are also listed.

REFERENCES

1. D'A. W. Thompson, *On Growth and Form*, abridged edition edited by J. T. Bonner, Cambridge, England: Cambridge University Press (1961)
2. H. Weyl, *Symmetry*, Princeton, N. J.: Princeton University Press (1962)
3. A. F. Wells, *The Third Dimension in Chemistry*, London: Oxford University Press (1962)
4. E. W. Sinnott, *The Problem of Organic Form*, New Haven, Conn.: Yale University Press (1963)
5. *Aspects of Form*, edited by L. L. Whyte, Bloomington, Ind.: Indiana University Press (1961)
6. a) C. S. Smith, "Some Elementary Principles of Polycrystalline Microstructure," *Met. Rev.* **9,** **33** (1964) 1–48
   b) ———, "Grain Shapes and Other Metallurgical Applications of Topology," in *Metal Interfaces*, Cleveland, O.: American Society for Metals (1952) 65–113
   c) ———, "Microstructure," *Trans. ASM* **45** (1953) 533; *Trans. Chalmers Univ. Tech.*, Gothenburg, Sweden, No. 152 (1954)
   d) ———, "The Shape of Things," *Sci. Am.* **190,** 1 (1954) 58–64
7. S. A. Saltykov, *Stereometric Metallography*, second edition, Moscow, *Metallurgizdat* (1958)
8. Grain Size Chart, Plate I, ASTM Stds., Pt. 3 (1958) 506, Philadelphia, Pa.: Amer. Soc. for Testing Materials
9. H. F. Fischmeister, "Apparative Hilfsmittel der Stereologie," *Quantitative Methods in Morphology*, edited by E. R. Weibel and H. Elias, Berlin, Springer-Verlag (1967) 221
10. E. Scheil and H. Wurst, "Statistische Gefügeuntersuchungen II. Messung der räumlichen Kristallgrösse," *Z. Metallk.* **28** (1936) 340
11. F. C. Hull and W. J. Houk, "Statistical Grain Structure Studies, Plane Distribution Curves of Regular Polyhedrons," *Trans. AIME* **197** (1953) 565
12. H. H. Hausner, "Characterization of the Powder Particle Shape," *Planseeberichte für Pulvermetallurgie* **14** (1966) 75
13. H. Hertel, *Structure, Form and Movement*, New York: Reinhold (1966)
14. H. Green, "The Effect of Nonuniformity and Particle Shape on Average Particle Size," *J. Franklin Inst.* **204** (1927) 713
15. C. S. Smith, private communication (see also page 58 of Ref. 39)
16. F. N. Rhines, "Measurement of Curvature and Number in Space by the Use of the Tangent Count in Random Plane Sectioning," Private communication (1958)
17. A. D. Aleksandrov, "Curves and Surfaces," in *Mathematics, Its Content, Methods, and Meaning*, Vol. 2, (translated by S. H. Gould), Cambridge, Mass.: MIT Press (1963) 57

18. D. J. Struik, *Classical Differential Geometry*, second edition, Reading, Mass.: Addison-Wesley (1961)
19. W. A. Granville, P. F. Smith, and W. R. Longley, *Elements of the Differential and Integral Calculus*, Boston: Ginn (1941)
20. See Ref. 18, page 19
21. F. T. Lewis, "A Further Study of the Polyhedral Shapes of Cells," *Proc. Amer. Acad. Arts Sci.* **61**, 1 (1925) 1; 22
22. See Ref. 17, Vol. 1, B. D. Delone, "Analytic Geometry," 183
23. *Ency. Britannica* (1954), Vol. 6. See articles on "Curve," 885, and "Curves, Special," 887
24. G. P. Tolstov, *Fourier Series*, Englewood Cliffs, N.J.: Prentice-Hall (1962)
25. See Ref. 17, Vol. 2, S. M. Nikolskii, "Approximations of Functions," 265
26. D. Greenhood, *Mapping*, Chicago: University of Chicago Press (1964)
27. See Ref. 18, page 169
28. F. B. Hildebrand, *Advanced Calculus for Engineers*, Englewood Cliffs, N.J.: Prentice-Hall, (1949) 322–346
29. P. M. Morse and H. Feshbach, *Methods of Theoretical Physics*, Part 1, New York: McGraw-Hill (1953) 501 *et seq.*
30. S. A. Schelkunoff, *Applied Mathematics for Engineers and Scientists*, second edition, New York: D. Van Nostrand (1965) 138
31. See Ref. 29, page 656 *et seq.*
32. L. P. Eisenhart, *A Treatise on the Differential Geometry of Curves and Surfaces*, New York: Dover (1960)
33. H. S. M. Coxeter, "Close Packing and Froth," *Ill. J. Math.* **2** (1958) 746
34. J. L. Meijering, "Interface Area, Edge Length, and Number of Vertices in Crystal Aggregates with Random Nucleation," *Philips Res. Rep.* **8** (1953) 270
35. J. W. Cahn, "The Significance of Average Mean Curvature and Its Determination by Quantitative Metallography," *Trans. AIME* **239** (1967) 610
36. R. T. DeHoff, "The Quantitative Estimation of Mean Surface Curvature," *Trans. AIME* **239** (1967) 617
37. R. T. DeHoff, "The Determination of the Geometric Properties of Aggregates of Constant Size Particles from Counting Measurements Made on Random Plane Sections," *Trans. AIME* **230** (1964) 764
38. G. Herdan, *Small Particle Statistics*, second edition, London: Butterworth's (1960)
39. F. Schückher, "Grain Size," *Acta Polytechnica Scand.*, Ch. 54, Stockholm (1966)
40. E. E. Underwood, unpublished research, Lockheed Missiles & Space Co., Palo Alto, Calif. (1963). See also U.S. Patent No. 3,337,334, dated 22 August 1967
41. J. E. Hilliard, "Applications of Quantitative Metallography in Recrystallization Studies," in *Recrystallization, Grain Growth, and Textures*, Cleveland, Ohio, ASM (1965) 267
42. E. J. Myers, "Quantitative Metallography of Cylinders, Cubes, and Other Polyhedrons," *Proc. First Int. Cong. for Stereology*, Vienna, Austria (1963), Paper 15
43. A. Braun, "Considerations sur la Micrographie Quantitative," *Rev. Métallurgie* **52**, 9 (1955) 676

44. G. Bach, "Schätzung charakteristischer Partikelgrössen im Gefüge durch einfache Auszählungen an zufälligen Anschliffebenen," *Z. Metallk.* **56** (1965) 376
45. R. L. Fullman, "Measurement of Particle Sizes in Opaque Bodies," *Trans. AIME* **197** (1953) 447
46. R. T. DeHoff, "The Estimation of Particle Size Distributions from Simple Counting Measurements Made on Random Plane Sections," *Trans. AIME* **233** (1965) 25
47. E. R. Weibel, "Principles and Methods for the Morphometric Study of the Lung and Other Organs," *Lab. Invest.* **12**, 2 (Feb. 1963) 131
48. G. Martin, *Trans. Brit. Cer. Soc.* **27** (1928) 247; 259
49. W. Stoeber, "Statistical Size Distribution Analysis," *Quantitative Electron Microscopy*, edited by G. F. Bahr and E. H. Zeitler, Baltimore, Md.: Williams and Wilkins (1965) 154/892
50. J. M. DallaValle, *Micromeritics, The Technology of Fine Particles*, second edition, New York: Pitman (1948)
51. G. A. Moore, "Applications of Computers to Quantitative Analysis of Microstructures," *Stereology*, Proc. Second Int. Cong. for Stereology, edited by H. Elias, New York: Springer-Verlag (1967) 281
52. H. F. Fischmeister, "Scanning Methods in Quantitative Metallography," in *Quantitative Microscopy*, edited by R. T. DeHoff and F. N. Rhines, New York: McGraw-Hill (1968) 336
53. *Stereology*, Proc. Second Int. Cong. for Stereology, edited by H. Elias, New York: Springer-Verlag (1967)
54. G. A. Moore, "Application of Computers to Quantitative Analysis of Microstructures," in *Ceramic Microstructures, Their Analysis, Significance, and Production*, edited by R. M. Fulrath and J. A. Pask, New York: John Wiley (1968) 71
55. *Proc. First Int. Cong. for Stereology*, Vienna, Austria, published by the Vienna Medical Academy (1963)
56. A. J. Sedriks and R. D. Garwood, "A Serial Sectioning Technique for the Determination of the Habit Planes of Widmanstätten Precipitates," *J. Inst. Metals* **91** (1962–63) 317
57. M. Gihr, "Rekonstruktion von Nervenzellen," *Proc. First Int. Cong. for Stereology*, Vienna, Austria (1963), Paper 37
58. H. Elias and J. C. Sherrick, *Morphology of the Liver*, New York: Academic Press (1968)
59. W. J. S. Krieg, *Connections of the Cerebral Cortex*, Evanston, Ill.: Brain Books (1965)
60. H. Elias and J. E. Pauly, *Human Microanatomy*, third edition, Philadelphia, Pa.: F. A. Davis (1966)
61. H. Klemm, "Die Gefügeuntersuchung in der dritten Dimension," *Metall* **5** (1951) 337
62. J. G. Hines and R. W. Hugill, in *Physical Metallurgy of Stress Corrosion Fracture*, edited by T. N. Rhodin, New York: Interscience (1959) 193
63. H. Elias, "The Renal Glomerulus by Light and Electron Microscopy," *Res. in Service of Medicine* **46** (1956) 6
64. E. E. Underwood, "Quantitative Metallography," *Metals Eng. Quart.* **1**, 3 (August 1961) 70
65. E. R. Weibel, *Morphometry of the Human Lung*, Berlin: Springer-Verlag (1963) 112

66. G. Masing, *Ternary Systems*, New York: Reinhold (1944)
67. F. N. Rhines, *Phase Diagrams in Metallurgy*, New York: McGraw-Hill (1956)
68. G. T. Hahn, "Seeing Through Stressed Steel," *Battelle Tech. Rev.* **14,** 12 (December 1965) 12
69. J. Kronsbein, L. J. Buteau, Jr., and R. T. DeHoff, "Measurement of Topological Parameters for Description of Two-Phase Structures with Special Reference to Sintering," *Trans. AIME* **233** (1965) 1961
70. L. D. Dyer, "The Actual Dislocation Distribution Around a Ball Indentation on the Cube Face of Copper," *Trans. ASM* **58** (1965) 620
71. E. E. Underwood and W. C. Coons, "The Role of Quantitative Stereology in Deformation Twinning," in *Deformation Twinning,* edited by R. E. Reed-Hill, J. P. Hirth, and H. C. Rogers, New York: Gordon and Breach (1964) 405
72. E. S. Hegre, "Serial Section Cinematography," *Stereology,* Proc. Second Int. Cong. for Stereology, edited by H. Elias, New York: Springer-Verlag (1967) 316
73. J. C. Kendrew, "The Three-Dimensional Structure of a Protein Molecule," *Sci. Am.* **208** (December 1961) 96
74. J. W. Cahn and R. J. Charles, "The Initial Stages of Phase Separation in Glasses," Schenectady, N.Y.: General Electric Res. Lab., Rep. No. 65-RL-3899M (March 1965)
75. R. H. Hopkins and R. W. Kraft, "A Rapid Technique for Observation of Three-Dimensional Microstructures," *Trans. AIME* **233** (1965) 1526
76. S. N. Postlethwait and R. Mills, "An Automated Device for Cinematography of Sequential Microscope Sections," *Stereology,* Proc. Second Int. Cong. for Stereology, edited by H. Elias, New York: Springer-Verlag (1967) 323
77. G. S. Tint and V. V. Damiano, *Rev. Sci. Instr.* **32,** 3 (1961) 325
78. V. V. Damiano and N. Brown, "Dislocation Sources at Inclusions in Zinc," *Trans. AIME* **230** (1964) 926
79. V. I. Shevchenko and E. N. Alpator, *Zavod. Lab.*, No. 9 (1963)
80. R. L. de Montebello, "The Synthalyzer for Optical Reconstruction and Dissection of Structures in Three Dimensions," *Stereology,* Proc. Second Int. Cong. for Stereology, edited by H. Elias, New York: Springer-Verlag (1967) 306
81. B. Langefors, "Automated Design," *Int. Sci. and Tech.*, 26 (February 1964) 90; also "Annual Computer Art Contest," in *Computers and Automation* **16,** 8 (1967) 8
82. D. L. Flanagan and O. V. Hefner, "Surface Molding—New Tool for the Engineer," *Astro. and Aeron.* **5,** 4 (1967) 58
83. E. N. Leith and J. Upatnieks, "Photography by Laser," *Sci. Am.* **212** (June 1965) 24
84. E. Scheil, "Statistiche Gefügeuntersuchungen, I," *Z. Metallk.* **27** (1935) 199
85. J. H. Steele, Jr., "Serial Section Analysis of Grain Structures," *Stereology,* Proc. Second Int. Cong. for Stereology, edited by H. Elias, New York: Springer-Verlag (1967) 74
86. J. Kronsbein and J. H. Steele, Jr., "The Topological Properties of Structural Components," *Stereology,* Proc. Second Int. Cong. for Stereology, edited by H. Elias, New York: Springer-Verlag (1967) 252
87. C. H. Desch, *J. Inst. Met.* **11** (1914) 57

88. C. H. Desch, "The Solidification of Metals from the Liquid State," *J. Inst. Met.* **22** (1919) 241

89. F. T. Lewis, "Geometric Accounting for Diverse Shapes of 14-hedral Cells: Transition from Dodecahedra to Tetrakaidecahedra," *Am. J. Botany* **30** (1943) 74

90. E. B. Matzke and J. Nestler, "Volume-Shape Relationships in Variant Foams. A Further Study of the Role of Surface Forces in Three-Dimensional Cell Shape Determination," *Am. J. Botany* **33** (1946) 130

91. R. L. Hulbary, "The Influence of Air Spaces on the Three-Dimensional Shapes of Cells in Elodea Stems, and a Comparison with Pith Cells of Ailanthus," *Am. J. Botany* **31** (1944) 561

92. W. M. Williams and C. S. Smith, "A Study of Grain Shape in an Aluminum Alloy and Other Applications of Stereoscopic Microradiography," *Trans. AIME* **194** (1952) 755

93. E. B. Matzke, "The Three-Dimensional Shape of Bubbles in a Foam—An Analysis of the Role of Surface Forces in Three-Dimensional Cell Shape Determination," *Am. J. Botany* **33** (1946) 58

94. G. Lupacciolu, "Sur la Representation Troidimensionnelle de l'image Roentgen sur la Toile," *Proc. First Int. Cong. for Stereology*, Vienna Medical Academy, Vienna (1963) Paper 35/1

95. J. D. Bernal, "A Geometrical Approach to the Structure of Liquids," *Nature* **183** (January 17, 1959) 141

96. F. N. Rhines and W. Timpe, "Gefügebeobachtungen in drei Dimensionen an eutektischen Legierungen," *Z. Metallk.* **48** (1957) 109

97. J. W. Marvin, "The Shape of Compressed Lead Shot and Its Relation to Cell Shape," *Am. J. Botany* **26** (1939) 280

98. N. A. Valyus, *Stereoscopy*, translated into English by H. Asher, London: Focal Press (1966)

99. V. E. Cosslett and W. C. Nixon, *X-Ray Microscopy*, Cambridge, England: Cambridge University Press (1960) 45

100. A. R. Lang, "Studies of Individual Dislocations in Crystals by X-ray Diffraction Microradiography," *J. App. Phys.* **30** (1959) 1748

101. V. Gerold and F. Meier, "Der röntgenographische Nachweis von Versetzungen in Germanium," *Z. Physik* **155** (1959) 387

102. W. M. Williams, "Stereoscopic Microradiography," *Metallurgia* **63** (1961) 95

103. See the proceedings of the various international conferences on electron microscopy

104. *Quantitative Electron Microscopy*, edited by G. F. Bahr and E. H. Zeitler, Baltimore, Md.: Williams and Wilkins (1965)

105. J. G. Helmcke, "Determination of the Third Dimension of Objects by Stereoscopy," see Ref. 104, page 195/933

106. G. Pohlmann, "Ein einfaches Verfahren zur Auswertung elektronenmikroskopischer Stereoaufnahmen," *Proc. Fourth Int. Conf. on Electron Microscopy* (1960) 152

107. Z. S. Basinski, "Stereo-Electron Micrography of Dislocation Networks," *Fifth Int. Cong. for Electron Microscopy*, edited by S. S. Breese, Jr., New York: Academic Press (1962) B-13

108. *Manual of Photogrammetry*, third edition, edited by M. M. Thompson, Am. Soc. of Photogrammetry (1966); see Chapter II, "Basic Mathematics of Photogrammetry"

109. A. Hennig, "Formbestimmung von Körpern aus ebenen Schnitten," *Stereology*, Proc. Second Int. Cong. for Stereology, edited by H. Elias, New York: Springer-Verlag (1967) 83
110. F. N. Rhines, "Measurement of Topological Parameters," *ibid.*, 235
111. R. T. DeHoff, "The Relationship Between Mean Surface Curvature and the Stereologic Counting Measurements," *ibid.*, 95
112. S. H. Chasen and R. N. Seitz, "On-Line Systems and Man-Computer Graphics," *Computers and Automation* **16** (November 1967) 23
113. S. A. Coons, "Surfaces for Computer-Aided Design of Space Forms," Cambridge, Mass.: MAC-TR-41, June 1967, Mass. Inst. of Technology
114. *The Encyclopedia of Microscopy*, edited by G. L. Clark, New York: Reinhold (1961) 464
115. L. K. Barrett and C. S. Yust, "Progressive Shape Changes of the Void During Sintering," *Trans. AIME* **239** (1967) 1172
116. G. F. Jenks and D. A. Brown, "Three-Dimensional Map Construction," *Science* **154** (18 November 1966) 857
117. B. Meltzer, N. H. Searle, and R. Brown, "Numerical Specification of Biological Form," *Nature* **216** (October 7, 1967) 32; **217** (March 30, 1968) 1289.
118. G. Matheron, *Éléments pour une théorie des Milieux Poreux*, Paris: Masson (1967)
119. G. Matheron, *Les Variables Régionalisées et leur Estimation*, Paris: Masson (1965)
120. A. Haas, G. Matheron, and J. Serra, "Morphologie Mathématique et Granulométries en Place," *Ann. des Mines* (1967), No. 11, page 735; No. 12, page 767
121. J. Serra, "Buts et Réalisation de l'analyseur de Textures," *Revue de l'Industrie Minérale* **49**, 9 (1967)
122. J. Serra, "Remarques sur une lame mince de minerai Lorrain," extrait du *Bull. du B.R.G.M.*, No. 6 (1966)
123. W. Thompson (Lord Kelvin), "On the Division of Space with Minimum Partitional Area," in *Mathematical and Physical Papers*, **5** (1911) 297, Cambridge, England: Cambridge University Press
124. R. E. Williams, "Space-Filling Polyhedron: Its Relation to Aggregates of Soap Bubbles, Plant Cells, and Metal Crystallites," *Science* **161** (July 19, 1968) 276

PROBLEMS

7.1. The periodic function in Fig. 7.25 represents a violin note. How well can you reproduce this curve using Eq. (7.19) with $k = 1$ to 3?

7.2. Calculate the frequency distribution versus $A/A_{\max}$ curve for a sphere. How does the area frequency curve compare with the diameter frequency curve (curve number 4 in Fig. 7.15b) and the chord frequency curve (Fig. 5.9)?

Fig. 7.25. Sound of a violin.

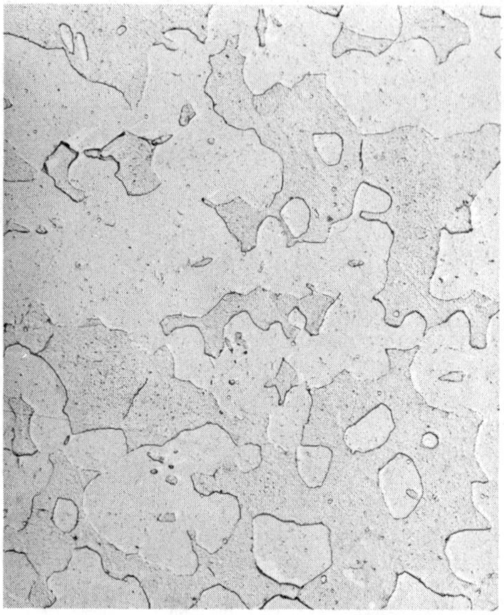

Fig. 7.26. Concavities and convexities of dark phase in a light matrix.

7.3. Calculate the form factor ϕ (number 9 under "Shape Indices" in Section 7.6.1) for the particles in Fig. 1.2(b). Consider only the particles and portions of particles lying within the inner circle.

7.4. Calculate the grain shape index I_s (number 3 under "Shape Indices" in Section 7.6.1) for the particles in Fig. 1.2(b) (as in Problem 7.3) and in Fig. 4.5(b).

7.5. Determine the average mean curvature $\bar{K}_m$ of the interfaces in Fig. 7.26 with respect to the dark phase, using (a) the tangent count (Eq. 7.36), and (b) the + and − particle density count (Eq. 7.43). Would contiguity concepts (as described in Ref. 35 and Section 4.5) be possible with this microstructure?

7.6. According to Eq. (7.48), the average mean curvature $\bar{K}_m$ is proportional to N_A/N_L. What is the value of $\bar{K}_m$ for a system of polyhedrons with flat sides? If the latter system, and a system of convex particles with curved sides, have the same value of N_A/N_L, is $\bar{K}_m$ the same? (See Note at end of Ref. 35.)

7.7. Derive the equations in (a) Table 7.3 and (b) Table 7.4.

7.8. Calculate the value of the mean number of edges per face, $\bar{n} = \sum nF_n/F$, for the following isolated polyhedrons:*

	F_3	F_4	F_6	F_8	C	E
Cube	–	6	–	–	8	12
Cuboctahedron	8	6	–	–	12	24
Truncated cube	8	–	–	6	24	36
Truncated octahedron	–	6	8	–	24	36
Truncated cuboctahedron	–	12	8	6	48	72

*For illustrations of these polyhedrons, see Cundy and Rollett, *Mathematical Models*, second edition, London: Oxford University Press (1961).

INDEXES

AUTHOR INDEX

*Page numbers in italics refer to references listed at end of chapters.

SUBJECT INDEX

ABCDE79876543210